Recent Titles in This Series

- 555 **Tomás Feder,** Stable networks and product graphs, 1995
- 554 **Mauro C. Beltrametti, Michael Schneider, and Andrew J. Sommese,** Some special properties of the adjunction theory for 3-folds in $\mathbb{P}^5$, 1995
- 553 **Carlos Andradas and Jesús M. Ruiz,** Algebraic and analytic geometry of fans, 1995
- 552 **C. Krattenthaler,** The major counting of nonintersecting lattice paths and generating functions for tableaux, 1995
- 551 **Christian Ballot,** Density of prime divisors of linear recurrences, 1995
- 550 **Huaxin Lin,** C^*-algebra extensions of $C(X)$, 1995
- 549 **Edwin Perkins,** On the martingale problem for interactive measure-valued branching diffusions, 1995
- 548 **I-Chiau Huang,** Pseudofunctors on modules with zero dimensional support, 1995
- 547 **Hongbing Su,** On the classification of C^*-algebras of real rank zero: Inductive limits of matrix algebras over non-Hausdorff graphs, 1995
- 546 **Masakazu Nasu,** Textile systems for endomorphisms and automorphisms of the shift, 1995
- 545 **John L. Lewis and Margaret A. M. Murray,** The method of layer potentials for the heat equation on time-varying domains, 1995
- 544 **Hans-Otto Walther,** The 2-dimensional attractor of $x'(t) = -\mu x(t) + f(x(t-1))$, 1995
- 543 **J. P. C. Greenlees and J. P. May,** Generalized Tate cohomology, 1995
- 542 **Alouf Jirari,** Second-order Sturm-Liouville difference equations and orthogonal polynomials, 1995
- 541 **Peter Cholak,** Automorphisms of the lattice of recursively enumerable sets, 1995
- 540 **Vladimir Ya. Lin and Yehuda Pinchover,** Manifolds with group actions and elliptic operators, 1994
- 539 **Lynne M. Butler,** Subgroup lattices and symmetric functions, 1994
- 538 **P. D. T. A. Elliott,** On the correlation of multiplicative and the sum of additive arithmetic functions, 1994
- 537 **I. V. Evstigneev and P. E. Greenwood,** Markov fields over countable partially ordered sets: Extrema and splitting, 1994
- 536 **George A. Hagedorn,** Molecular propagation through electron energy level crossings, 1994
- 535 **A. L. Levin and D. S. Lubinsky,** Christoffel functions and orthogonal polynomials for exponential weights on [-1,1], 1994
- 534 **Svante Janson,** Orthogonal decompositions and functional limit theorems for random graph statistics, 1994
- 533 **Rainer Buckdahn,** Anticipative Girsanov transformations and Skorohod stochastic differential equations, 1994
- 532 **Hans Plesner Jakobsen,** The full set of unitarizable highest weight modules of basic classical Lie superalgebras, 1994
- 531 **Alessandro Figà-Talamanca and Tim Steger,** Harmonic analysis for anisotropic random walks on homogeneous trees, 1994
- 530 **Y. S. Han and E. T. Sawyer,** Littlewood-Paley theory on spaces of homogeneous type and the classical function spaces, 1994
- 529 **Eric M. Friedlander and Barry Mazur,** Filtrations on the homology of algebraic varieties, 1994
- 528 **J. F. Jardine,** Higher spinor classes, 1994
- 527 **Giora Dula and Reinhard Schultz,** Diagram cohomology and isovariant homotopy theory, 1994

(Continued in the back of this publication)

Memoirs
of the
American Mathematical Society

Number 555

Stable Networks
and Product Graphs

Tomás Feder

1991 *Mathematics Subject Classification.*
Primary 68Q05, 05C12.

Library of Congress Cataloging-in-Publication Data
Feder, Tomás.
   Stable networks and product graphs / Tomás Feder.
     p. cm. – (Memoirs of the American Mathematical Society, ISSN 0065-9266; no. 555)
   Includes bibliographical references (p.   ).
   ISBN 0-8218-0347-6
   1. Machine theory. 2. Graph theory. 3. System analysis. I. Title. II. Series.
QA3.A57 no. 555
[QA267]
510 s—dc20   95-15926
[511.3]   CIP

Memoirs of the American Mathematical Society

This journal is devoted entirely to research in pure and applied mathematics.

Subscription information. The 1995 subscription begins with Number 541 and consists of six mailings, each containing one or more numbers. Subscription prices for 1995 are $369 list, $295 institutional member. A late charge of 10% of the subscription price will be imposed on orders received from nonmembers after January 1 of the subscription year. Subscribers outside the United States and India must pay a postage surcharge of $25; subscribers in India must pay a postage surcharge of $43. Expedited delivery to destinations in North America $30; elsewhere $92. Each number may be ordered separately; *please specify number* when ordering an individual number. For prices and titles of recently released numbers, see the New Publications sections of the *Notices of the American Mathematical Society.*

Back number information. For back issues see the *AMS Catalog.*

Subscriptions and orders should be addressed to the American Mathematical Society, P. O. Box 5904, Boston, MA 02206-5904. *All orders must be accompanied by payment.* Other correspondence should be addressed to P. O. Box 6248, Providence, RI 02940-6248.

Copying and reprinting. Individual readers of this publication, and nonprofit libraries acting for them, are permitted to make fair use of the material, such as to copy a chapter for use in teaching or research. Permission is granted to quote brief passages from this publication in reviews, provided the customary acknowldgement of the source is given.

Republication, systematic copying, or multiple reproduction of any material in this publication (including abstracts) is permitted only under license from the American Mathematical Society. Requests for such permission should be addressed to the Manager of Editorial Services, American Mathematical Society, P. O. Box 6248, Providence, RI 02940-6248. Requests can also be made by e-mail to reprint-permission@math.ams.org.

The owner consents to copying beyond that permitted by Sections 107 or 108 of the U.S. Copyright Law, provided that a fee of $1.00 plus $.25 per page for each copy be paid directly to the Copyright Clearance Center, Inc., 222 Rosewood Dr., Danvers, MA 01923. When paying this fee please use the code 0065-9266/95 to refer to this publication. This consent does not extend to other kinds of copying, such as copying for general distribution, for advertising or promotion purposes, for creating new collective works, or for resale.

Memoirs of the American Mathematical Society is published bimonthly (each volume consisting usually of more than one number) by the American Mathematical Society at 201 Charles Street, Providence, RI 02904-2213. Second-class postage paid at Providence, Rhode Island. Postmaster: Send address changes to Memoirs, American Mathematical Society, P. O. Box 6248, Providence, RI 02940-6248.

© Copyright 1995, American Mathematical Society. All rights reserved.
Printed in the United States of America.
This volume was printed directly from author-prepared copy.
∞ The paper used in this book is acid-free and falls within the guidelines
established to ensure permanence and durability.
♻ Printed on recycled paper.

10 9 8 7 6 5 4 3 2 1   00 99 98 97 96 95

Contents

Abstract   x

Acknowledgements   xii

1 Introduction   1
    1.1 Organization   7

2 Preliminaries   8
    2.1 Gates, Circuits, and Networks   8
         2.1.1 Modifying a Network   11
    2.2 Evaluation, Convergence, and Stability   15
    2.3 Classification of Bases   15
         2.3.1 Universal Bases   16
         2.3.2 Linear Bases   18
         2.3.3 Monotone Bases   18
    2.4 Nonexpansive Bases   20
         2.4.1 No Finite Basis for Nonexpansive Gates   23
         2.4.2 Circuits and Networks with Variable Inputs   25
    2.5 Complexity Assumptions   28

3 Stability in Nonexpansive Networks   30
    3.1 Hamming Distance and Convergent Networks   30
         3.1.1 Convergent Networks   32
         3.1.2 Properties of Convergent Networks   34
         3.1.3 Algorithms for Convergent Networks   39
         3.1.4 Representation by Convergent Networks   42
         3.1.5 Algorithms for Stability   45
         3.1.6 Randomized Algorithms   47
         3.1.7 Lower Bounds   52
    3.2 Fixed Points and Retracts   58
         3.2.1 Fixed Points and Medians   58
         3.2.2 Median Sets and 2SAT   60
         3.2.3 Representation by Circuits   65

3.3 Periodic Points and Isomorphisms 68
    3.3.1 Distances and Periodic Points 68
    3.3.2 Efficient Representation of Stable Configurations 71
    3.3.3 Efficient Representation in Scatter-Free Networks 77
3.4 Unstable Networks and Fixed Cubes 79
    3.4.1 Minimal Fixed Cubes 81
    3.4.2 Algorithms for Fixed Cubes 84
    3.4.3 From Monte Carlo to Las Vegas 84
    3.4.4 A Small Certificate for Mappings without Fixed Point 85
    3.4.5 The Convergent Network Representation is Decidable 89
    3.4.6 Convergent Network Evaluation and Linear Programming 94
3.5 Non-Periodic Points and Iterates 97
    3.5.1 The Scatter-Free Case 97
    3.5.2 The Nonexpansive Case 100
    3.5.3 Evaluation, Stability and Convergence 104
3.6 Discussion 105

4 Optimization and Enumeration   108
4.1 Uncapacitated Flow 108
    4.1.1 Flow Problems and Width 109
    4.1.2 Maximum Flow when the Width is Small 114
    4.1.3 Maximum Flow when the Optimum is Small 115
    4.1.4 Uncapacitated Blocking Flow 121
4.2 Optimization on 2SAT instances 123
    4.2.1 The Non-Bipartite Case 124
    4.2.2 The Bipartite Case 126
4.3 Enumeration on 2SAT instances 128
4.4 Discussion 131

5 Stable Matching   133
5.1 The Stable Arrangement Problem 133
    5.1.1 Stable Arrangements 134
    5.1.2 Reduction to Network Stability 135
    5.1.3 Instabilities and Local Search 138
    5.1.4 Structural Properties 140
    5.1.5 Complexity of Several Arrangement Problems 144
5.2 Optimization and Enumeration in Stable Matching 148
    5.2.1 Size and Width 149
    5.2.2 Optimization 152
    5.2.3 Enumeration and Partial Arrangements 155
5.3 Discussion 157

CONTENTS

6 Metric Networks and Product Graphs — **158**
- 6.1 Metric Spaces — 159
- 6.2 Representations — 162
- 6.3 Isometric Representation — 165
- 6.4 Prime Factorization — 168
- 6.5 The 2-Isometric Representation — 171
  - 6.5.1 Closed Sets and the Imprint Function — 171
  - 6.5.2 2-Isometric Subspaces — 174
  - 6.5.3 2-Isometric Representations — 179
- 6.6 Retracts — 183
  - 6.6.1 Subspaces without Holes and the Distance Center — 184
  - 6.6.2 Partial Mappings and Projections — 185
  - 6.6.3 Canonical Retract Representation — 189
- 6.7 Dynamic Search in Graphs — 194
- 6.8 Network Stability — 196
  - 6.8.1 Convergent Networks — 197
  - 6.8.2 Fixed Points and Retracts — 198
  - 6.8.3 Periodic Points and Isomorphisms — 200
  - 6.8.4 Unstable Mappings and Fixed Products — 202
  - 6.8.5 Non-Periodic Points and Iterates — 205
- 6.9 Discussion — 211

Bibliography — **213**

Index — **221**

List of Tables

1.1 Complexity of six problems on bases with fanout. 3
5.1 The preference lists in a stable arrangements instance. 134

List of Figures

2.1 The network N with transition function f. 10
2.2 The network (N, S) with transition function (f, S). 12
2.3 The network (N, k) with transition function (f, k). 13
2.4 The network N' and the corresponding circuit $K = (N', L(N))$. 14

3.1 The n-cube for $n = 3$. 31
3.2 The gate (f_S, S) obtained by convergence from (N, S), and the projection f_S. 34
3.3 The network N_k that converges to X_k 43
3.4 The median function. 59

4.1 Providing flow at supply vertices. 113
4.2 Representation of the layered graph. 118

5.1 Network representation of a stable arrangement instance. 137
5.2 Breaking a path. 145
5.3 Breaking a path in the monotone case. 145

Abstract

A network is a collection of gates, each with many inputs and many ouputs, where links join individual outputs to individual inputs of gates; the unlinked inputs and outputs of gates are viewed as inputs and outputs of the network. A stable configuration assigns values to inputs, outputs, and links in a network, so as to ensure that the gate equations are satisfied.

The problem of finding stable configurations in a network is computationally hard, even when all values are boolean and all input values are specified in advance; in general, the difficulty of a stability problem seems to depend on the kinds of gates present in the network. The study can be restricted to gates that satisfy a nonexpansiveness condition requiring small perturbations at the inputs of a gate to have only a small effect at the outputs of the gate. The stability question on the class of networks satisfying this local nonexpansiveness condition contains stable matching as a main example, and defines the boundary between tractable and intractable versions of network stability.

The structural and algorithmic study of stability in nonexpansive networks is based on a representation of the possible assignments of boolean values for a network as vertices in a boolean hypercube under the associated Hamming metric. This global view takes advantage of the median properties of the hypercube, and extends to metric networks, where individual values are now chosen from finite metric spaces and combined by means of an additive product operation. The relationship between products of metric spaces and products of graphs then establishes a connection between isometric representations in graphs and nonexpansiveness in metric networks.

1991 Mathematics subject classification: 68Q05, 05C12.

Key words: network stability, nonexpansive mappings, satisfiability, stable matching, network flow, graph distance, median graphs, cartesian product graphs, isometric embeddings, retracts.

*A mis padres
y a mi hermana*

Acknowledgements

I am grateful to Don Knuth, my dissertation advisor.

Ashok Subramanian introduced me to the framework on which this work is based, and encouraged me to explore it further.

This work benefitted greatly from conversations with Yossi Azar, Fan Chung, Joan Feigenbaum, Ron Graham, Dan Gusfield, David Johnson, Nati Linial, Nimrod Megiddo, Rajeev Motwani, Moni Naor, Christos Papadimitriou, Shaibal Roy, Alex Schäffer, Peter Winkler, and Mihalis Yannakakis.

The material presented here consists of my doctoral dissertation at Stanford University, with a few additions and modifications.

Chapter 1

Introduction

Machine models of computation are characterized by a collection of possible configurations, together with a transition function that maps each configuration to another configuration. The computation of a machine starts with an initial configuration, traverses a sequence of configurations by iteratively applying the transition function, and terminates if it reaches a *stable configuration*, one that is left unchanged by the transition function. In order to consider different computations of the same machine, each configuration is subdivided into two separate parts, the input part and the internal part, and all initial configurations agree in the internal part, so that the initial configuration is determined by the input part. Configurations, in particular stable ones, may also have an output part.

The notion of *uniformity* distinguishes two kinds of model: a model is uniform or nonuniform depending on whether the machine operates on inputs of possibly different sizes or of a single size. The Turing machine is a uniform machine model, where the input part is given by the contents of an input tape, while the internal part is given by the state of a finite control, the contents of an internal tape, and the position of heads that are used to access the two tapes. The question of whether computation on a given input terminates, known as the halting problem, is undecidable for Turing machines. When the resources available to the machine are limited, the problem becomes decidable, and this leads to the definition of computational complexity classes. The class $\mathcal{PSPACE}$ contains the decision problems that can be solved by a Turing machine with space limited to a polynomial in the size of the input. The class $\mathcal{P}$ contains the decision problems that can be solved with time limited to a polynomial in the size of the input. The class $\mathcal{NP}$ contains the decision problems that can be solved in polynomial time by a nondeterministic Turing machine. The class $\#\mathcal{P}$ is the class of counting problems that can be described as the problem of counting accepting computations on a given input for a nondeterministic Turing machine running in polynomial time.

Networks constitute a nonuniform model. In this model, the configurations are given by words of fixed length over a finite alphabet; the transition function is described by a

---

Received by the editor March 15, 1992.

computational circuit. The circuit takes as inputs the input and internal parts of a configuration, and produces as outputs the internal and output parts of the next configuration; the input part is left unchanged from each configuration to the next. The fact that the internal outputs obtained at each iteration are used as internal inputs for the next iteration can be represented by feedback arcs from each internal output to the corresponding internal input. Such a circuit with feedback arcs is called a *network*.

Four basic questions that can be posed on networks are: convergence, evalutation, stability, and counting. These four questions are in close correspondence with the four complexity classes defined above for a Turing machine, and in particular, they define complete problems for the four corresponding classes. The *evaluation question* is the problem of evaluating a computational circuit. The *convergence question* asks whether some configuration eventually reached by iterating the transition function of a network, starting with a given initial configuration, is a stable configuration. The *stability question* asks whether given the input part of a configuration, the network has a stable configuration consistent with this input part. The *counting problem* asks for the number of stable configurations consistent with a given input part. There are two other problems that can also be studied on networks, namely the enumeration and optimization problems. The *enumeration problem* asks to list all stable configurations. The *optimization problem* asks for a stable configuration that maximizes or minimizes some given objective function.

Circuits and networks are built with gates. The study of the complexity of various questions on networks in terms of the kinds of gates used to build the networks was initiated by Mayr and Subramanian [78, 109]. They showed that in the case of boolean gates, there are essentially three kinds of restrictions on the types of gates that can sometimes reduce the complexity of network problems. The first two, monotonicity and linearity, limit the individual output functions allowed at each gate. It is not hard in general to determine how the complexity of the problems under consideration is affected by these restrictions.

At this stage, the fanout of gates is unrestricted, and the ability to produce multiple copies of the output values produced by a gate and feed these copies as inputs to different gates can be represented by a COPY gate. If in addition, AND and NOT gates are available, then every boolean function can be computed. If only AND and OR gates are available, then precisely the *monotone* functions can be computed. If only XOR gates are available, then the *linear* functions can be computed. The complexity of the six problems on networks defined above, as a function of the gates available, but assuming that fanout in the form of a COPY gate is available, is summarized in Table 1.1. (The * indicates that for monotone networks, the answer to the stability question is always 'yes', and the associated search problem has complexity comparable to that of the evaluation question for the same gates.)

The third possible kind of restriction is meaningful only when gates that can produce several possibly different outputs are considered: it does not restrict individual outputs of gates, but rather how the various outputs are related. The imposed restriction requires that a small perturbation on the inputs (affecting only a few inputs to the gate) produce only

Basis	Evaluation	Stability	Convergence
{AND, NOT, COPY}	$\mathcal{P}$-complete	$\mathcal{NP}$-complete	$\mathcal{PSPACE}$-complete
{AND, OR, COPY}	$\mathcal{P}$-complete	*	$\mathcal{PSPACE}$-complete
{AND, COPY}; {OR, COPY}	$\mathcal{NL}$-complete	*	in $\mathcal{P}$
{XOR, COPY}	in $\mathcal{NC}^2$	in $\mathcal{NC}^2$	in $\mathcal{P}$
{NOT, COPY}	$\mathcal{L}$-complete	$\mathcal{L}$-complete	in $\mathcal{P}$
{COPY}	$\mathcal{L}$-complete	*	in $\mathcal{P}$

Basis	Enumeration	Counting	Optimization
{AND, NOT, COPY}	first sol. $\mathcal{NP}$-hard	$\#\mathcal{P}$-complete	$\mathcal{NP}$-hard
{AND, OR, COPY}	third sol. $\mathcal{NP}$-hard	$\#\mathcal{P}$-complete	$\mathcal{NP}$-hard
{AND, COPY}; {OR, COPY}	polynomial per sol.	$\#\mathcal{P}$-complete	$\mathcal{NP}$-hard
{XOR, COPY}	polynomial per sol.	in $\mathcal{P}$	$\mathcal{NP}$-hard
{NOT, COPY}	polynomial per sol.	in $\mathcal{P}$	in $\mathcal{P}$
{COPY}	polynomial per sol.	in $\mathcal{P}$	in $\mathcal{P}$

Table 1.1: Complexity of six problems on bases with fanout.

a small perturbation on the outputs (affecting only a few outputs), and defines the class of *nonexpansive* gates (termed 'adjacency-preserving' in [78, 109]). More precisely, in the boolean case, changing the value of one of the input bits of the gate can only change the value of one of the output bits. The condition can be viewed as a limitation on the fanout of gates, since it forbids gates that make multiple copies of the same input; such a gate would multiply the effect of a perturbation on the inputs. Circuits of nonexpansive gates provide an intermediate model between general circuits and formulas, since in the case of formulas all gates have fanout 1 and are hence nonexpansive.

The work of Mayr and Subramanian gave special emphasis to an important family of nonexpansive gates, the *scatter-free* gates: an example of a scatter-free gate is the *comparator*, which outputs the 'min' and the 'max' of two given inputs. Here we continue this study and extend it to the general nonexpansive case. The configurations of a network are first restricted to words over a boolean alphabet; they can then be viewed as vertices in a boolean hypercube, and the nonexpansiveness condition can then be stated in terms of the associated Hamming metric. More precisely, if two configurations are at Hamming distance d, then the two configurations obtained by applying the transition function of the network are at distance at most d.

The effect of nonexpansiveness on the complexity of the computational questions considered is not as clear as for the first two types of restriction. To begin the study of nonexpansive gates, we show that the study cannot be reduced to a finite number of possible gates by means of simple simulations, as was the case when the COPY gate was available, in which case it was only necessary to consider seven different sets of gates. We also show that when the input part is not fixed ahead of time, the stability question can become $\mathcal{NP}$-complete for some nonexpansive gates; here the lack of fanout at the gates seems to be compensated by taking advantage of variable inputs. The proof also shows that it is $\mathcal{NP}$-hard to distinguish sorters from comparator circuits that on certain inputs produce outputs far from sorted.

The first basic result derived from the metric characterization of nonexpansiveness is the fact that given an assignment to the inputs of a network, the resulting assignment to the outputs of the network is independent of the choice of a stable configuration, or even a periodic configuration (a periodic point of the transition function). The network is then called *convergent*, and it is said to converge to the nonexpansive gate that maps input assignments to output assignments (under an arbitrary choice of a periodic configuration). This leads to a polynomial time algorithm for the stability question on networks of nonexpansive gates, extending the known scatter-free results [78, 109]. In fact, we show that the stability question has the same parallel complexity as the evaluation question, for networks of nonexpansive gates, under $\mathcal{NC}^1$ reductions. For the evaluation question, on scatter-free circuits, we give an $\Omega(m^{1/3})$ lower bound on the parallel time complexity, up to logarithmic factors, to be compared with the $O(m^{1/2})$ upper bound of [78], when the number of processors is polynomial.

An important property of the hypercube that we use next is that any three points have a unique *median*, a point that simultaneously lies on shortest paths between each pair from the three points (i.e., the hypercube is a *median graph*). The bits of the median are obtained by taking bitwise majority of the corresponding bits of the three given points. A basic result here, derived from the metric properties, is that the median of three stable configurations is a stable configuration, in a network of nonexpansive gates. It follows that the set of stable configurations can be characterized as the set of solutions to a 2SAT instance. The stability question for networks of nonexpansive gates as a search problem is then also reduced to the evaluation quetion, under $\mathcal{NC}^1$ reductions; we also show from this characterization that the counting problem for networks of comparators is $\#\mathcal{P}$-complete, and that the evaluation question is hard for nondeterministic logspace.

When a nonexpansive mapping on the hypercube is restricted to the periodic points, it is an isomorphism on the subgraph induced by the periodic points. We show that this isomorphism can be extended to an isomorphism on the entire hypercube; such an isomorphim consists simply of permutation and negation of bits. This leads to an efficient algorithm for finding the 2SAT instance that characterizes the stable configurations in a network of nonexpansive gates.

A theorem of Bandelt and Vel [11] states that every nonexpansive mapping on a median graph, in particular on the hypercube, has a *fixed cube*, i.e., a subcube such that the mapping restricted to it is an isomorphism. We give a new proof of this result. The theorem leads to finding a small certificate for networks of nonexpansive gates that do not have a stable configuration. It also leads to showing decidability for whether a given nonexpansive mapping can be defined as the mapping to which a network using a given set of nonexpansive gates converges. We also give here a linear programming formulation for the evaluation of the gate to which a nonexpansive network converges. From this formulation we obtain, with Nimrod Megiddo and Serge Plotkin, a parallel algorithm for network stability on networks of nonexpansive gates of constant size that runs in time $O(\sqrt{m})$ up to logarithmic factors, with a polynomial number of processors.

The 2SAT characterization of stable configurations and the characterization of the isomorphism on periodic points are then used to show that a nonexpansive mapping need only be iterated a polynomial number of times before reaching a periodic point, thus reducing the convergence question for networks of nonexpansive gates to the evaluation question. Therefore the evaluation, stability, and convergence questions, which have significantly different complexities in the general case, have essentially the same complexity in the nonexpansive case, while the counting problem remains #$\mathcal{P}$-complete as in the general case.

There are thus essentially three polynomially solvable cases of the stability question for networks of boolean gates, namely monotone, linear, and nonexpansive networks. It is worth noting here the similarity with Schaefer's classification of polynomially solvable cases of boolean satisfiability [104], namely Horn clauses, linear equations modulo 2, and 2SAT, in close correspondence with the above three cases.

The 2SAT characterization of stable configurations leads to efficient algorithms for the optimization and enumeration problems on networks of nonexpansive gates. We introduce a notion of *width*, which arises in applications of nonexpansive network stability, and give an efficient algorithm for the optimization problem via a flow algorithm that runs in $O(wm \log K)$ time for flow problems with m arcs, of width w and optimum flow K. In some applications, the optimum flow K is not too large, and the running time of an alternative algorithm is bounded by $O(m\sqrt{K})$. This solves the optimization problem in polynomial time for monotone nonexpansive networks because the associated 2SAT instance is then bipartite, while in the nonmonotone case the 2SAT instance is non-bipartite, obtaining an exact optimum is $\mathcal{NP}$-complete, and a factor-of-two approximation can be obtained in $O(m\log(w^2/m+2))$ time. We also obtain an efficient algorithm for enumerating solutions to a 2SAT instance, where the time spent per solution is proportional to the maximum degree when the instance is viewed as a graph (the degree is bounded from above by the width).

Stable matching problems have a natural formulation in the boolean nonexpansive network model. This formulation, discovered by Subramanian [108], establishes a correspondence between stable matching problems and the stability question for networks of certain scatter-free gates, and provides a new approach to the study of stable matching. The stable marriage problem is thus essentially the same as the stability question for networks of comparators, while the stable roommates problem is essentially the stability question for networks of X gates (a combination of comparators and negation). For instance, in terms of parallel complexity, stable marriage, comparator circuit evaluation, and lex-first maximal matching, all reduce to each other [78, 109]. Using this approach, we give new results on stable matching, as well as new and often simpler proofs of old results. We study stable matching in the context of a new more general problem, the stable arrangement problem. One of the structural results that follows from the structural properties of nonexpansive networks is that the median of three stable arrangements (obtained by giving the intermediate choice, from the three in the given arrangements, to each person) is also a stable arrangement. We prove that several different versions of stable matching, both as a decision and as a search problem, are also equivalent to comparator circuit evaluation under $\mathcal{NC}^1$ reductions. The results on optimization and enumeration for 2SAT instances apply to stable matching, with w and m corresponding to the number of people and the total length of the preference lists respectively. Thus, for instance, the egalitarian stable roommates problem is $\mathcal{NP}$-complete but can be efficiently approximated within a factor of two. We give a parallel algorithm for stable matching running in essentially $O(\sqrt{m})$ time via linear programming.

The study of nonexpansiveness extends to the non-boolean case, where the hypercube is now replaced by the cartesian product of graphs, or more generally of finite metric spaces (instead of just two points 0 and 1 at distance 1). Distances in a cartesian product are computed by adding distances from each of the factors. The basis of the study is several unique representation theorems for finite metric spaces as subspaces of a cartesian product of metric spaces, where the distance function on the space is inherited by the subspace. The uniqueness of the representations implies that an isomorphism on the given subspace can be extended to an isomorphism on the cartesian product, a property that was used earlier in the boolean case, when examinining the periodic points of a nonexpansive mapping in the hypercube. Our proof of Bandelt and Vel's fixed cube theorem extends to the case of cartesian products of metric spaces.

We consider theorems on four different unique representations, the *isometric*, *2-isometric*, *retract* representations, and the *prime factorization*. Here the existence of a unique isometric representation was established by Graham and Winkler [46, 116], the unique prime factorization was shown by Sabidussi [103], and the other two unique representation theorems are new. In the case where the metric spaces being represented are graphs, the running time is $O(nm)$ for n vertices and m edges, except for the retract representation, where finding the unique representation is $\mathcal{NP}$-hard. The starting point of gaining efficient in the algorithms is the replacement of a relation θ used by Graham and Winkler in the isometric embedding representation with a smaller relation θ_1. While we are especially interested in the retract representation (the space of periodic points of a nonexpansive mapping is a retract), we consider also the weaker 2-isometric representation because of its computational tractability, and because it provides an appropriate generalization to medians from the boolean case, which coincides in the case of products of cliques with the notion of *imprint* considered by Chung, Graham, and Saks [20, 21] in their study of dynamic search in graphs. The notion of windex defined in their study can be efficiently computed using the 2-isometric representation defined here.

1.1 Organization

Chapter 2 gives the classification of the six basic question for bases with fanout, and begins the study of nonexpansiveness by showing that nonexpansive gates cannot be built with a finite basis of gates, and studying networks whose input values are not fixed ahead of time.

Chapter 3 studies nonexpansive networks under a fixed input assignment. Such networks exhibit a rich structure that can be used to obtain efficient algorithms. The study is based on the distance properties of nonexpansive mappings in the hypercube, leading to the notion of a convergent network; the median structure of stable configurations, giving a characterization of the solutions to network stability in terms of a 2-satisfiability instance; and the properties of certain fixed cubes of Bandelt and Vel [11] and of certain isomorphisms for nonexpansive mappings.

Chapter 4 exploits the characterization of the set of stable configurations to give algorithms for the optimization problem and the enumeration problem, for nonexpansive networks. The enumeration algorithm is efficient when the characterization of the solution space is a 2-satisfiability instance of small degree. The basic approach for the optimization problem is a formulation in terms of a maximum flow problem, and several flow algorithms are introduced, some of them using a new notion of width to achieve greater efficiency in some special cases that arise in applications of network stability.

Chapter 5 studies stable matching as the main application of nonexpansive network stability. The study of Chapters 3 and 4 is used to obtain structural and algorithmic results in stable matching. These include the median structure of stable matchings and a link between convergent networks and the linear program associated with stable matching; efficient algorithms for the stable marriage optimization problem and for the approximate stable roommates optimization problem, whose exact version is shown to be $\mathcal{NP}$-hard; and algorithms for enumerating complete solutions and recognizing partial solutions.

Chapter 6 studies nonexpansive networks in a more general set-up, where each value can now be any point from a given finite metric space. The underlying graph structure of these *metric networks* is that of cartesian products, of which the boolean hypercube is the simplest example. We give an efficient implementation of the isometric representation theorem of Graham and Winkler [46, 116], and show that the theorem still holds when the notion of 'isometric' is replaced by 'retract' or by a new intermediate notion of '2-isometric'. These representation results are used to study structural and algorithmic aspects of dynamic search [20, 21] and metric networks.

Many of the chapters can be read independently of one another to a large extent. Chapter 3 requires mainly the definition of the model from section 2.1. In Chapter 4, the algorithms on 2-satisfiability use some definitions from section 3.2.2. Chapter 5 is based on results from Chapters 3 and 4. The study of products in Chapter 6 is presented independently of earlier chapters; the application to network stability omits the proofs that carry over directly from Chapter 3.

Chapter 2

Preliminaries

In section 2.1, we introduce the network model and some of the basic operations on networks. Section 2.2 defines three basic questions on networks, namely evaluation, stability, and convergence, as well as three auxiliary questions on the space of stable configurations of a network, namely optimization, enumeration, and counting. Section 2.3 summarizes the classification of the computational complexity of these six questions in terms of the kinds of boolean gates used in the networks; at this point, we assume that the ability to make multiple copies of values is available in circuits and networks. The remaining case for boolean gates is that of nonexpansive gates. We begin the study of nonexpansive gates in section 2.4, where we show that the study cannot be reduced to a finite number of gates by means of simple simulations, as was done when copying was available. We also show that the stability question for nonexpansive gates can become computationally hard if the inputs to the network are not given in advance; here the lack of fanout at the gates seems to be compensated by taking advantage of variable inputs. The proof also shows that it is hard to distinguish sorters from comparator circuits that on certain inputs produce outputs far from sorted. The model of complexity that is used in Chapter 3 for algorithms on networks is defined in section 2.5.

2.1 Gates, Circuits, and Networks

A *(boolean) assignment* on a set S is a mapping $x : S \to \{0,1\}$ from the set $S = S(x)$ to the set of boolean values 0 and 1. An element $i \in S(x)$ is a *coordinate* of x, and the image $x(i)$ is its *value*. Given a set of coordinates $T \subseteq S(x)$, we denote by x_T the restriction of x to the set T. If $T = \{i\}$, where i is a coordinate in $S(x)$, then x_T is denoted by x_i. The empty assignment $x_\emptyset$ is denoted by ϵ. If x and y are assignments with $S(x) \cap S(y) = \emptyset$, then xy denotes the union of the two assignments, with $S(xy) = S(x) \cup S(y)$. In particular, if $S(x) = \{1, 2, \ldots, n\}$, then $x = x_1 x_2 \ldots x_n$. With a slight abuse of notation, we shall identify each x_i with its value $x(i)$. For example, the statement $x = x_1 x_2 x_3 = 011$ indicates that $S(x) = \{1,2,3\}$ and $(x(1), x(2), x(3)) = (0, 1, 1)$. Two assignments x and y are *consistent* if $x_i = y_i$ for all $i \in S(x) \cap S(y)$.

A *gate* is a mapping $g : (I \to \{0,1\}) \longrightarrow (O \to \{0,1\})$ from assignments on the input set $I = I(g)$ to assignments on the output set $O = O(g)$. The coordinates in $I(g)$ and $O(g)$ are called *inputs* and *outputs* of g respectively. The gate g is a k-input, l-output gate if $|I(g)| = k$ and $|O(g)| = l$. Given a gate g, an assignment x with $S(x) \subseteq I(g)$ and a coordinate set $T \subseteq O(g)$, the *restriction* $g_{x,T}$ of the gate g is the gate g' obtained from g by discarding the outputs not in T and discarding the inputs in $S(x)$ after assigning to them the values given by x. More formally, the gate g' has inputs $I(g') = I(g) \backslash S(x)$, outputs $O(g') = O(g) \cap T$, and satisfies $g'(y) = g(xy)_T$.

A *basis* is a set of gates. We shall often need to use gates from a given basis but with different names given to their inputs and outputs. A *relabelling map* on a set of coordinates S is a one-to-one correspondence $\rho : S \to S'$ between coordinate sets. Given $T \subseteq S = S(\rho)$, we write $\rho(T) = \{\rho(i) : i \in T\}$. If x is an assignment on T, then $\rho(x)$ denotes the assignment y on $\rho(T)$ defined by $y_{\rho(i)} = x_i$. A *relabelling* is a pair $\rho = (\rho^I, \rho^O)$ of relabelling maps, called the input and output relabelling maps respectively. Given a gate g and a relabelling ρ with $I(g) \subseteq S(\rho^I)$ and $O(g) \subseteq S(\rho^O)$, we let $g' = \rho(g)$ denote the gate with $I(g') = \rho^I(I(g))$ and $O(g') = \rho^O(O(g))$ defined by $g'(\rho^I(x)) = \rho^O(g(x))$. Given gates g, g', we say that g' is a g gate if $g' = \rho(g)$ for some relabelling ρ; if Ω is a basis, then we say that g' is an Ω gate if g' is a g gate for some $g \in \Omega$.

We define a few useful gates g with $I(g) = \{1, 2, \ldots, k\}$ and $O(g) = \{1, 2, \ldots, l\}$. Four particularly simple gates are the 1-input, 1-output *identity* gate defined by $\mathrm{ID}(x_1) = x_1$; the 0-input, 1-output *constant* gates K_0 and K_1 defined by $\mathrm{K}_0(\epsilon) = 0$ and $\mathrm{K}_1(\epsilon) = 1$; and the 1-input, 0-output *absorption* gate A defined by $\mathrm{A}(x_1) = \epsilon$. We shall implicitly assume that all bases contain at least these four gates. A basis may, in addition, contain other gates; here are a few examples. The 1-input, 2-output COPY gate is defined by $\mathrm{COPY}(x_1) = x_1 x_1$ (i.e., $\mathrm{COPY}(x_1) = y_1 y_2$ with $(y(1), y(2)) = (x(1), x(1))$). The 1-input, 1-output NOT gate computes $\mathrm{NOT}(x_1) = \overline{x_1}$. The 2-input, 1-output AND gate computes $\mathrm{AND}(x_1 x_2) = x_1 \wedge x_2$. The 2-input, 1-output OR gate computes $\mathrm{OR}(x_1 x_2) = x_1 \vee x_2$. The 2-input, 1-output XOR gate computes $\mathrm{XOR}(x_1 x_2) = x_1 \oplus x_2$. The operations $\wedge$, $\vee$ and $\oplus$ are commutative and associative, and can therefore be applied to sets of boolean values. When these operations are applied to a set of values indexed by S, we denote them by $\bigwedge_{i \in S} b_i$, $\bigvee_{i \in S} b_i$ and $\bigoplus_{i \in S} b_i$ respectively.

A *network* is a set of gates that do not share any inputs and do not share any outputs. This means that if N is a network and g, g' are distinct gates in N, then $I(g) \cap I(g') = \emptyset$ and $O(g) \cap O(g') = \emptyset$. On the other hand, given two (not necessarily distinct) gates g, g' in N, it may happen that an output of g is also an input of g'. If $i \in O(g) \cap I(g')$, then we say that output i of gate g and input i of gate g' are *linked*. By the disjointness property, every input is linked to at most one output, and every output is linked to at most one input. These links give a topology to the network that can be described by a directed multigraph (a directed graph with loops and parallel edges allowed) on the gates of the network, with a directed edge from g to g' for every output of g linked to an input of g'. If the underlying directed multigraph of a network is acyclic, the network is called a *circuit*. A network (circuit) is

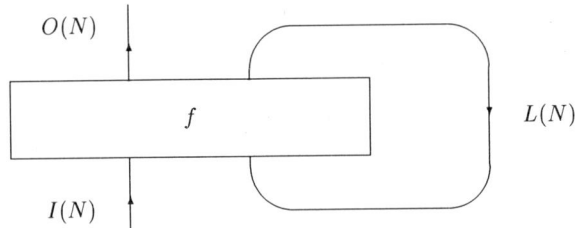

Figure 2.1: The network N with transition function f.

said to be *over* a basis Ω if every gate in it is an Ω gate. Fig. 3.3 illustrates a network N_k over $\Omega = \{C, X\}$; Fig. 5.1 gives a network over $\Omega = \{X_2, X_3\}$.

The *transition function* of a network N is a single gate f that describes all the gates in N. The gate f has $I(f) = \bigcup_{g \in N} I(g)$ and $O(f) = \bigcup_{g \in N} O(g)$, and satisfies $y = f(x)$ if and only if $y_{O(g)} = g(x_{I(g)})$ for all gates $g \in N$. Note that if f is the transition function of N, then the networks N and $N' = \{f\}$ have the same transition function; we shall see that, for many purposes, they can actually be treated as the same network. Given a network N with transition function f, the set $R(N) = I(f) \cup O(f)$ is the set of *coordinates* of the network N, and consists of three parts: the set of *links* $L(N) = I(f) \cap O(f)$, the set of *inputs* $I(N) = I(f) \setminus O(f)$, and the set of *outputs* $O(N) = O(f) \setminus I(f)$ of the network. A convenient pictorial representation for a network N with transition function f together with the three associated coordinate sets is given in Fig. 2.1.

A *configuration* of a network N is an assignment t on the coordinate set $R(N)$, and consists of an *input assignment* $t_{I(N)}$, an *output assignment* $t_{O(N)}$, and an *internal assignment* $t_{L(N)}$. A network N can be used to define an associated mapping on the configurations of N. Given two configurations x and y of a network N with transition function f, we write $y = N(x)$ if $y_{I(N)} = x_{I(N)}$ and $y_{O(N) \cup L(N)} = f(x_{I(N) \cup L(N)})$. In other words, all gates are evaluated using the values assigned to their inputs by x as inputs, thus obtaining at their outputs the values for the configuration y; the inputs to the network are not outputs of any gate, and thus keep their value from x. A configuration x is *stable* if $N(x) = x$. A configuration x is therefore stable if it satisfies $f(x_{I(f)}) = x_{O(f)}$ for the transition function f, or equivalently $g(x_{I(g)}) = x_{O(g)}$ for each gate $g \in N$, i.e., if it satisfies all the gate equations.

In a circuit, given an input assignment, these gate equations determine unique internal and output assignments such that the resulting configuration is a stable configuration consistent with the given input assignment. The circuit is then said to *compute* the corresponding mapping from input assignments to output assignments. A basis Ω can *simulate* a gate g if there is a circuit over Ω that computes g; it can simulate a basis Ω' if it can simulate every gate in Ω'. Note that the basis $\Omega = \{g\}$ can simulate all the restrictions of g by means of the constant and absorption gates that are always implicitly present in Ω.

The *kth iterate* of a mapping τ on a set U is the mapping $\tau^{(k)}$ defined inductively by letting $\tau^{(0)}(z) = z$ and $\tau^{(k+1)}(z) = \tau(\tau^{(k)}(z))$ for all $z \in U$. A *periodic point* of τ is a z such that $\tau^{(p)}(z) = z$ for some $p \geq 1$. The least such p is the *period* of z. A *fixed point* of τ is a periodic point of period 1. We are particularly interested in the iterates and periodic points of the mapping associated with a network N. It will sometimes be useful to look at the iterates $N^{(k)}$ in terms of the transition function f of the network. For this purpose, we define two restrictions of f given an input assignment for the network. Given an assignment x on $I(N)$, the *output function* of the network is the mapping $g_x = f_{x,O(N)}$, and the *internal function* of the network is the mapping $h_x = f_{x,L(N)}$, so that if z is an assignment on $L(N)$, then $f(xz) = g_x(z)h_x(z)$. If y is an assignment on $O(N)$, then $N(xyz) = xg_x(z)h_x(z)$, and $N^{(k+1)}(xyz) = x\, g_x(h_x^{(k)}(z))\, h_x^{(k+1)}(z)$ for all $k \geq 0$. The periodic points of the mapping associated with N are called *periodic configurations*; the fixed points are precisely the stable configurations. The periodic configurations xyz consistent with an input assignment x are determined by the choice of a periodic point z of the internal function h_x. For if z has period p and $z' = h_x^{(p-1)}(z)$, then the periodic configuration must have $z = h_x(z')$ and $y = g_x(z')$. The periodic configurations are therefore the configurations $xg_x(z')h_x(z')$ with z' a periodic point of h_x.

2.1.1 Modifying a Network

We shall often need to break or create links in a network. This is achieved by applying a relabelling to the gates of the network. If N is a network and $\rho = (\rho^I, \rho^O)$ is a relabelling, we define a network $\rho(N) = \{\rho(g) : g \in N\}$. If $i \in L(N)$ and $\rho^I(i) \neq \rho^O(i)$, then ρ has the effect of breaking the link i, because if i was an input to g and an output to g' in N, then $\rho^I(i)$ will be an input to $\rho(g)$ while $\rho^O(i)$ will be an output to $\rho(g')$ in $\rho(N)$, hence the common input and output for the two gates have become an input and an output that are different. A special kind of relabelling will be particularly useful. Given a set of coordinates S and two tags α, β, we define a relabelling map ρ by $\rho(i) = (i, \alpha)$ if $i \in S$ and $\rho(i) = (i, \beta)$ otherwise, and denote this map on coordinates by $\rho(i) = (i, S, \alpha, \beta)$. A set of coordinates T can be tagged by letting $(T, \alpha) = \{(i, \alpha) : i \in T\}$. If x is an assignment on T, then the assignment $y = (x, \alpha)$ on (T, α) is defined by $y_{(i,\alpha)} = x_i$.

Given a network N, a set of coordinates $S \subseteq L(N)$, and three tags α, β, γ, we define a relabelling $\rho = (\rho^I, \rho^O)$ such that $\rho^I(i) = (i, S \cup I(N), \alpha, \gamma)$ and $\rho^O(i) = (i, S \cup O(N), \beta, \gamma)$, and denote this relabelling on N by $\rho(N) = (N, S, \alpha, \beta, \gamma)$. If $\alpha \neq \beta$, then this relabelling has the effect of breaking the links in S, adding tag α to the inputs (including the inputs

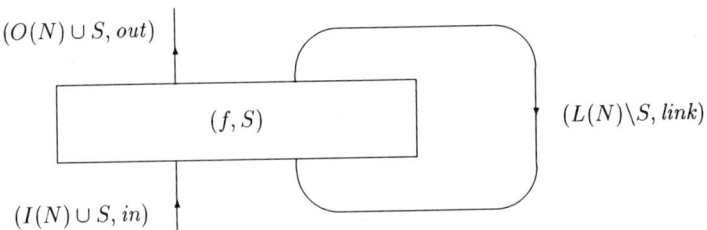

Figure 2.2: The network (N, S) with transition function (f, S).

created by breaking the links in S), tag β to the outputs (including the outputs created by breaking the links in S), and tag γ to the links (only for those links that are not in S and where therefore not broken). In many cases the choice of tags will not be important. We introduce three special tags in, out, and $link$, and define $(N, S) = (N, S, in, out, link)$. When S consists of all the links in N, we use the notation $(N, \alpha, \beta) = (N, L(N), \alpha, \beta, link)$. (Note that the tag $link$ does not get used in this case.) If f is the transition function of N, then $(f, S, \alpha, \beta, \gamma)$, (f, S), and (f, α, β) denote the transition functions of $(N, S, \alpha, \beta, \gamma)$, (N, S), and (N, α, β) respectively. The network (N, S) is illustrated in Fig. 2.2; note how S becomes part of the sets of input and output coordinates, since the links in $S \subseteq L(N)$ are broken, and then the relabelling (tagging with in or with out) ensures that the coordinates from S have different names as inputs and as outputs to the network.

The relabelling $(N, S, \alpha, \beta, \gamma)$ can also be used to define circuits and unravel networks into circuits. Given k networks $N_1, \ldots, N_k$ with the same set of links $L = L(N_i)$, we define the *composition* of the N_i as the circuit $(N_1, \ldots, N_k) = \bigcup_{1 \leq i \leq k}(N_i, i-1, i)$. This circuits breaks the links of each N_i and then uses those links to connect consecutive levels N_i and N_{i+1}; the reason here is that link $j \in L$ gets tagged as (j, i) as an output to a gate in the relabelled N_i, and gets also tagged as (j, i) as an input to a gate in the relabelled N_{i+1} (namely $(N_i, i-1, i)$ and $(N_{i+1}, i, i+1)$). Note that if N_i has transition function f_i, then $(N_i, i-1, i)$ has transition function $f' = (f_i, i-1, i)$, where all inputs of f' are tagged by $i-1$ and all outputs of f' are tagged by i. When the networks N_i are all the same network N, we introduce the notation $(N, k) = (N, N, \ldots, N)$ (k copies), and denote by (f, k) the transition function of this circuit, where f is the transition function of N. The circuit (N, k) is illustrated in Fig. 2.3.

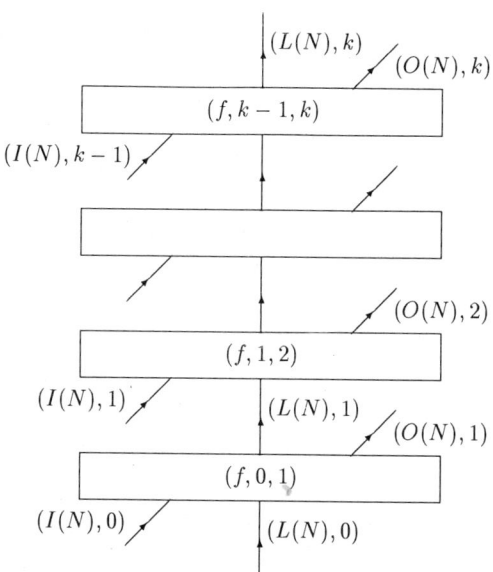

Figure 2.3: The network (N, k) with transition function (f, k).

Given a configuration xyz on N, where x, y, z are input, output, and internal assignments on N respectively, consider the tagged input assignment $(z, 0)(x, 0)(x, 1)\ldots(x, k-1)$ on (N, k) (e.g., $(x, 2)$ provides values for the inputs in $(I(N), 2)$, see Fig. 2.3), and let t be the unique stable configuration of the circuit (N, k) consistent with this input assignment. Then the levels within t give $t_{(L(N),i)} = (N^{(i)}(xyz)_{L(N)}, i)$ and $t_{(O(N),i)} = (N^{(i)}(xyz)_{O(N)}, i)$, so the circuit (N, k) obtained by unravelling N can be used to obtain iterates of N.

We have assumed so far that the transition function of a network N consists of a single level of gates. Assume now instead that we have a network $N = \{f\}$, where the transition function f is described by a circuit K over Ω. In order to distinguish inputs and outputs of f, assume that K computes the function $(f, L(N))$ obtained by breaking all the links of N. We would now like to view N directly as a network over Ω, rather than a network over $\{f\}$. For this purpose, we replace f by K in N. This is achieved by establishing links between inputs and outputs of K as in N, thus obtaining a network N'. More precisely, we can assume that K is of the form $K = (N', L(N))$ for some network N'. See Fig. 2.4.

A configuration x of N can be mapped to a configuration x' of N' as follows. Note that $I(N') = I(N)$, $O(N') = O(N)$, $L(N') \supseteq L(N)$, and $(L(N')\backslash L(N), link) = L(K)$. Therefore $S(x') = R(N) \cup (L(N')\backslash L(N))$. We let $x'_{R(N)} = x$. For the remaining coordinates

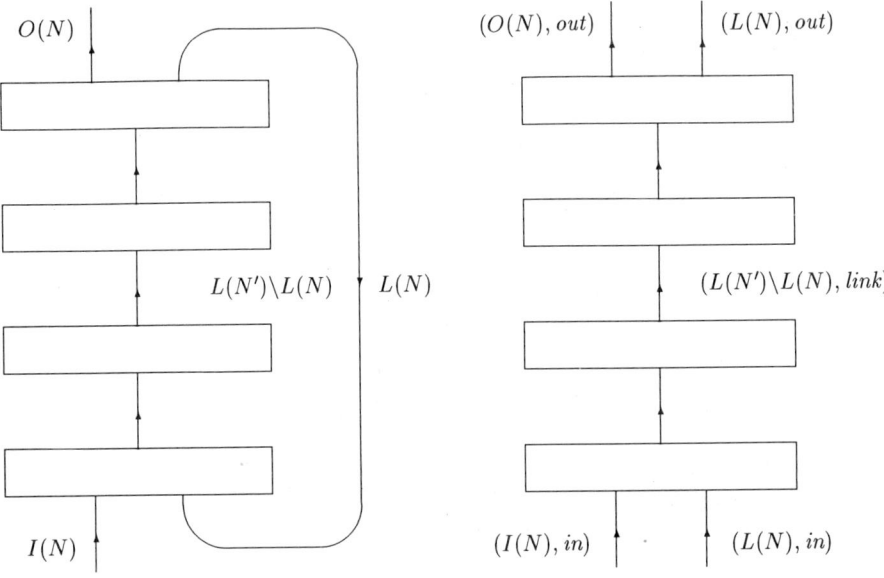

Figure 2.4: The network N' and the corresponding circuit $K = (N', L(N))$.

$L(N')\backslash L(N)$, we use $(x_{I(N) \cup L(N)}, in)$ as an input assignment to K, find the unique stable configuration t of K consistent with this input assignment, and set $(x'_{L(N')\backslash L(N)}, link) = t_{L(K)}$. Suppose that K consists of d levels $1, 2, \ldots, d$ of gates, where the outputs of gates at level $i-1$ are inputs of gates at level i, for $1 < i \le d$. It can then be shown, by a pipelining argument, that if $N(x) = y$, then $N'^{(d)}(x') = y'$, so that one iteration of N corresponds to d iterations of N'; furthermore, the stable configurations x of N correspond to the stable configurations x' of N'. The fact that N' consists of d contiguous levels of gates is important only for the simulation of the transition function of N with the transition function of N'; the correspondence of stable configurations holds even if N' is not of this form. In any case, the format consisting of contiguous levels can always be achieved by stretching certain links of N' and inserting an ID gate in them. Formally, we say that N'' is obtained from N' by *stretching* coordinate i if $N'' = (N', \{i\} \cap L(N')) \cup \{\rho(\text{ID})\}$, where $\rho^I(1) = (i, out)$ and $\rho^O(1) = (i, in)$. By stretching enough links, we can ensure that N' has the appropriate level structure, completing the transformation.

2.2 Evaluation, Convergence, and Stability

We shall study three basic questions for circuits and networks. If K is a circuit and x is an input assignment to K, then the *evaluation question* for K and x is the problem of finding the output of the function computed by K on input x. The decision problem associated with this search problem asks for the value of some particular boolean coordinate of the output.

If N is a network and t is a configuration of N, then the *convergence question* for N and t is the problem of finding a periodic configuration of the form $N^{(k)}(t)$ for some $k \geq 0$; such a periodic configuration always exists since the set of possible configurations is finite. The decision problem associated with this search problem asks whether this periodic configuration is a stable configuration. If N is a circuit, then the answer to this decision problem is always 'yes', and the stable configuration obtained provides the answer to the evaluation question.

If N is a network and x is an input assignment to N, then the *stability question* for N and x is the problem of finding a stable configuration of N consistent with the input assignment x, if one exists. The decision problem associated with this search problem asks whether a stable configuration exists. Here again, if N is a circuit, then the answer to this decision problem is always 'yes', and the stable configuration obtained gives the answer to the evaluation question.

There are several other problems associated with the stability question. The *enumeration problem* is the task of listing all the stable configurations consistent with a given input assignment. The *counting problem* is the task of counting all such stable configurations. The *optimization problem* is the task of selecting the best such stable configuration under some optimization criterion. For instance, we can associate a nonnegative weight with each condition of the form '$t_i = 0$' or '$t_i = 1$', where i is a coordinate, and ask for a stable configuration t consistent with the input assignment x that minimizes the sum of the weights of the conditions that hold for t.

In the following sections, we shall examine the complexity of some of these questions in terms of the types of gates used to build the networks.

2.3 Classification of Bases

We are interested in the complexity of the evaluation, convergence, and stability questions for circuits and networks over a given basis Ω. If a basis Ω can simulate a basis Ω', then the evaluation and stability questions for circuits and networks over Ω' can be viewed as questions for circuits and networks over Ω, simply by substituting in the appropriate simulating circuits. A similar observation holds for the convergence question, provided that the simulating circuits have an appropriate level structure (see section 2.1.1) and the number of levels is the same for all the simulating circuits. Therefore, in the study of the complexity of the various questions, one can select one representative from each equivalence class of bases that can simulate each other.

The limitation in the model requiring each output to be linked to at most one input is an uncommon one, and was introduced in order to study the role of 'fanout' in circuits and networks [78, 109]. This limitation can be circumvented by only considering bases Ω that contain the COPY gate, and we shall do so throughout this section. Section 2.4 will begin the study of bases that do not contain, and cannot simulate, the COPY gate.

Mayr and Subramanian showed that there are precisely seven bases containing the COPY gate, up to simulations. A basis is *universal* if it can simulate all bases. One such universal basis is the basis {AND, NOT, COPY}. If a basis containing the COPY gate is not universal, it must be either unable to simulate AND or unable to simulate NOT. The bases that cannot simulate NOT are precisely the *monotone* bases; the basis {AND, OR, COPY} is a monotone basis that can simulate all monotone bases. A monotone basis containing COPY that cannot simulate all monotone bases must be either unable to simulate AND, or unable to simulate OR; all such bases can be simulated either by {OR, COPY} or by {AND, COPY}. The only monotone basis containing COPY that cannot simulate either of these two bases is the basis {$COPY$}. This gives the only four monotone bases (up to simulations). Every non-monotone basis can simulate NOT; if it cannot simulate AND, then it must be *linear*. All linear bases can be simulated by the linear basis {XOR, COPY}. The only non-monotone linear basis containing COPY that cannot simulate all linear bases is {NOT, COPY}.

This completes the list of seven bases containing COPY. The complexity of several questions for these bases is given in Table 1.1. The evaluation, stability, and convergence questions are viewed as decision problems in the table; for monotone bases, the answer to the stability question (indicated by a * in the table) is always 'yes', and the associated search problem has complexity comparable to that of the evaluation question for the same basis. The classification for evaluation and stability is from [43, 44, 78, 109]. For the remaining questions, we have not attempted in general to give the tightest possible complexity result, but rather to distinguish the bases for which the question is as hard as for universal bases from those for which it is significantly easier. The optimization result for {XOR, COPY} is based on an observation of Beigel [14]. The remainder of this section sketches the proofs of some of these complexity facts.

2.3.1 Universal Bases

A polynomial space Turing machine can be described by a network $N = \{f\}$, where f describes the transition function of the machine. The transition function f can in turn be described by a fairly simple circuit K over a universal basis, for instance the basis $\Omega = $ {AND, NOT, COPY}. From section 2.1.1, we know that the network N can be represented by a network N' over Ω, so that one iteration of N corresponds to some number d of iterations of N'. As a result, the computation of k steps of the machine is described by an evaluation of the circuit (N', kd), and the computation of a polynomial time Turing machine is described by a polynomial size circuit. Conversely, a circuit can always be evaluated in polynomial time. This gives the following result of Ladner [77].

Theorem 2.1 *The evaluation question for circuits over* $\{\text{AND}, \text{NOT}, \text{COPY}\}$ *is $\mathcal{P}$-complete.*

A polynomial space Turing machine computes by iterating its transition function, accepting its input if it reaches a stable configuration. If the machine is represented by a network N' over a universal basis as before, then k iterations of the machine correspond to kd iterations of N'; the machine accepts if and only if a stable configuration is reached when iterating N'. Conversely, the iterations of an arbitrary network N' can be performed by a polynomial space Turing machine. This gives:

Theorem 2.2 *The convergence question for networks over* $\{AND, \text{NOT}, \text{COPY}\}$ *is $\mathcal{PSPACE}$-complete.*

A nondeterministic polynomial time Turing machine can be described as a machine that operates on a given input by guessing a polynomial size certificate and then performing a deterministic polynomial time computation to verify the certificate on the given input. This polynomial time computation can viewed as the transition function of a network $N = \{f\}$, where f is described by a polynomial size circuit over a universal basis Ω. If $x_{I(N)}$ describes the input of the machine, and $x_{L(N)}$ describes the certificate, then f can be designed so that $y = f(x)$ satisfies $y_{L(N)} = x_{L(N)}$ if and only if $x_{L(N)}$ is a valid certificate for the given input. As a result, N has a stable configuration consistent with its input if and only if there exists a valid certificate for the input to the machine. The network N can in turn be represented by a network N' over Ω as before, with a one-to-one correspondence between the stable configurations of both networks. Conversely, the existence of a stable configuration consistent with a given input can be verified for a given network N by a nondeterministic polynomial time Turing machine by guessing and verifying a stable configuration. This gives the following result of Mayr and Subramanian [78, 109].

Theorem 2.3 *The stability question for networks over* $\{\text{AND}, \text{NOT}, \text{COPY}\}$ *is $\mathcal{NP}$-complete.*

The stable configurations of the network N' in the preceding construction are precisely the certificates of the corresponding $\mathcal{NP}$ problem, so the number of stable configurations is the same as the number of certificates. This reduces every counting problem in $\#\mathcal{P}$ to the counting problem for networks.

Theorem 2.4 *The counting problem for networks over* $\{\text{AND}, \text{NOT}, \text{COPY}\}$ *is $\#\mathcal{P}$-complete.*

2.3.2 Linear Bases

Given that universal bases make computational problems on circuits or networks hard (complete for some complexity class), it is natural to consider less powerful bases. A basis is *linear* if the outputs of each gate in it are sums of some subset of the inputs, up to negation, where addition is over GF(2). If a basis can simulate NOT but cannot simulate AND, then it must be linear; If a basis is linear, then it can be simulated by the linear basis {XOR, COPY}.

The transition function f of a network over a linear basis can be represented by a boolean matrix A so that $f(x) = y$ if and only if $Ax = y$, where sums and products for matrix multiplication are performed over GF(2) [109]. The evaluation and stability questions reduce then to solving a system of linear equations over GF(2), and this can be done in $\mathcal{NC}^2$ by Borodin [17] and Mulmuley [82]. (The classes $\mathcal{NC}^k$ contain problems that can be solved by circuits of size polynomial in n and depth $O(\log^k n)$, where n is the size of the input, and copying is free.) If the system of equations has r degrees of freedom, then it has 2^r solutions, so the counting and enumeration problems also have relatively low complexity. The convergence question requires iterating the function f at most 2^m times, for networks of size m. This computation can be performed in polynomial time from the matrix representation, by repeated squaring starting from A.

This discussion indicates that questions on circuits and networks over a linear basis are relatively easy from a computational point of view. The problems have even lower complexity if more restrictions are imposed (e.g., for the basis {NOT, COPY} [78, 109]).

2.3.3 Monotone Bases

Given two assignments x, y on $S = S(x) = S(y)$, we write $x \leq y$ if $x_i \leq y_i$ for all $i \in S$. A gate g is *monotone* if $x \leq y$ implies $g(x) \leq g(y)$ for every x, y. A basis is *monotone* if every gate in it is monotone. If a basis cannot simulate NOT, then it must be monotone; If a basis is monotone, then it can be simulated by the monotone basis {AND, OR, COPY}.

For monotone bases, the evaluation and the convergence questions remain complete for $\mathcal{P}$ (see Goldschlager [43]) and $\mathcal{PSPACE}$ respectively. The basic idea in the reductions considers the circuit or network from the non-monotone case, and eliminates negations by replacing each boolean coordinate x_i with two coordinates that take values $x_i, \overline{x_i}$. This idea does not work directly for the stability question, because it is hard to enforce the condition that the two values corresponding to a given coordinate be complementary values. (For the other two questions, this can be enforced directly at the starting input or internal values.) In fact, if we denote by 0 and by 1 the all-zero and the all-one internal assignments respectively, then a monotone internal function h will give $0 \leq h(0)$ $(1 \geq h(1))$ and by induction $h^{(n)}(0) \leq h^{(n+1)}(0)$ (resp. $h^{(n)}(1) \geq h^{(n+1)}(1)$). This can be used to show that the two sequences $h^{(n)}(0)$ and $h^{(n)}(1)$ converge to the internal assignments of stable configurations, namely the smallest and the largest stable configurations (with respect to $\leq$) that are consistent with the input, within $|L(N)|$ iterations [78, 109]. Therefore these

CHAPTER 2. PRELIMINARIES

two stable configurations can be found in polynomial time. On the other hand, we have the following.

Theorem 2.5 *The question of whether a network over $\{\text{AND}, \text{OR}, \text{COPY}\}$ has any stable configurations consistent with a given input other than the smallest and the largest such stable configurations is $\mathcal{NP}$-complete.*

Proof. Consider an instance of 3SAT, where each clause is of the form $u \vee v \vee w$, and each of u, v, w is a literal (either a variable x_i or its negation $\overline{x_i}$). Include also the trivial clauses $x_i \vee \overline{x_i}$. Construct a circuit with inputs $x = x_1 x_2 \ldots x_n x'_1 x'_2 \ldots x'_n$, where each x_i corresponds to a variable in the 3SAT instance and each x'_i to its negation. The circuit makes multiple copies of each input using COPY gates, feeds a copy of each pair x_i, x'_i into an AND gate, and feeds the outputs of all these AND gates into a large OR gate. The output α of this OR gate will be 0 if and only if at least one of each pair x_i, x'_i is 0. Feed copies of the inputs (literals) involved in each clause into an OR gate, and let β be the AND of the outputs of these OR gates. Thus β equals 1 if and only if the x_i, x'_i satisfy all the clauses (in particular, at least one of each pair x_i, x'_i is 1). Therefore $\alpha\beta = 01$ if and only if the input x characterizes a solution to the 3SAT instance. Now, pick a copy of each input in x, take its OR with α and take the AND of the result with β. The resulting output $y = y_1 y_2 \ldots y_n y'_1 y'_2 \ldots y'_n$ will satisfy $y = x$ when $\alpha\beta = 01$, i.e., when x characterizes a 3SAT solution. Otherwise either $\alpha = 1$ or $\beta = 0$, and so y is either all zeros or all ones, in which case the equation $y = x$ is satisfied if and only if x is either all zeros or all ones. If we link each output in y to the corresponding input in x, we obtain a network with two trivial stable configurations, namely the assignments consisting of all zeros or all ones. The remaining stable configurations describe the solutions of the 3SAT instance. □

This complexity is reduced if we restrict monotone bases even further. If a monotone basis Ω cannot simulate AND (the case where OR cannot be simulated is identical), then it can be shown that Ω can be simulated by the basis $\{\text{OR}, \text{COPY}\}$. In this case, it is possible to find more stable configurations for the network, and in fact all stable configurations can be enumerated in polynomial time per configuration. This is made possible by the fact that now, given an assignment of boolean values to some subset of the coordinates, we can test whether this partial assignment can be extended to a stable configuration by giving values to the remaining coordinates. This is done by considering all the coordinates in the network that were assigned the value 0; if such a coordinate is the output of an OR gate, then the inputs of the gate must also have value 0; if both inputs of an OR gate have value 0, then the output must have value 0. For COPY gates, if an input or an output has value 0, then the input and both outputs must have value 0. This process assigns value 0 to a subset of the coordinates in the network, and fails if it attempts to assign value 0 to a coordinate that had been assigned the value 1, in which case the network has no stable configuration consistent with the given partial assignment. If this process does not fail, then a stable configuration can be obtained by assigning value 1 to the remaining unassigned coordinates. To see this, note that the output of an OR gate is assigned value 0 if and only if both inputs are assigned value 0, so the equation for the OR gate holds; similarly, an output of a COPY gate is assigned value 0 if and only if the input is assigned value 0, so the equation for the COPY gate also holds. A recursive traversal on the space of partial assignments that extend to stable configurations can then be used to enumerate all stable configurations.

The evaluation and convergence questions become easier in the $\{\text{OR}, \text{COPY}\}$ case because the transition function can again be represented by a boolean matrix as in the $\{\text{XOR}, \text{COPY}\}$ case, where addition and multiplication correspond now to the boolean operations $\vee$ and $\wedge$ respectively, and repeated squaring can once again be used (actually, evaluation is slightly easier than in the XOR case, see [78, 109]). On the other hand, the counting problem remains $\#\mathcal{P}$-complete, even in the $\{\text{OR}, \text{COPY}\}$ case. The reduction is from the problem of counting ideals of a partial order, which was proved $\#\mathcal{P}$-complete by Provan and Ball [95], and consists of constructing a network by assigning an OR gate (actually, an OR gate with multiple copies of its output value) to each element of the partial order, and linking an ouput of an OR gate g to an input of an OR gate g' if $g \leq g'$ in the partial order (in particular, the output of g feeds back as an input to g). In a stable configuration, if the output of g has value 1, then the output of g' has value 1, whenever $g \leq g'$; furthermore, this is the only condition that the outputs of gates must satisfy to ensure stability, so the outputs of value 1 in stable configurations correspond to ideals of the partial order.

2.4 Nonexpansive Bases

The classification of bases from the last section considered bases that could not simulate NOT or could not simulate AND. An important third case was left out, namely that of bases that cannot simulate COPY. We begin here the study of such bases.

Two assignments x, y on $S = S(x) = S(y)$ are *adjacent* if they differ in precisely one coordinate, i.e., if $x_i \neq y_i$ for some i in S and $x_j = y_j$ for all $j \neq i$ in S. We denote by e^i the assignment z with $z_i = 1$ and $z_j = 0$ for $j \neq i$. Given assignments x, y, z on $S = S(x) = S(y) = S(z)$, we write $y = x \oplus z$ if $y_i = x_i \oplus z_i$ for all $i \in S$. The assignments adjacent to x are then those of the form $x \oplus e^i$. A gate g is *nonexpansive* if it maps adjacent assignments x, y to assignments $g(x), g(y)$ that are either equal or adjacent. A basis is *nonexpansive* if every gate in it is nonexpansive. We have opted here for the term 'nonexpansive', instead of the term 'adjacency-preserving' from [78, 109], because it captures the notion that a small perturbation on the inputs to a gate can only produce a small perturbation on its outputs. A more precise formulation of this notion, expressed in terms of a metric on the space of assignments, will be used extensively in Chapter 3. In fact, it is this metric point of view that will lead to the generalization to metric spaces in Chapter 6.

Mayr and Subramanian found the following simple characterization of bases with limited fanout [78, 109].

Lemma 2.6 *The bases that cannot simulate the* COPY *gate are precisely the nonexpansive bases.*

Proof. A circuit of nonexpansive gates computes a nonexpansive mapping. (This will follow from a more general result in section 3.1.1.) Since the COPY gate is *expansive* (the opposite of nonexpansive), it follows that it cannot be simulated by a nonexpansive basis. Conversely, suppose that a basis is expansive, and therefore contains an expansive gate g. There are therefore two input assignments x, x' that differ in only one coordinate, say $x_1 \neq x'_1$ and $z = x_S = x'_S$ for $S = I(g)\setminus\{1\}$, whose images $y = g(x)$ and $y' = g(x')$ differ in at least two coordinates, say $y_1 \neq y'_1$ and $y_2 \neq y'_2$. We can then build a circuit that computes a COPY gate as follows. Consider the 1-input, 2-output restriction $h = g_{z,\{1,2\}}$ of the gate g. The gate h produces two output bits that depend on the single input bit, and are therefore equal to the input bit or its negation. In the latter case, one can obtain a NOT gate as a restriction of h and use it to negate the negated outputs of h. Therefore g can simulate a 2-output gate that produces two copies of its single input, i.e., g can simulate a COPY gate. □

An important class of nonexpansive gates was defined by Mayr and Subramanian, namely the scatter-free gates. Say that an output i of a gate h is *constant* if all input assignments x on $I(h)$ give the same output value $h(x)_i$. A restriction $h = g_{x,T}$ of a gate g is a *proper restriction* if it has no constant outputs. A gate g is *scatter-free* if g and all its proper restrictions have no more outputs than inputs, i.e., if every proper restriction $h = g_{x,T}$ has $|O(h)| \leq |I(h)|$. Scatter-free gates are nonexpansive: Given two adjacent input assignments x, y that differ only in coordinate i, with $z = x_S = y_S$ for $S = I(g)\setminus\{i\}$, the 1-input restriction $g_{z,O(g)}$ can have at most one output j that is not constant by the definition of scatter-freedom, so $g(x)$ and $g(y)$ must be either adjacent or equal. The converse of this statement does not hold: The 3-input, 4-output gate g defined by $g(100) = 1000$, $g(010) = 0100$, $g(001) = 0010$, $g(111) = 0001$, and $g(x) = 0000$ elsewhere, is nonexpansive but not scatter-free (it has *scatter*). In practice, many of the interesting nonexpansive gates turn out to be scatter-free as well. Gates computed by circuits of scatter-free gates are in turn scatter-free [78, 109].

The following scatter-free gates will often be used. The *comparator* is a two-input, two-output gate C defined by $C(x_1 x_2) = y_1 y_2$, where $y_1 = x_1 \wedge x_2$ and $y_2 = x_1 \vee x_2$. More generally, the *sorter* is defined by $\text{SORT}(x_1 x_2 \cdots x_k) = y_1 y_2 \cdots y_k$, where the y_i are the x_i in sorted order (i.e., the y_i are a permutation of the x_i such that $y_i \leq y_j$ for $i \leq j$). Thus the comparator is a 2-input, 2-output sorter; sorters, in turn, can be simulated using comparators. If U is a set of assignments on $\{1, 2, \ldots, k\}$ that does not contain the all-zero or the all-one assignments, and has the property that no two assignments from U are adjacent, then the *near-sorter* associated with U is a gate g with $I(g) = O(g) = \{1, 2, \ldots, k\}$ defined as follows: If $x \notin U$, then $g(x) = \text{SORT}(x)$; If $x \in U$, and i is such that $y_{i-1} = 0$ and $y_i = 1$ for $y = \text{SORT}(x)$, then $z = g(x)$ is given by $z_{i-1} = 1$, $z_i = 0$, and $z_j = y_j$ for $j \neq i-1, i$. In other words, on inputs from U, the gate g takes the outputs from SORT and swaps two consecutive but different bits. One can verify directly that each of these gates is monotone and scatter-free. In fact, by the following lemma of Mayr and Subramanian [78, 109], it is sufficient to verify that these gates are monotone and nonexpansive.

Lemma 2.7 *All monotone nonexpansive gates are scatter-free.*

Proof. Given a nonexpansive gate g, consider an assignment z of boolean values to all but k of the inputs to g. Consider the two complete assignments x^0 and x^1 of input values for g obtained from this partial assignment by respectively assigning all zeros and all ones to the remaining k inputs. Since g maps adjacent assignments to adjacent or equal assignments, it must map assignments that differ in k bits to assignments that differ in at most k bits, so $g(x^0)$ and $g(x^1)$ differ in at most k bits. Let S be the set of bits where $g(x^0)$ and $g(x^1)$ differ. Every complete input assignment y consistent with z satisfies $x^0 \leq y \leq x^1$, and therefore $g(x^0) \leq g(y) \leq g(x^1)$ since g is monotone. It follows that the assignments $g(y)$ agree in all bits outside S, and hence in all but at most k bits, proving that g is scatter-free. □

Among non-monotone scatter-free gates, one of the simplest is the X gate, defined by $X(x_1 x_2) = y_1 y_2$, where $y_1 = x_1 \wedge \overline{x_2}$ and $y_2 = x_2 \wedge \overline{x_1}$. The basis $\{X\}$ can simulate NOT since $X(x_1 1) = 0\overline{x_1}$, and can simulate a comparator because if $X(x_1 \overline{x_2}) = y_1 \overline{y_2}$, then $C(x_1 x_2) = y_1 y_2$. These equations also show that X is computed by a small circuit of comparators and negations, so the bases $\{X\}$ and $\{\text{NOT}, C\}$ can simulate each other. The k-variate X gate is a generalization of the X gate, defined by $X_k(x_1 x_2 \ldots x_k) = y_1 y_2 \ldots y_k$, where $y_i = x_i \wedge (\bigwedge_{j \neq i} \overline{x_j})$. These gates are all scatter-free. A family of nonexpansive gates that are not necessarily scatter-free is provided by the following special kind of *decoder*. Given a set U of assignments on $\{1, 2, \ldots, k\}$ with the property that no two assignments from U are adjacent, define the decoder associated with U as a k-input gate that has an output associated with each element $x \in U$. Such an output has value 1 precisely when the input assignment is equal to x. For instance, if U is the set of assignments that contain a single 1, then the corresponding decoder is the X_k gate. If U consists of the *odd* assignments (those that contain an odd number of 1s), then the corresponding decoder is a k-input, 2^{k-1}-output nonexpansive gate. Mayr and Subramanian [78, 109] have shown that the increase from k inputs to 2^{k-1} nonconstant outputs is worst possible for nonexpansive gates. This increase indicates that most decoders are not scatter-free.

In general, there does not seem to be a simple way to describe the possible nonexpansive or scatter-free gates. The following lemma estimates the number of such gates. For comparison, note that the total number of k-input, k-output gates (on given input and output sets) is 2^{k2^k}.

Lemma 2.8 *There are $2^{\Theta(k2^k)}$ k-input nonexpansive gates with no constant outputs; $k^{\Theta(2^k)}$ k-input, k-output nonexpansive gates; $2^{\Theta(2^k)}$ k-input, k-output scatter-free gates; and $2^{\Theta(2^k)}$ k-input, k-output monotone scatter-free gates.*

Proof. Consider a nonexpansive gate g with k inputs and l outputs. The gate g can be defined for each of the 2^k input assignments x in turn, by changing one coordinate of x at a time, starting with the all-zero assignment. There are 2^l possible choices for $g(0)$, but for each assignment x subsequently considered there are at most $l + 1$ choices for $g(x)$, namely the $l + 1$ assignments adjacent or equal to $g(x')$, where x' is the assignment listed before x. This gives $2^l(l+1)^{2^k-1}$ as an upper bound on the number of possible gates. The k-output bound is obtained by letting $l = k$; the bound in the case where the number of outputs is not fixed is obtained from the fact that the number of non-constant outputs of a nonexpansive gate is at most $l = 2^{k-1}$. To obtain the lower bound, we consider gates with $g(x) = 0$ (the all-zero assignment) for all even assignments x, and where $g(x)$ is an arbitrary assignment adjacent or equal to 0 for all odd assignments x. All such mappings are nonexpansive. There are 2^{k-1} odd assignments, and there are $l + 1$ possible choices for their images. If we fix the images of the first l odd assignments to be the l assignments adjacent to 0, then the gate has no constant outputs. The number of possible such mappings is still at least $(l+1)^{2^{k-1}-l}$, and the two lower bounds are obtained with $l = 2^{k-2}$ and $l = k$ respectively.

In the scatter-free case, if we assign values to $k - r$ of the inputs, then at least $k - r$ of the outputs must become constants. We can describe g by a complete binary tree of depth k, with 2^k leaves. The choice of the ith edge traversed on a path from the root to a leaf determines the value (0 or 1) of the ith input. Thus the leaves correspond to all possible input assignments. Each choice for an input assigns a constant value to an output. When the value of the ith input is chosen, $i-1$ outputs have already been assigned, so there are $k-i+1$ possible choices for the ith output to be assigned, giving $2(k - i + 1)$ possible assignments since each output can be assigned either 0 or 1. There are therefore $2(k - i + 1)$ possible output choices for each of the 2^i edges that occur as the ith edge on a path from the root to a leaf. This gives $\prod_{1 \leq i \leq k} (2(k - i + 1))^{2^i}$ as an upper bound on the number of possible scatter-free gates. The logarithm of this product is $\sum_i 2^i \log(2(k - i + 1)) = O(2^k)$, proving the stated upper bound. The $2^{\Omega(2^k)}$ lower bound on the number of monotone scatter-free gates is obtained by counting the number of near-sorters, which are always monotone and scatter-free. A near-sorter is defined by a set U of assignments on $\{1, 2, \ldots, k\}$ with no adjacent assignments (and excluding the all-zero and all-one assignments). The number of such sets consisting of odd assignments alone is at least $2^{2^{k-1}-1}$, proving the lower bound. □

2.4.1 No Finite Basis for Nonexpansive Gates

Subramanian [109] gave some examples of nonexpansive bases that cannot simulate certain nonexpansive gates. For example, the basis $\{X\}$ cannot simulate the scatter-free Δ_4 gate with $I(\Delta_4) = O(\Delta_4) = \{0, 1, 2, 3\}$ defined (with a slight change in the naming convention for inputs and outputs of [109]) by $[\Delta_4(x)]_i = \overline{x_{i+1}} \wedge x_i \wedge (\overline{x_{i-1}} \vee x_{i-2})$, where the arithmetic on indices is performed modulo 4. In this section, we show that the scatter-free X_k gate cannot be simulated by any nonexpansive basis that does not contain a gate with at least k inputs and k outputs. It then follows that no finite basis can simulate all X_k gates.

Say that h is a *restriction of g in the weak sense* if h can be obtained from a restriction of g by negating some subset of the inputs and outputs, and relabelling inputs and outputs.

CHAPTER 2. PRELIMINARIES 24

Lemma 2.9 *Every nonexpansive basis that can simulate* X_k *must have a gate with at least k inputs and k outputs. Every scatter-free basis that can simulate X_k must have a gate g such that X_k is a restriction of g in the weak sense.*

Proof. The simulation provides a circuit K over a nonexpansive basis Ω that computes X_k. We assume that every restriction (in the weak sense) of a gate in Ω is also in Ω. We also assume that K is a circuit over Ω that computes X_k with the smallest possible number of gates. Let g be a lowest gate in the circuit, i.e., a gate all of whose inputs are inputs of the circuit K. We can view g as defined on the entire input set $I(K) = \{1, 2, \ldots, k\}$; the inputs of K that are not inputs to g are simply left unchanged by g. Therefore g is a gate from assignments to $I(K)$ to assignments to some set S; if $g(x) = y$, then the output $X_k(x)$ of the circuit K can only depend on y. By introducing negations at the outputs of g, we can normalize g so that it maps the all-zero assignment to the all-zero assignment, i.e., $g(0) = 0$ (these negations can be absorbed into gates occurring above g in the circuit without increasing the size of the circuit). The gate g must also map the assignment 0 and all the assignments x adjacent to 0 to distinct assignments y, because the images of these assignments under X_k are all distinct. This means that assignments x with a single bit x_i equal to 1 must map to assignments y with a single bit y_j equal to 1 under g, and that different i must correspond to different j. We can then rename the outputs of g so that $j = i$. Let g' be the gate obtained from g by considering only the first k outputs of g; by the preceding argument, we have that $g'(x) = x$ for all x that have at most one coordinate equal to 1.

Let e^i be the word whose only 1 is the ith bit, and let e^{ij} be the word whose only 1s are the ith and the jth bits, with $i \neq j$. Then $g'(e^i) = e^i$ for all i, and by adjacency we must have either (1) $g'(e^{ij}) = e^{ij}$ or (2) $g'(e^{ij}) = 0$, for each $i \neq j$; furthermore, the outputs of g that are not outputs of g' must be 0s in $g(e^{ij})$. Suppose that case (1) holds for some i, j, and case (2) holds for some i', j'. We can further assume that $j = j'$ (otherwise, if case (1) holds for j, j' we can replace i, j by j, j', and if case (2) holds we can replace i', j' by j', j). Thus $g'(e^{ij}) = e^{ij}$ and $g'(e^{i'j}) = 0$. Let $e^{ii'j}$ be the assignment whose only 1s are the ith, i'th and jth bits. Then, by adjacency, we must have $g'(e^{ii'j}) = e^{j''}$, where $j'' = i$ or $j'' = j$; furthermore, the outputs of g that are not outputs of g' must be 0s in $g(e^{ii'j})$. But this means that $g(e^{ii'j}) = g(e^{j''})$, and this is not possible because these two input assignments map to different outputs under X_k.

Therefore either case (1) holds for all i, j or case (2) holds for all i, j. If case (1) holds for all i, j, then $g'(x) = x$ for all x with at most two bits equal to 1. This implies, as we shall later prove in Corollary 3.35, that $g'(x) = x$ for *all* x. By adjacency, it follows that the outputs of g that are not outputs of g' must be 0s for all inputs x. The gate g can then be removed from the circuit K, decreasing the total number of gates in the circuit and contradicting minimality.

The only remaining possibility is that case (2) holds for all i, j, i.e., $g'(e^{ij}) = 0$ for all i, j. This means that each output of g' depends on each input of g': Output i depends on input i because $g'(0) = 0$ and $g'(e^i) = e^i$; output i depends on input j for $j \neq i$ because $g'(e^i) = e^i$ and $g'(e^{ij}) = 0$. Therefore the circuit K has a gate with at least k inputs and k outputs, as claimed.

In the scatter-free case, no more than k outputs of g can depend on the k inputs, so the outputs of g that are not outputs of g' must be 0s for all inputs x. We can therefore assume that $g = g'$. Recall that the inputs x with two 1s map to 0 under g. If we consider the inputs with r 1s for $r = 3, 4, \ldots$ in turn until some x with $g(x) \neq 0$ is found, then the first such x must satisfy $g(x) = e^i$ for some i, by adjacency. But then $g(x) = g(e^i)$, and this is not possible because these two input assignments map to different outputs under X_k. Therefore $g(x) = 0$ for all x with two or more 1s, and the gate g is then an X_k gate. □

It follows from this lemma that a nonexpansive basis capable of simulating all X_k must necessarily have gates with at least k inputs and k outputs for all k, and cannot be finite.

Theorem 2.10 *There is no finite nonexpansive basis that can simulate all nonexpansive gates, or all scatter-free gates.*

2.4.2 Circuits and Networks with Variable Inputs

We shall usually be concerned with circuits and networks under a fixed input assignment. In this section we show that this limitation is important, because several questions on nonexpansive networks become computationally hard when the input assignment is not fixed in advance.

Given n and a pair $\{i, j\}$ of distinct integers between 1 and n, define the gate $g = g^{\{i,j\}}$ with $I(g) = O(g) = \{1, \ldots, n\}$ by letting $g(x) = y$ with $y_i y_j = X(x_i x_j)$ and $y_{i'} = x_{i'}$ for $i' \neq i, j$. Thus the gate $g^{\{i,j\}}$ applies an X gate to the ith and jth inputs, leaving the remaining inputs unchanged. In what follows, we shall consider circuits obtained by composing gates $g^{\{i,j\}}$, i.e., circuits of the form $K = (\{g^{\{i_1,j_1\}}\}, \{g^{\{i_2,j_2\}}\}, \ldots, \{g^{\{i_m,j_m\}}\})$, with $I(K) = \{(1,0), (2,0), \ldots, (n,0)\}$ and $O(K) = \{(1,m), (2,m), \ldots, (n,m)\}$, using the notation from section 2.1.1. In words, first an X gate is applied to inputs i_1, j_1, (setting them both to 0 if they both equal 1), then an X gate is applied in the resulting assignment to inputs i_2, j_2, and so on; the second component (tag) of pairs indicates the number of times coordinates have been updated, hence equals 0 for inputs and m for outputs. These circuits can be viewed as a special kind of circuit over $\{X\}$.

Lemma 2.11 *The question of whether a given output of a given circuit over $\{X\}$ takes the value 1 for some input assignment is $\mathcal{NP}$-complete.*

Proof. The reduction is from *independent set*. Given a undirected graph G with n vertices $V(G) = \{1, 2, \ldots, n\}$ and m edges $E(G) = \{\{i_1, j_1\}, \{i_2, j_2\}, \ldots, \{i_m, j_m\}\}$, construct the circuit K with n inputs, n outputs, and m gates $g^{\{i,j\}}$, as described above.

Given an output assignment y for K determined by the choice of an input assignment for K, let $S_y = \{i : y_{(i,m)} = 1\}$. We shall show that the sets of the form S_y for some such output assignment y are precisely the independent sets of G. As a result, the circuit K can produce r or more outputs equal to 1 for some input assignment if and only if G has an independent set of size r. Therefore, if we sort the outputs of K, then the question of whether G has an independent set of size r reduces to the question of whether the rth largest output equals 1 for some input assignment, showing that the problem of testing whether a chosen output equals 1 for some input assignment is $\mathcal{NP}$-hard for circuits over $\{X, \text{SORT}\}$, and hence for circuits over $\{X\}$ because $\{\text{SORT}\}$ can be simulated by $\{C\}$ and in turn by $\{X\}$.

It remains to prove the claim relating the possible output values for K to the independent sets of G. If S is an independent set of G, let y be the assignment on $\{1, 2, \ldots, n\}$ defined by $y_i = 1$ if and only if $i \in S$. Then $g^{\{i,j\}}(y) = y$ for all edges $\{i,j\} \in E(G)$, because $X(y_i y_j) = y_i y_j$ unless $y_i = y_j = 1$, and this possibility is excluded by the fact that S is an independent set of G. Therefore the circuit K, given the input assignment $(y, 0)$, returns the output assignment (y, m), and this output assignment satisfies $S_{(y,m)} = S$.

In the other direction, let t be the stable configuration for the circuit determined by some input assignment, and choose an edge $\{i,j\} = \{i_k, j_k\} \in E(G)$. Then either $t_{(i,k)} = 0$ or $t_{(j,k)} = 0$, because the coordinates (i,k) and (j,k) are the outputs of the X gate of $(g^{\{i,j\}}, k-1, k)$ inside K. It follows that either $t_{(i,m)} = 0$ or $t_{(j,m)} = 0$, because each of the gates g in K satisfies $[g(x)]_i = 0$ whenever $x_i = 0$. Therefore the set S_y for $y = t_{O(K)}$ is an independent set of G. □

A 1-forcer is a 1-input, 0-output network that has a stable configuration if and only if the value of its single input is 1. A 1-forcer can be constructed with XOR gates or with X gates [78, 109]. By feeding the output tested in the preceding construction into a 1-forcer, we obtain the following:

Corollary 2.12 *The question of whether a given network over $\{X\}$ has an input assignment for which a consistent stable configuration exists is $\mathcal{NP}$-complete.*

Simple modifications of the above constructions give hardness results for comparator circuits as well. This time the gates $g = g^{\{i,j\}}$ with $i < j$ are defined using comparators instead of X gates, by letting $g(x) = y$ with $y_i y_j = C(x_i x_j)$ and $y_{i'} = x_{i'}$ for $i' \neq i, j$, and the circuit K is constructed by composing $g^{\{i,j\}}$ gates as before. This special kind of comparator circuit is often called a *sorting network*. A sorting network *sorts* if it computes the function SORT (more precisely, if it computes (SORT, 0, m), where m is the number of gates in the sorting network); otherwise, the sorting network is a *non-sorter*. The $\mathcal{NP}$-completeness of testing whether a comparator circuit is a non-sorter was first proved by Rabin [68]. This result was generalized to a broad class of problems on comparator circuits by Chung and Ravikumar [22]. One of these problems is the following:

Lemma 2.13 *Given a sorting network and two selected outputs of this circuit, the question of whether there is some value of the inputs for which the first selected output has value 1 and the second one has value 0 is $\mathcal{NP}$-complete.*

Proof. The *balanced bipartite independent set* problem asks whether a given bipartite graph $G = (V_1, V_2, E)$ has an independent set with r vertices from V_1 and r vertices from V_2. This problem was shown to be $\mathcal{NP}$-complete by Garey and Johnson [39, 69]. We write $V_1 = \{1, 2, \ldots, n'\}$ and $V_2 = \{n'+1, n'+2, \ldots, n\}$. The proof of the previous lemma used a circuit K of X gates to generate all possible independent sets in a graph; if we replace the second input and the second output of every X gate by its negation, then the X gate becomes a comparator. This modification gives a comparator circuit that generates the independent sets S of G, with membership in S represented by an output value 1 for vertices in V_1 and by an output value 0 for vertices in V_2. Thus the balanced independent set problem reduces to the question of whether it is possible to generate an output word with at least r 1s in outputs corresponding to vertices in V_1 and at least r 0s in outputs corresponding to vertices in V_2. If we sort these two sets of outputs separately, then it is sufficient to test whether the rth largest output corresponding to vertices in V_1 (output $n' - r + 1$) has value 1 and the rth smallest output corresponding to vertices in V_2 (output $n' + r$) has value 0. □

This construction can also be used to obtain Rabin's result:

Lemma 2.14 *The question of whether a given sorting network is a non-sorter is $\mathcal{NP}$-complete.*

Proof. Let K be the circuit used in the previous construction, which produces an output assignment on $\{1, 2, \ldots, n', n'+1, \ldots, n\}$ with at least r 1s in the first n' coordinates and at least r 0s in the last $n - n'$ coordinates if and only if a given bipartite graph G has a balanced independent set (with at least r elements in each of the two vertex sets). We compose K with a circuit K' that sorts precisely those input assignments on $\{1, 2, \ldots, n\}$ for which the minimum k of the number k_1 of 1s in the first n' coordinates and the number k_0 of 0s in the last $n - n'$ coordinates is smaller than r. As a result, the input assignments to K' (coming from output assignments to K) that do not get sorted by K' are precisely those corresponding to balanced independent sets, proving that testing whether the composition is a non-sorter is $\mathcal{NP}$-complete.

The circuit K' is the composition of three sorting steps: (1) Sort the first n' coordinates and the last $n - n'$ coordinates separately; (2) Sort the $2(r-1)$ intermediate coordinates $\{n' - r + 2, \ldots, n' + r - 1\}$; (3) Again, sort the first n' coordinates and the last $n - n'$ coordinates separately. If $k_1, k_0 \geq r$, then after step (1) the coordinate $n' - r + 1$ (the rth largest one among the first n') has value 1, and the coordinate $n' + r$ (the rth smallest one among the last $n - n'$) has value 0; Step (2) leaves these two coordinates unchanged; After step (3), the first n' coordinates have at least one 1 somewhere, and the last $n - n'$ coordinates have at least one 0 somewhere, so the output is not sorted. On the other hand, if $k = \min(k_1, k_0) < r$, say $k = k_1$ so that $k_1 \leq k_0$ (the case $k = k_0$ is similar), then after step (1) the first $n' - r + 1$ coordinates are all 0s, and there are at most as many 1s among the $r - 1$ coordinates $\{n' - r + 2, \ldots, n'\}$ as there are 0s among the $r - 1$ coordinates $\{n' + 1, \ldots, n' + r - 1\}$; After step (2), the $r - 1$ coordinates $\{n' - r + 2, \ldots, n'\}$ are all 0s, so the first n' coordinates are all 0s; After step (3) the coordinates $\{1, \ldots, n\}$ are sorted. □

Say that a sorting network is a k-sorter if it produces output assignments that differ from sorted outputs in at most k coordinates. It turns out that even the question of distinguishing sorters from sorting networks that are not even ϵn-sorters is $\mathcal{NP}$-complete, for some constant $\epsilon > 0$. The proof uses the result of Arora, Lund, Motwani, Sudan, and Szegedy [4] that it is $\mathcal{NP}$-complete to distinguish instances of 3SAT that have a solution from those that can satisfy at most $1 - \epsilon$ of the clauses. We can create a bipartite graph where the three literals in each clause c are represented by sets S_1, S_2 of three vertices in each side, then join vertices representing complementary literals in the two sides by an edge, and also join vertices representing different literals from the same clause c in S_1 and S_2. We now ask for an independent set. If the 3SAT instance has a solution, then there is an independent set containing at least one element from each set S_1 or S_2; furthermore, if an independent set contains elements both from S_1 and S_2 for $1 - \epsilon$ of the clauses c, then a satisfying assignment for $1 - \epsilon$ of the clauses is obtained. As in the case of balanced independent sets, it is not hard to design sorting networks to test the minimum of the number of S_1 and the number of S_2 that an independent set intersects, and this completes the reduction. We do not know whether the balanced independent set problem itself is hard to approximate. It is only known how to distinguish sorters from sorting networks that are only $(\frac{n}{2} - O(\sqrt{n \log n}))$-sorters, because such k-sorters fail to sort at least an inverse polynomial fraction of the input assignments.

2.5 Complexity Assumptions

Several parameters will be used to analyze the complexity of algorithms on networks. The *size* of a gate g is the total number of inputs and outputs, $|I(g)| + |O(g)|$; its *gatewidth* is the minimum of the number of inputs and outputs, $\min(I(g), O(g))$. For a network N, the size is the total number of coordinates $|R(N)|$; the gatewidth is the maximum gatewidth of the gates in it, $\max_{g \in N} \min(I(g), O(g))$.

Given a gate f, we shall treat f as an oracle (a black box) and assume that the time needed to evaluate f on an input x is proportional to the size of f, namely $|I(f)| + |O(f)|$. This assumption is consistent with the actual complexity of evaluating f when f consists of a collection of gates with a constant number of inputs and outputs, i.e., when f is the transition function of a network $N = \{g^1, g^2, \ldots, g^k\}$, so that $y_{O(g^i)} = g^i(x_{I(g^i)})$ for $y = f(x)$, since in that case the evaluation of each g^i takes constant time.

If f is nonexpansive, then a different kind of query is also needed. If a new query x' is obtained from a previous query x by changing the value of only one coordinate, then $f(x')$ and $f(x)$ differ in the value of at most one coordinate. We assume that f can identify this coordinate in constant time. This assumption is again justified if f consists of a collection of constant size gates g^i, because in that case the coordinate that is changed is an input of some g^i, and so the output coordinate that changes as a result is an output of g^i that can be determined in constant time by inspecting g^i.

If f is scatter-free, then we assume that f can be restricted to a new gate f' by assigning a value to one of its input coordinates, in constant time; we also assume that if f has more outputs than inputs, then a constant output (known to exist by the definition of scatter-freedom) can be determined in constant time. Once again, this is justified when f consists of constant size gates g^i: if we restrict f by fixing one input, then we are just restricting some g^i by fixing one of its inputs; if f has more outputs than inputs, then some g^i must have more outputs than inputs, and a constant output of g^i can be found in constant time (provided that we keep track of the g^i that have more outputs than inputs).

Chapter 3

Stability in Nonexpansive Networks

The fanout available in networks can be limited by requiring that all the gates present be nonexpansive. When only nonexpansive gates are allowed, the resulting networks exhibit a rich structure. Section 3.1 begins the study of this structure by viewing the space of configurations as the set of vertices in a hypercube under the Hamming metric. This leads to an important notion, that of a convergent network. Nonexpansive networks are shown to be convergent, and this results in a polynomial time algorithm for the stability question, as well as a reduction from stability, as a decision problem, to evaluation. Section 3.2 exploits the median structure of the hypercube to give a succint characterization of the set of stable configurations in terms of the solution set of a 2-satisfiability instance. This characterization is then used to reduce stability, as a search problem, to evaluation, and to determine the complexity of the counting problem. Section 3.3 examines the behavior of nonexpansive networks on the space of periodic configurations, and relates it to the isomorphisms of the hypercube. This relationship is used to give an algorithm for the simulation of gates by a given basis, and an efficient algorithm that finds the succint characterization for the stable configurations of nonexpansive networks. Section 3.4 presents Bandelt and Vel's fixed cube theorem. This result is then used to study the distinction between networks that have stable configurations and those that do not, and to give an algorithm for the convergent network representation. Section 3.5 studies the iterative behavior of networks on non-periodic configurations, and derives a polynomial-time solution for the convergence question, as well as a reduction from the convergence question to the evaluation question. Section 3.6 summarizes the results and presents some open questions.

3.1 Hamming Distance and Convergent Networks

Given a coordinate set S, the *hypercube* on S is the graph whose vertices are the assignments on S and whose edges join adjacent assignments. If $|S| = n$, then the hypercube is called an n-cube. We shall often assume $S = \{1, 2, \ldots, n\}$ for simplicity. The 3-cube is illustrated

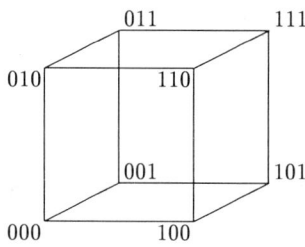

Figure 3.1: The n-cube for $n = 3$.

in Fig. 3.1.

Nonexpansive mappings, when viewed as mappings from a hypercube to a hypercube, map vertices to vertices so that an edge joining two vertices is mapped to an edge or to a single vertex. This means that the nonexpansive mappings are the (weak) homomorphisms from hypercubes to hypercubes, where the adjective 'weak' permits mappings that collapse edges into single vertices. If the mapping is one-to-one and onto, then the homomorphism is an *isomorphism*.

The distance between two vertices in a graph is the minimum number of edges along a path joining those two vertices. In the case of the hypercube, the distance $d(x, y)$ between two vertices x and y is simply the Hamming distance, i.e., the number of coordinates in which x and y differ. A homomorphism maps paths to paths, and therefore cannot increase distances. This gives the following characterization of nonexpansive gates:

Lemma 3.1 *A gate g is nonexpansive if and only if $d(g(x), g(y)) \leq d(x, y)$ for all input assignments x, y.*

Nonexpansive gates cannot increase distances. The following lemma states that for periodic points, they cannot decrease distances either.

Lemma 3.2 *If f is a nonexpansive mapping on a hypercube, and x, y are periodic points of f, then $d(f(x), f(y)) = d(x, y)$.*

Proof. Let $k \geq 1$ be a common multiple of the periods of x and y. Then $d(x, y) = d(f^{(k)}(x), f^{(k)}(y)) \leq d(f(x), f(y)) \leq d(x, y)$ by the previous lemma, so equality holds throughout. □

Let N be a network of nonexpansive gates with transition function f, so that $f(x) = y$ if and only if $g(x_{I(g)}) = y_{O(g)}$ for each gate g in N. Given two input assignments x, x' for f with $y = f(x)$ and $y' = f(x')$, we have

$$d(y, y') = \sum_{g \in N} d(y_{O(g)}, y'_{O(g)}) \leq \sum_{g \in N} d(x_{I(g)}, x'_{I(g)}) = d(x, x'),$$

so f is nonexpansive. Furthermore, if $d(y, y') = d(x, x')$, then equality holds throughout and so $d(y_{O(g)}, y'_{O(g)}) = d(x_{I(g)}, x'_{I(g)})$ for each gate $g \in N$.

Given an input assignment x for a network N of nonexpansive gates, the periodic configurations consistent with x are determined by the choice of a periodic point z of the internal function $h = h_x$ (see section 2.1). Note that this internal function is the transition function of the network N' consisting of the restrictions $g' = g_{x',L'}$ of gates $g \in N$, where $x' = x_{I(g)}$ and $L' = L(N) \cap O(g)$, i.e., the network obtained by eliminating the inputs to the network using the values x and eliminating the outputs to the network. Any two periodic points z, z' of h must satisfy $d(h(z), h(z')) = d(z, z')$, and therefore $d(g'(z_{I(g')}), g'(z'_{I(g')})) = d(z_{I(g')}, z'_{I(g')})$ for every gate $g' \in N'$. In particular, if the two assignments in the left-hand side coincide, then so do the two assignments in the right-hand side. This gives:

Lemma 3.3 *Given a network N of nonexpansive gates, an input assignment x for N and a gate $g \in N$, let t vary over the periodic configurations for N consistent with x. Then the values of the internal coordinates of N in $g(t_{I(g)})$ uniquely determine $t_{I(g)}$. In particular, if t varies over the stable configurations for N consistent with x, then $t_{O(g) \cap L(N)}$ uniquely determines $t_{I(g)}$.*

3.1.1 Convergent Networks

A network N is said to be *convergent* if for every input assignment x there exists an output assingment y such that every configuration consistent with x maps to a configuration consistent with y under sufficiently many iterations of N. More formally, for every configuration t consistent with x, there must exist an integer k_0 such that $N^{(k)}(t)$ is consistent with y for all $k \geq k_0$. Since every configuration maps to a periodic configuration for sufficiently large k, and every periodic configuration maps to itself for infinitely many values of k, the condition of convergence is equivalent to the requirement that every periodic configuration consistent with x must also be consistent with y. Recall that the periodic configurations of N are the configurations $xg_x(z)h_x(z)$, where z is a periodic point of h_x and the mappings g_x, h_x are the output and internal functions of N (see section 2.1). The condition defining convergent networks becomes then the statement that $g_x(z) = y$ for all periodic points z of h_x. If a network N is convergent, then for every input assignment x there is a unique corresponding output assignment y for N, and we say that N converges to the gate g with $I(g) = I(N)$ and $O(g) = O(N)$ that computes $g(x) = y$.

The notion of a convergent network evolved out of discussions with Ashok Subramanian, and was motivated by the following observation (see also [109]).

Theorem 3.4 *All networks of nonexpansive gates are convergent, and converge to a nonexpansive gate.*

Proof. Let f be the transition function of a network N of nonexpansive gates, given by $f(xz) = g_x(z)h_x(z)$, where g_x and h_x are the output and internal functions. Let z, z' be periodic points of h_x. Then

$$\begin{aligned} d(g_x(z), g_x(z')) + d(h_x(z), h_x(z')) &= d(f(xz), f(xz')) \\ &\leq d(xz, xz') = d(z, z') = d(h_x(z), h_x(z')) \end{aligned}$$

by Lemmas 3.1 and 3.2, so $d(g_x(z), g_x(z')) = 0$ and $g_x(z) = g_x(z')$. Therefore the output $y = g_x(z)$ depends only on x, and not on the choice of a periodic point z. This shows that the network is convergent, and converges to some gate g, where $g(x) = g_x(z)$ for all periodic points z of h_x.

Given two input assignments x, x', let z and z' be periodic points of h_x and $h_{x'}$ respectively that are closest to each other. In particular, $d(z, z') \leq d(h_x(z), h_{x'}(z'))$. Then

$$\begin{aligned} d(g_x(z), g_{x'}(z')) + d(h_x(z), h_{x'}(z')) &= d(f(xz), f(x'z')) \\ &\leq d(xz, x'z') = d(x, x') + d(z, z') \\ &\leq d(x, x') + d(h_x(z), h_{x'}(z')). \end{aligned}$$

Therefore $d(g(x), g(x')) = d(g_x(z), g_{x'}(z')) \leq d(x, x')$, so by Lemma 3.1 the gate g is nonexpansive. □

In fact, a basis Ω is nonexpansive if and only if all networks over Ω are convergent [109]. Note that if the network is a circuit K, then it converges to the gate computed by K. Convergence can therefore be viewed as a generalization of the notion of computation by circuits. The lemma gives as a special (and easier) case the fact that circuits of nonexpansive gates compute nonexpansive mappings.

There is an alternative and often useful characterization of the relationship between a network N and the gate g to which it converges. Given a gate h with $I(h) = O(h)$, we call z a *minimum distance point* of h if z minimizes $d(z, h(z))$ over all assignments to $I(h)$. Given a network N with internal function h_x, we let d_x denote the distance $d(z, h_x(z))$, where z is a minimum distance point of h_x.

Lemma 3.5 *Let N be a nonexpansive network with transition function f given by $f(xz) = g_x(z)h_x(z)$, so that g_x is the output function and h_x is the internal function of the network. Suppose that N converges to a gate g. Then $g(x) = g_x(z)$ for every minimum distance point z of h_x, so that the output of the gate to which N converges can be obtained by applying the output function to a minimum distance point of the internal function.*

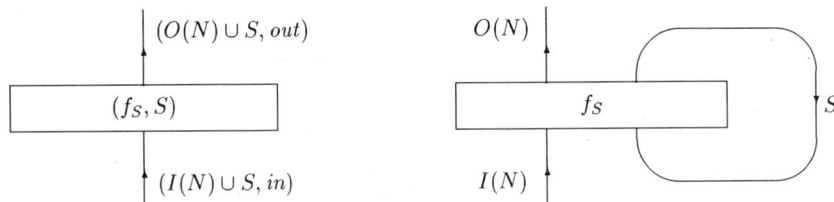

Figure 3.2: The gate (f_S, S) obtained by convergence from (N, S), and the projection f_S.

Proof. Given x, suppose that z is a minimum distance point of h_x, and let $z' = h_x(z)$. Then

$$d(g_x(z), g_x(z')) + d(h_x(z), h_x(z')) = d(f(xz), f(xz')) \leq d(xz, xz') = d(z, z') \leq d(z', h_x(z'))$$

by minimality, so $d(g_x(z), g_x(z')) = 0$ and $d(z', h_x(z')) = d(z, z')$. Therefore z' is also a minimum distance point and z' gives the same image under g_x as z does. By iterating this construction k times, it follows that $z^{(k)} = h_x^{(k)}(z)$ is also a minimum distance point and that $z^{(k)}$ gives the same image under g_x as z does. For k sufficiently large, $z^{(k)}$ is periodic, and so $g_x(z) = g_x(z^{(k)}) = g(x)$ by the definition of convergence. □

Given a network N with transition function f, and a set of links $S \subseteq L(N)$, there is an important function f_S associated with N and S, defined as follows. Suppose that we break the links in S to obtain a network (N, S). This network converges to a gate f'. If we restore the names of the original links, we obtain a gate f_S such that $(f_S, S) = f'$. See Figs. 2.1, 2.2 and 3.2. The gate f_S is called the *projection* of f under S. From the definition of convergent networks, this means that f_S is the gate with $I(f_S) = I(N) \cup S$, $O(f_S) = O(N) \cup S$, and $f_S(x) = f_{x,O(N) \cup S}(z)$ for every periodic point z of $f_{x,L(N) \setminus S}$. In particular, if $S = L(N)$, then $f_S = f$, and if $S = \emptyset$, then f_S is the gate to which N converges. If $S = \{i\}$, then f_S is denoted by f_i.

3.1.2 Properties of Convergent Networks

In order to better understand a nonexpansive network N as it converges to a gate g, it is useful to observe N iterate. Given an *input sequence* of input assignments $x^0, x^1, \ldots, x^{k-1}$, and an *initial internal assignment* z^0, we can iterate N and obtain an *output sequence* of output assignments $y^1, y^2, \ldots, y^k$ together with a *final internal assignment* z^k. In terms of the transition function f of the network N, this computation is expressed by the equations $f(x^0 z^0) = y^1 z^1$, $f(x^1 z^1) = y^2 z^2$, $f(x^2 z^2) = y^3 z^3$, $\ldots$, $f(x^{k-1} z^{k-1}) = y^k z^k$. If we consider the circuit $K = (N, k)$ (see section 2.1.1 and Fig. 2.3), and let f' be the function computed by K, then

$$f'((z^0, 0)(x^0, 0)(x^1, 1) \ldots (x^{k-1}, k-1)) = (y^1, 1)(y^2, 2) \ldots (y^k, k)(z^k, k).$$

If we replace the sequences x^i, y^i, z^i with sequences $\widehat{x^i}, \widehat{y^i}, \widehat{z^i}$ obtained in a similar fashion, we get a corresponding equation. Comparing both equations, and applying Lemma 3.1 to the nonexpansive mapping f', we obtain the following *basic equation*:

$$d(z^k, \widehat{z^k}) + \sum_{0 < i \leq k} d(y^i, \widehat{y^i}) \leq d(z^0, \widehat{z^0}) + \sum_{0 \leq i < k} d(x^i, \widehat{x^i}).$$

An important special case of this equation arises when we let $\widehat{x^i} = x^{i+1}$ for all i, and $\widehat{z^0} = z^1$, so that $\widehat{z^i} = z^{i+1}$ and $\widehat{y^i} = y^{i+1}$. This gives:

$$d(z^k, z^{k+1}) + \sum_{0 < i \leq k} d(y^i, y^{i+1}) \leq d(z^0, z^1) + \sum_{0 \leq i < k} d(x^i, x^{i+1}).$$

Another special case is obtained by chosing an input assignment x, setting $\widehat{x^i} = x$ for all i, and letting $\widehat{z^0}$ be a periodic point of the internal function h_x. Then each $\widehat{z^i}$ is a periodic point of h_x, so $\widehat{y^i} = g(x)$ for all i. The basic equation then becomes:

$$d(z^k, \widehat{z^k}) + \sum_{0 < i \leq k} d(y^i, g(x)) \leq d(z^0, \widehat{z^0}) + \sum_{0 \leq i < k} d(x^i, x).$$

We give two applications of these equations.

Lemma 3.6 *Given a nonexpansive network N that converges to a gate g, and an input assignment x, let $y^1, y^2, \ldots$ be the output sequence obtained by iterating N, using the input assignment x and an arbitrary initial internal assignment z^0, and let $\widehat{z^0}$ be a periodic internal assignment of the internal function for input x. Then*

$$\sum_{0 < i \leq k} d(y^i, g(x)) \leq d(z^0, \widehat{z^0}) - d(z^k, \widehat{z^k}) \leq |L(N)|,$$

where the z^i and the $\widehat{z^i}$ are the internal assignments obtained by iterating N on input x and initial internal assignments $z^0, \widehat{z^0}$.

Proof. In the second special case of the basic equation, set $x^i = x$ for all i, and use the fact that $d(z^0, \widehat{z^0}) \leq |L(N)|$. □

We know that the gate g to which a nonexpansive network N converges can be defined in terms of minimum distance points of the internal function h_x of the network. The following lemma exhibits a relationship between the corresponding minimum distances d_x.

Lemma 3.7 *Let N be a nonexpansive network that converges to a gate g, and let x, x' be two input assignments for g. Then*

$$|d_x - d_{x'}| \leq d(x, x') - d(g(x), g(x')).$$

Proof. By symmetry, it is sufficient to show $d_x - d_{x'} \leq d(x, x') - d(g(x), g(x'))$. In the first special case of the basic equation, let $x^0 = x'$ and let z^0 be a minimum distance point of the internal function $h_{x'}$, so that $d(z^0, z^1) = d_{x'}$ and $y^1 = g(x')$. Let $x^i = x$ for $i \geq 1$, and choose $k \geq 1$ sufficiently large so that $y^{k+1} = g(x)$. The four terms in the equation can be bounded as follows: (1) $d(z^k, z^{k+1}) \geq d_x$; (2) $\sum_{0 < i \leq k} d(y^i, y^{i+1}) \geq d(y^1, y^{k+1}) = d(g(x'), g(x))$; (3) $d(z^0, z^1) = d_{x'}$; and (4) $\sum_{0 \leq i < k} d(x^i, x^{i+1}) = d(x', x)$. The equation then becomes

$$d_x + d(g(x'), g(x)) \leq d_{x'} + d(x', x).$$

□

The following theorem shows that convergent networks are well-behaved with respect to the operation of breaking links.

Theorem 3.8 *Let N and N' be two nonexpansive networks, and suppose that $S \subseteq L(N) \cap L(N')$ is a coordinate set such that the networks (N, S) and (N', S) converge to the same gate. Then the networks N and N' also converge to the same gate.*

Proof. We shall show that if (N, S) converges to a gate (g, S), then the networks N and $\{g\}$ converge to the same gate h. As a result, if both (N, S) and (N', S) converge to (g, S), then both N and N' converge to h, and the theorem follows.

The internal assignments of N are assignments on $L(N)$; we write them as assignments zt, where z is an assignment on S and t is an assignment on $L(N) \backslash S$. The transition function f of the network N can be represented as an output function and an internal function, namely $f(xzt) = f_{x,O(N)}(zt) f_{x,L(N)}(zt)$, where x and zt are assignments to $I(N)$ and $L(N)$ respectively. Alternatively, we can write $f(xzt) = f_{xz,O(N) \cup S}(t) f_{xz,L(N) \backslash S}(t)$. The fact that (N, S) converges to (g, S) means that $f_S = g$, i.e., $f_{xz,O(N) \cup S}(t) = g(xz)$ for every periodic point t of $f_{xz,L(N) \backslash S}$. The mapping g can be written as $g(xz) = g_{x,O(N)}(z) g_{x,S}(z)$. The fact that $\{g\}$ converges to h means that $g_{x,O(N)}(z) = h(x)$ for every periodic point z of $g_{x,S}$.

The strategy of the proof will be the following. Given an assignment x on $I(N)$, we shall prove the existence of a minimum distance point zt of $f_{x,L(N)}$ such that t is a periodic point of $f_{xz,L(N) \backslash S}$ and z is a periodic point of $g_{x,S}$. As a result, we will have $f_{xz,O(N) \cup S}(t) = g(xz)$ and $g_{x,O(N)}(z) = h(x)$, so that $f_{x,O(N)}(zt) = h(x)$. Since x was arbitrary, it will follow from Lemma 3.5 that N also converges to h.

The proof of the existence of a minimum distance point satisfying the two periodicity conditions is a two-step argument. The first step shows that if zt is a minimum distance point of $f_{x,L(N)}$, then there exists a minimum distance point zt'' such that t'' is a periodic point of $f_{xz,L(N)\backslash S}$. The second step shows that if zt is a minimum distance point of $f_{x,L(N)}$, then there exists a minimum distance point $z't'$ such that z' is a periodic point of $g_{x,S}$. From this point $z't'$, by the result of the first step, there must then exist a minimum distance point $z't''$ such that t'' is a periodic point of $f_{xz,L(N)\backslash S}$ (while z' is still a periodic point of $g_{x,S}$), so the point $z't''$ has all the properties needed to complete the proof.

To establish the first step, it is sufficient to show that if zt is a minimum distance point of $f_{x,L(N)}$, then so is $z\tilde{t}$ for $\tilde{t} = f_{xz,L(N)\backslash S}(t)$; we can then repeatedly replace t by the corresponding $\tilde{t}$ until a periodic point t'' of $f_{xz,L(N)\backslash S}$ is reached. The minimum distance property of $z\tilde{t}$ is inherited from that of zt by writing $f_{x,L(N)}(zt) = \tilde{z}\tilde{t}$ and observing that

$$\begin{aligned}d(z\tilde{t}, f_{x,L(N)}(z\tilde{t})) &\leq d(z\tilde{t}, \tilde{z}\tilde{t}) + d(\tilde{z}\tilde{t}, f_{x,L(N)}(z\tilde{t})) \\ &= d(zt, \tilde{z}\tilde{t}) - d(zt, z\tilde{t}) + d(f_{x,L(N)}(zt), f_{x,L(N)}(z\tilde{t})) \leq d(zt, \tilde{z}\tilde{t}).\end{aligned}$$

This completes the first step.

To establish the second step, it is sufficient to show that if zt is a minimum distance point of $f_{x,L(N)}$, then there is a minimum distance point $\tilde{z}\tilde{t}$ with $\tilde{z} = g_{x,S}(z)$; we can then repeatedly replace z by the corresponding $\tilde{z}$ (also updating t along the way) until a periodic point z' of $g_{x,S}$ is obtained. To achieve this, we first obtain a minimum distance point zt'' with t'' a periodic point of $f_{xz,L(N)\backslash S}$, as guaranteed by the result of the first step, and let $\tilde{z}\tilde{t} = f_{x,L(N)}(zt'')$ so that $\tilde{z} = f_{xz,S}(t'') = g_{x,S}(z)$ by the periodicity of t''. Then $d(\tilde{z}\tilde{t}, f_{x,L(N)}(\tilde{z}\tilde{t})) = d(f_{x,L(N)}(zt''), f_{x,L(N)}(\tilde{z}\tilde{t})) \leq d(zt'', \tilde{z}\tilde{t})$, so the minimum distance property of $\tilde{z}\tilde{t}$ is inherited from that of zt''. This proves the second step and completes the proof of the theorem. □

A consequence of this theorem, and essentially a restatement of the theorem, is the following *substitutivity property*: Given a nonexpansive network N, if we replace a gate g in N by a nonexpansive network N_g that converges to g, then the resulting network N' also converges to g. To make this statement precise, we first make sure that the inputs and outputs of g have different names by considering the gate $g' = (g, L(\{g\}))$, we perform the same relabelling on N_g to obtain $N'_g = (N_g, L(\{g\}))$; we also assume that N_g shares no coordinates with N other than those from g, with $I(N_g) = I(g)$, $O(N_g) = O(g)$, and $L(N_g) \cap R(N) = \emptyset$. Finally, we assume that N'_g converges to g'. We then claim that the network N and the network $N' = (N\backslash\{g\}) \cup N_g$ resulting from the substitution converge to the same gate. To see this, let $S = R(\{g\}) \cap L(N)$. Then the networks (N, S) and (N', S) each consist of two disjoint parts (two parts that do not share any coordinates): The first part of the networks consists of $\{g'\}$ and N'_g respectively, and converges to the same gate g' in both cases; the second part is the same for both networks, and therefore also converges to the same gate in both cases. By the disjointness of the two parts, the two networks (N, S) and (N', S) must then converge to the same gate, and so must the networks N and N', by the theorem.

A special case of this substitutivity property is a result due to Subramanian [109]. Construct from a given network N a network N' by streching a coordinate of N and inserting an ID gate in it (see section 2.1.1). Then N and N' must converge to the same gate. This follows by observing that the operation of stretching a coordinate can be viewed as the replacement of a gate g involving that coordinate by the same gate combined with an ID gate, and the resulting extension N_g of the gate g converges to g. The operation of stretching a coordinate 'disrupts the timing' of the network, and can significantly alter its periodic configurations; this explains why an approach not involving periodicity, such as the use of the minimum distance property, was useful in the proof of the theorem.

The substitutivity property also gives the following essentially equivalent property. Given a nonexpansive network N with transition function f, suppose that $S \subseteq T \subseteq L(N)$. Then $[f_T]_S = f_S$. To see this, let $g = f_T$, so that (N,T) converges to (g,T), with $T = L(\{g\})$, and $((N,S),T')$ converges to $((g,S),T')$ for $T' = L(\{(g,S)\})$. We can then use the substitutivity property to replace $\{(g,S)\}$ with (N,S). The network $\{(g,S)\}$ converges to (g_S, S). By substitutivity, (N,S) also converges to (g_S, S), so $f_S = g_S$.

We conclude the presentation of basic properties of convergent networks with an application of the two preceding results. The following property was observed by Mayr and Subramanian [78, 109] in the scatter-free case.

Theorem 3.9 *Let $N = \{f\}$ be a nonexpansive network, and consider two sets $S, T \subseteq L(N)$. Then $\{f_{S \cup T}\}$ has a stable configuration consistent with a given input assignment if and only if both $\{f_S\}$ and $\{f_T\}$ have such a stable configuration. If these equivalent conditions hold, then the stable configurations of $\{f_S\}$ are precisely the configurations consistent with stable configurations of $\{f_{S \cup T}\}$.*

Proof. First note that if $N = \{f_U\}$ has a stable configuration u for some $U \subseteq L(N)$, and $V \subseteq U$, then $\{f_V\}$ has a stable configuration consistent with u. To see this, let $f' = f_U$. A stable configuration $xyzt$ of N with $x \in I(N)$, $y \in O(N)$, $z \in V$ and $t \in U \backslash V$, must satisfy $f'(xzt) = yzt$, so t is a periodic point (a fixed point) of $f'_{xz,U \backslash V}$ and therefore $f_V(xz) = f'_V(xz) = f'_{xz,O(N) \cup V}(t) = yz$, giving a stable configuration xyz for $\{f_V\}$.

It is therefore sufficient to show that if both $\{f_S\}$ and $\{f_T\}$ have a stable configuration consistent with a given input assignment, and if z is a such a stable configuration for $\{f_S\}$, then there is a stable configuration of $\{f_{S \cup T}\}$ consistent with z. We first reduce the general case to the interesting case where $S \cup T = L(N)$, $S \cap T = \emptyset$, and $I(N) = O(N) = \emptyset$. We can assume that $S \cup T = L(N)$ because if we know that the statement holds in this case, then the fact that the network $N' = \{f'\}$ with $f' = f_{S \cup T}$ has $S \cup T = L(N')$ implies that the statement holds for f', and therefore for f because $f_{S \cup T} = f'_{S \cup T}$, $f_S = f'_S$, and $f_T = f'_T$. We can assume that $S \cap T = \emptyset$, since we can replace T by $T' = T \backslash S$ and use the fact that if $\{f_T\}$ has a stable configuration, then so does $\{f_{T'}\}$. We can assume that $I(N) = O(N) = \emptyset$ by considering the network $N' = \{f'\}$ with no inputs or outputs, where $f' = f_{x,L(N)}$ and $x = z_{I(N)}$. The resulting stable configuration z' for N' will be consistent with z, and will give a stable configuration for N consistent with z in the remaining coordinates $I(N)$ (by inspection) and $O(N)$ as well, because $\{f\}$ and $\{f_S\}$ converge to the same gate $f_\emptyset = [f_S]_\emptyset$ that maps $z_{I(N)}$ to $z_{O(N)}$.

Suppose then that $S \cup T = L(N)$ and $S \cap T = \emptyset$, so that $L(N)\setminus S = T$ and $L(N)\setminus T = S$, with $I(N) = O(N) = \emptyset$. Since the networks have no inputs or outputs, the stable configurations are the same as fixed points of the corresponding transition functions. By assumption, we have $f_S(z) = z$ and $f_T(t) = t$ for some assignments z, t on S and T respectively. Let z' be a periodic point of $f_{t,S}$ closest to z, and let $z'' = f_{t,S}(z')$, so that $f(z't) = z''t$ and $d(z, z') \leq d(z, z'')$. Then t is a periodic point (a fixed point) of $f_{z',T}$, and therefore $f_S(z') = z''$. By Lemma 3.7, since (N, S) converges to (f_S, S), we have

$$|d_z - d_{z'}| \leq d(z, z') - d(f_S(z), f_S(z')) = d(z, z') - d(z, z'') \leq 0,$$

so $d_z = d_{z'} = 0$. This means that $f_{z,T}$ has a fixed point t', and therefore $f(zt') = zt'$, giving a fixed point for f that is consistent with z. □

This theorem implies, in particular, that $N = \{f\}$ has a stable configuration consistent with a given input assignment if and only if $\{f_i\}$ has such a stable configuration for each link $i \in L(N)$.

Note that, in the main case of the above proof, given $z = f_S(z)$ and $t = f_T(t)$, the fixed point $zt' = f(zt')$ can be found simply by iterating $(\{f\}, S)$ using z as input and t as the initial internal assignment (but without explicit knowledge of the auxiliary points z', z'' used in the proof) until z appears at the output. This follows from the construction from Lemma 3.7 (used above to prove the existence of zt'), provided that we start the iteration in that construction at level $i = 1$.

3.1.3 Algorithms for Convergent Networks

Given a nonexpansive network N with transition function f and converging to a gate g, we would like to evaluate $g(x)$ for some input x. If we iterate N on input x, starting with some arbitrary internal assignment, we obtain a sequence of outputs y^i satisfying

$$\sum_{0 < i \leq k} d(y^i, g(x)) \leq |L(N)|$$

for all k, by Lemma 3.6. This means that $y^i \neq g(x)$ for at most $l = |L(N)|$ values of i. Therefore $y^i = g(x)$ will appear at least once by the time k reaches $l + 1$, and will appear at least $l + 1$ times by the time k reaches $2l + 1$. Once a value of y^i appears $l + 1$ times, we know that this value must be $g(x)$. In fact, once a value for coordinate j of y^i appears $l + 1$ times, we know that this value must be $g(x)_j$. Therefore $g(x)$ is just the majority (or coordinate-wise majority) of the y^i, over $0 < i \leq k = 2l + 1$. The computation requires $O(m)$ time for the first query to f, where $m = |R(N)|$; subsequent queries only require $O(l)$ time, since they only differ from each other in coordinates from $L(N)$. This gives:

Lemma 3.10 *Suppose that the nonexpansive network N converges to g, and let $m = |R(N)|$, $l = |L(N)|$. Then the evaluation of $g(x)$ on some input x can be performed with at most $2l+1$ queries to the transition function of N, where the first query takes $O(m)$ time and the remaining queries take $O(l)$ time.*

The complexity can be reduced if a small *feedback arc set* is known for the network, namely a set $S \subseteq L(N)$ such that (N, S) is a circuit K. In that case, we can replace the circuit K with the gate g that it computes, by the substitutivity property. After the substitution, the network will only have $s = |S|$ links, so only $2s+1$ queries to g are needed. (Their complexity is still $O(m)$ for the first query and $O(l)$ for the remaining queries, because a query to g requires an evaluation of the circuit K.) The notion of a feedback arc set depends on the topology of the network; we can give instead a construction that achieves the same speed-up in a more general context. Suppose that we assign an arbitrary value to the s coordinates of (S, in) in $K = (N, S)$, and find a stable configuration z for K (by circuit evaluation) in $O(m)$ time. Back in N, with the links in S restored, the configuration z (using the values from (S, in), not those from (S, out), to give values to S) will only differ from $f(z)$ in coordinates from S, i.e., the distance from z to $f(z)$ in coordinates from $L(N)$ is at most s. In general, if the distance in $L(N)$ between a configuration z and its image $f(z)$ is d (regardless of how such z was obtained), then the bound in the lemma can be improved to $O(d)$ per query, rather than $O(l)$ time per query, since at most d bits change from each query to the next. We shall always aim for such formulations in terms of distances, rather than in terms of the topology of the network.

Another case where a speedup can be achieved is the scatter-free case. The following algorithm is due to Mayr and Subramanian [78, 109]. Given the input assignment x on $I(N)$, restrict f to the mapping $f_{x,O(f)}$ from assignments to $L(N)$ to assignments to $O(N) \cup L(N)$. This mapping has $|O(N)|$ more outputs than inputs. As long as the number of outputs is larger than the number of inputs, repeat the following operation. Find a constant output i (guaranteed to exist by the definition of scatter-freedom). If output i has value b, and $i \in L(N)$, then also assign value b to the input i, so that the difference between the number of unassigned outputs and the number of unassigned inputs does not change. As a result, this difference can only decrease when an output $i \in O(N)$ is assigned a value, and so the difference reaches 0 when all the outputs in $O(N)$ have been assigned a value. Note that the algorithm assigns values to some subset $S \subseteq L(N)$, as well as to the outputs in $O(N)$, and runs in time $O(m')$ for $m' = |I(N) \cup O(N) \cup S|$.

Lemma 3.11 *Let N be a scatter-free network with transition function f, and let x be an input assignment on N. Then the scatter-free algorithm finds a set $S \subseteq L(N)$ and a stable configuration w for $\{f_S\}$ such that w is the unique stable configuration of $\{f_S\}$ consistent with x. The running time is $O(m')$, where $m' = |I(N) \cup O(N) \cup S|$.*

Proof. Let $w = xyz$ be the configuration for $\{f_S\}$ found by the algorithm, where x, y, z are assignments on $I(N)$, $O(N)$ and S respectively. Recall that when the subset $I(N) \cup S$ of the inputs to f was assigned value xz, the outputs in the subset $O(N) \cup S$ had value yz.

Therefore $f_{xz,O(N)\cup S}(t) = yz$ for all assignments t on $L(N)\backslash S$, proving that $f_S(xz) = yz$ and therefore that w is a stable configuration of $\{f_S\}$. Uniqueness follows inductively by observing that every output fixed by the algorithm is forced by the inputs already fixed for f, and therefore for f_S. □

If N converges to g, then the stable configuration w obtained gives in particular the output value $g(x) = w_{O(N)}$, since $g = f_\emptyset = [f_S]_\emptyset$.

The scatter-free algorithm can be run even if only a partial input assignment x' to all but k of the inputs is given, in which case it will only be able to determine the value of a partial output assignment y' to all but at most k of the outputs. This shows that scatter-free networks converge to scatter-free gates [109].

A problem is in $\mathcal{NC}^1$ if there is a circuit over $\{\text{AND}, \text{NOT}\}$ with unbounded fanout at the gates that computes a solution from an input instance of size n, where the circuit has size polynomial in n and depth proportional to $\log n$. A problem B is $\mathcal{NC}^1$-reducible to A if there is an $\mathcal{NC}^1$ circuit for B that uses, in addition to the gates AND and NOT, an oracle gate for A. The depth of the circuit is in this case computed by assigning depth $\log k$ to each occurrence of the gate for A with at most k inputs and k outputs. One usually assumes some kind of uniformity for the different circuits corresponding to different values of n. We shall assume logspace uniformity, so that the n-input circuit in the family can be found in $O(\log n)$ space.

We can relate the complexity of evaluating convergent networks to that of evaluating circuits, extending a result of Subramanian [109] in the scatter-free case to the general nonexpansive case.

Lemma 3.12 *The problem of evaluating the gate defined by a convergent network over Ω reduces to the evaluation question for circuits over Ω, for all nonexpansive bases Ω.*

Proof. Given a network N with $|L(N)| = l$ that converges to g, and an input assignment x, the reduction constructs a circuit (N, k) with $k = 2l + 1$, sets all the inputs in the input sequence to x, and chooses an arbitrary initial internal assignment. After evaluation of the circuit, the majority of the y^i obtained as an output sequence will give the value of $g(x)$, as shown above, and majority can be computed in $\mathcal{NC}^1$. This reduction views the evaluation question as a search problem. The search problem can in turn be reduced to the decision problem by asking the value of each output bit of (N, k) separately. □

In the scatter-free case, majority is not needed for this reduction, because the above scatter-free algorithm shows that in fact the output assignments y^i will remain at $g(x)$ once $i \geq l$.

3.1.4 Representation by Convergent Networks

In Chapter 2, we considered the simulation of gates g by means of circuits over a basis Ω. The notion of a convergent network makes it possible to view this as a special case of the representation of gates g by means of convergent networks over a basis Ω. Is this new form of representation more powerful? Subramanian gave a positive answer to this question, by showing that the gate Δ_4 (see section 2.4.1), which cannot be simulated by a circuit over $\{X\}$, can be represented by means of a network over $\{X\}$ that converges to it [109]. In section 2.4.1, we saw that the gates X_k could not be decomposed into simpler gates (gates with fewer than k inputs or fewer than k outputs) by means of simulating circuits, and used this fact to show that the nonexpansive gates (and the scatter-free gates) could not be described with a finite basis. We now show, by contrast, that there exists for all k a network over $\{X\}$ that converges to X_k.

The network $N = N_k$, illustrated in Fig. 3.3 for the case $k = 5$, has $k(k+1)$ coordinates $ij0$ with $0 \leq i < k$ and $0 \leq j \leq k$, and k^2 coordinates $ij1$ with $0 \leq i, j < k$. These coordinates are partitioned into $I(N) = \{ij0 : j = 0\}$, $O(N) = \{ij0 : j = k\}$, and $L(N) = \{ij0 : 0 < j < k\} \cup \{ij1\}$. Coordinate j in $ij0$ is treated as an integer; coordinate i in $ij0$ and coordinates i, j in $ij1$ are treated as integers *modulo k*. The network N consists of k^2 gates g^{ij}, with $0 \leq i, j < k$. The gates g^{ij} with $j \neq 0$ are X gates; the gates g^{ij} with $j = 0$ are comparators (recall that comparators can be simulated with X gates). The gates are defined by $I(g^{ij}) = \{ij0, ij1\}$, $O(g^{ij}) = \{i(j+1)0, (i+1)(j+1)1\}$, and $g^{ij}(x) = y$ if $h(x_{ij0}x_{ij1}) = y_{i(j+1)0}y_{(i+1)j1}$, where h is either C or X depending on whether $i = 0$ or $i \neq 0$. In the associated directed graph, the coordinates $ij0$ form k paths, one for each value of i, going through the gates g^{ij} with the same value of i; the coordinates $ij1$ for k cycles, one for each value of $(i - j) \bmod k$, going through the gates g^{ij} with the same value of $(i - 1) \bmod k$.

We now show that N_k converges to a gate g with $g(x) = y$ if $X_k(x_{000}x_{100}\cdots x_{(k-1)00}) = y_{0k0}y_{1k0}\cdots y_{(k-1)k0}$.

Lemma 3.13 *The network N_k converges to the gate X_k.*

Proof. We use the scatter-free algorithm of Subramanian to perform the evaluation of the gate g to which N_k converges. If input $i00$ has value 0, then the algorithm assigns value 0 to coordinates $i10, i20, \ldots, ik0$ in that order, using gates $g^{i0}, g^{i1}, \ldots, g^{i(k-1)}$, because if the first input of a comparator or X gate has value 0, then so does its first output. Therefore, if the ith input of gate g has value 0, then the ith output of g has value 0. In particular $g(0) = 0 = X_k(0)$ for the all-zero input assignment.

Suppose that we have a list of distinct values $i', i' + 1, i' + 2, \ldots, i'' - 1, i''$ modulo k, such that input $i'00$ has value 1, input $i''00$ has value 1, and input $i00$ has value 0 for the intermediate i in the list. Suppose also that $i' \neq i''$, so that the list consists of at least 2 and at most k values. We show that the algorithm assigns value 0 to output $i''k0$. As a result, if the input assignment x for g has at least two coordinates set to 1, and i'' is one

CHAPTER 3. STABILITY IN NONEXPANSIVE NETWORKS

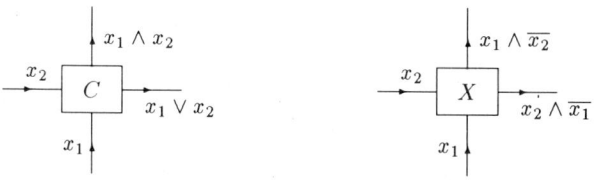

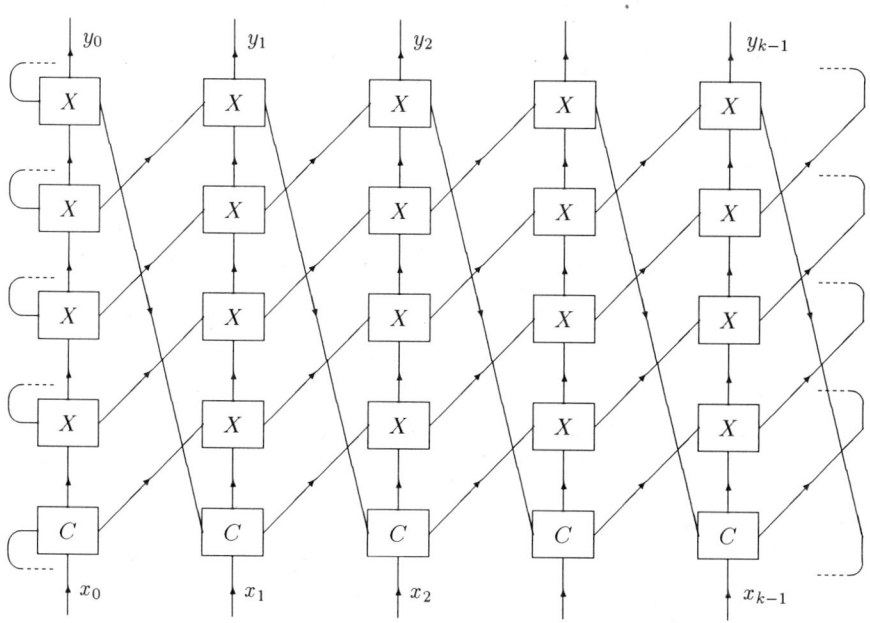

Figure 3.3: The network N_k that converges to X_k

such coordinate, we can let i' be another such coordinate immediately preceding it (modulo k), so that the intermediate coordinates $i'+1, i'+2, \ldots, i''-1$ are set to 0, and conclude that the corresponding output i'' in $g(x)$ has value 0. On the other hand, if an input i in x is set to 0, then we know that the corresponding ouput i in $g(x)$ has value 0. Therefore $g(x) = 0 = X_k(x)$ for all x with at least two 1s. To prove the claim, we first note that, the algorithm assigns value 0 to all $ij0$ for which input $i00$ has value 0, hence for all the intermediate i in the list. The algorithm assigns value 1 to the coordinate $(i'+1)11$, the second output of comparator $g^{i'0}$ with input $i'00$ of value 1. The algorithm then assigns value 1 to coordinates $(i'+2)21, (i'+3)31, \ldots, i''(i''-i')1$ in that order, because if coordinate $i(i-i')1$ for some intermediate i has value 1, then as the second input of the X gate $g^{i(i-i')}$ whose first input has value 0, it must give value 1 to the second output $(i+1)(i+1-i')1$. The algorithm assigns value 0 to coordinate $i''(i''-i'+1)0$, the first output of an X gate with first input $i''(i''-i')1$ of value 1. Finally, the algorithm assigns value 0 to coordinates $i''(i''-i'+2)0, i''(i''-i'+3)0, \ldots, i''k0$ in that order, by going through the X gates on the path i'', so the output $i''k0$ does indeed have value 0.

It remains to consider the case where the input assignment x has a single coordinate i_0 set to 1. For this case, it is easier to give a stable configuration of the network explicitly: Set coordinates of the form $i_0 j 0$ and of the form $(i_0+j)j1$ to 1, and all other coordinates to 0. It can be verified that this configuration is stable (satisfies all the gate equations), and therefore gives the correct input/output behavior for g. The only input set to 1 is $i_0 00$, and the only output set to 1 is $i_0 k 0$, proving that $g(x) = x = X_k(x)$ in this case as well. □

It is in general hard to decide whether a given basis can represent a given gate by means of convergent networks. In some cases, such a result follows from the fact that convergent networks of nonexpansive, scatter-free, or monotone gates must converge to nonexpansive, scatter-free, and monotone gates respectively [109]. The existence of a finite nonexpansive basis Ω that could be used to represent arbitrary nonexpansive gates via convergent networks over Ω cannot be ruled out at present. We do know, however, that there is a monotone scatter-free gate that cannot be represented with networks over $\{X\}$. This gate will be presented in section 3.4.5, in connection with a general decision procedure for deciding whether a given gate can be represented using a given basis.

The simulation of one basis Ω' with another basis Ω was used in Chapter 2 to reduce computational questions over Ω' to the same question over Ω. Can such a reduction be carried out if the representation of Ω' by Ω is given by convergent networks, rather than circuits? We address this question for the evaluation question and give a positive answer. This result was previously known in the scatter-free case, and also in the nonexpansive case if the networks representing gates are of constant size [109].

Lemma 3.14 *The evaluation question for circuits whose gates are defined by convergent networks over Ω reduces to the evaluation question for circuits over Ω.*

Proof. Let K be a circuit that computes a function f, and suppose that each gate $g \in K$ is described by a network N_g that converges to g. If we replace each gate in K by the corresponding network, we obtain a network N that converges to f, by the substitutivity property. The problem of evaluating f from N reduces to the problem of evaluating a circuit (N, k) by Lemma 3.12. The value k in Lemma 3.12 was proportional to the number of links in N, but can in fact be reduced to a value proportional to the number of links coming from the N_g (i.e., we need not count links from K), because such links form a feedback arc set for N (see section 3.1.3).] ☐

In particular, the evaluation question for circuits of X_k gates is no harder than the evaluation question for X gates, up to $\mathcal{NC}^1$ reductions, a result that could not be obtained by simple simulation in view of Lemma 2.9.

3.1.5 Algorithms for Stability

Let N be a nonexpansive network, f its transition function, and x an input assignment for N. A stable configuration for N consistent with x is a configuration $t = xyz$, where y and z are output and internal assignments respectively, such that $f(xz) = yz$. This means that z must satisfy $h_x(z) = z$, where h is the internal function of the network, so the problem of finding a stable configuration reduces to the problem of finding a fixed point of the internal function.

Suppose then that we are given a nonexpansive mapping f with $I(f) = O(f) = L$, and we wish to find a fixed point $x = f(x)$. Suppose that f has a fixed point. By Theorem 3.9, given a coordinate $i \in L$, the fixed points for f_i are precisely the coordinates x_i of fixed points x for f. Suppose that a is a fixed point of f_i. Then there must exist an assignment z on $L\setminus\{i\}$ such that $f(az) = az$, and therefore $f'(z) = z$ for the restriction $f' = f_{a,L\setminus\{i\}}$. Conversely, if z satisfies $f'(z) = z$, then $f_{a,\{i\}}(z) = f_i(a) = a$ from the definition of f_i, so $f(az) = az$. The problem of finding a fixed point of f reduces then to finding a fixed point a for f_i and then a fixed point z for $f' = f_{a,L\setminus\{i\}}$, giving a fixed point az for f.

In order to find a fixed point a for f_i, it is sufficient to test whether $f_i(a) = a$ for the two possible values $a = 0, 1$. The evaluation of f_i on input a is just the evaluation of the convergent network (N, S) with $N = \{f\}$ and $S = \{i\}$, and can be performed by Lemma 3.10 with $O(m)$ queries to f, each taking $O(m)$ time, where $m = |L|$. Once a fixed point a for f_i has been found, we are left with the subproblem of finding a fixed point of $f' = f_{a,L\setminus\{i\}}$, where $I(f') = O(f') = L\setminus\{i\}$ has one less coordinate than L. We solve this subproblem recursively, and the resulting fixed point z for f' gives a fixed point az for f. If the algorithm ever determines that a is not a fixed point of some f_i for both values $a = 0, 1$, then it can conclude that there is no fixed point for f. Since each recursive call reduces the number of coordinates by 1, the total number of recursive calls is m, each using $O(m)$ queries with $O(m)$ time per query.

Theorem 3.15 *Given a network N of size $|R(N)| = m$, and an input assignment x for N, one can find a stable configuration for N, if one exists, with $O(m^2)$ queries to the transition function of N, in $O(m^3)$ time, using $O(m)$ space.*

We seek a more efficient algorithm. The basic idea to achieve efficiency consists of the observation that most queries to f are close to earlier queries, i.e., require changing only a few coordinates; the complexity of these queries is then proportional to the number of coordinates changed, therefore smaller than m. We assume that for some x, the $2m+1$ points $f^{(j)}(x)$ with $0 \leq j \leq 2m$ have been previously obtained. This requires $O(m)$ queries that can be performed in $O(m^2)$ time. The evaluation of $f_i(a)$ using Lemma 3.10 computes $2m+1$ points $f'^{(j)}(z)$ with $0 \leq i \leq 2m$, where $f' = f_{a,L\setminus\{i\}}$ and $z = x_{L\setminus\{i\}}$, and therefore provides directly the $2m+1$ points for the recursive call on f'.

More generally, if the successive values for coordinates $1, 2, \ldots, m$ (where $m = |L|$) found recursively to obtain a fixed point a are $a_1, a_2, \ldots, a_m$, then the functions applied iteratively are $f, f_{a_1,L}, f_{a_1 a_2, L}, f_{a_1 a_2 \cdots a_m, L}$. At the first step of the respective iterations of these l functions, the values used as inputs to f are $x, a_1 x_{L\setminus\{1\}}, a_1 a_2 x_{L\setminus\{1,2\}}, \ldots, a_1 a_2 \cdots a_m = a$. If we denote these values by $x^{10}, x^{11}, x^{12}, \ldots, x^{1m}$, then

$$\sum_{1 \leq i \leq l} d(x^{1(i-1)}, x^{1i}) \leq d(x, a) \leq m,$$

because these points form a shortest path from x to a. Therefore the first step of all iterations, assuming that the a_i are guessed correctly, takes $O(m)$ time for all of them together. Then this inequality holds also for the outputs

$$\sum_{1 \leq i \leq l} d(f(x^{1(i-1)}), f(x^{1i})) \leq d(x, a) \leq m,$$

and therefore

$$\sum_{1 \leq i \leq l} d(x^{2(i-1)}, x^{2i}) \leq d(x, a) \leq m,$$

where $x^{20}, x^{21}, x^{22}, \ldots, x^{2m}$ are the inputs to the second iteration, because x^{2i} is obtained from $f(x^{1i})$ by setting coordinates $1, 2, \ldots, i$ to $a_1, a_2, \ldots, a_i$ respectively. The same $O(m)$ time bound holds then for each iteration step for all functions considered together, and since $2m+1$ steps of iteration are used, the total time bound is $O(m^2)$.

The assumption that each a_i in turn is guessed correctly can be justified by the fact that both values $a_i = 0, 1$ can be tried simultaneously and then the one that gives a correct answer first (a value that can be extended to a fixed point a) is kept. The time bound can be reduced from $O(m^2)$ to $O(m + d^2)$ if the starting point x is at distance at most d from some fixed point a.

Theorem 3.16 *Given a network N of size $|R(N)| = m$, and an input assignment x for N, one can find a stable configuration for N, if one exists, with $O(m^2)$ queries to the transition function of N, in $O(m^2)$ time, using $O(m^2)$ space.*

The scatter-free algorithm of Mayr and Subramanian [78, 109] can be viewed as an efficient implementation of the first fixed point algorithm above. In the scatter-free case,

CHAPTER 3. STABILITY IN NONEXPANSIVE NETWORKS

convergent networks can be evaluated in $O(m)$, rather than $O(m^2)$ time, so this already reduces the time complexity from $O(m^3)$ to $O(m^2)$. Moreover, the running time for evaluating $f_i(a)$ is actually $O(|S|)$, where $\{i\} \subseteq S \subseteq L$, and the algorithm gives a z such that $f_S(az) = f_i(a)z$ (see Lemma 3.11). Therefore, if $f_i(a) = a$, then $f_S(az) = az$, and we only need to find a fixed point of $f' = f_{az, L\setminus S}$ to obtain a fixed point for f. By trying the two values $a = 0, 1$ in parallel, one can find an a such that $f_i(a) = a$ in $O(|S|)$ time, if one exists; this time is well spent, because it reduces the problem to a fixed point problem on $|S|$ fewer coordinates, namely the problem for $f_{az, L\setminus S}$. This ensures linear-time complexity.

Theorem 3.17 *A stable configuration for a scatter-free network N of size $|R(N)| = m$ can be found in $O(m)$ time.*

We conclude this section by relating the complexity of the stability question as a *decision problem* to that of the evaluation question, extending a known result for the scatter-free case [78, 109]. The complexity of the search problem will be examined in section 3.2.2.

Theorem 3.18 *The stability question for networks over a nonexpansive basis Ω, as a decision problem, reduces to the evaluation question for circuits over Ω.*

Proof. The stability question asks whether the internal function f of the network has a fixed point. By Theorem 3.9, the mapping f has a fixed point if and only if each f_i has a fixed point, so the question reduces to the evaluation of each f_i on the two values $a = 0, 1$. This evaluation is a convergent network computation, which reduces to the evaluation question by Lemma 3.12. □

Note that the reduction requires the evaluation of $2m$ circuits over Ω (two for each f_i), each of depth $O(m)$ and size $O(m^2)$.

3.1.6 Randomized Algorithms

In this section we present a simple, randomized $O(m^2)$ algorithm for finding a fixed point of a nonexpansive mapping f with $I(f) = O(f) = L$ and $|L| = m$. The algorithm uses Monte Carlo randomization with one-sided error. This means that if a fixed point exists, then the algorithm will find it with high probability; if no fixed point exists, then the algorithm will fail to find a fixed point but, unlike the deterministic algorithm of the previous section, it will not provide a proof that no fixed exists. The first polynomial time fixed point algorithm [30] used the idea of repeatedly decreasing the pair $(d(x, f(x)), d(x, a))$ in lexicographic order, where x is a current point and a is some fixed point. We shall use both parameters, distance to the image and to a fixed point, in the algorithms below.

The algorithm is as follows. We start from an arbitrary point x and $y = f(x)$. A bit i is chosen uniformly at random from all the bits i such that $x_i \neq y_i$. The point x is then replaced by $x \oplus e^i$, its image y is updated accordingly, and the entire process is repeated. If, within $3cm^2$ such updates, the algorithm has not reached a fixed point, then it stops and announces that no fixed point was found.

Let $x^0, x^1, x^2, \ldots$ be the sequence of points generated by the algorithm. If f has a fixed point a, then as long as x^j is not a fixed point, we have $d(x^{j+1}, a) = d(x^j, a) - Z_j$, where $Z_j = 1$ if the bit i in which x^j and x^{j+1} differ satisfies $x_i^j \neq a_i$, and $Z_j = -1$ if the bit i satisfies $x_i^j = a_i$. Since the bit i is chosen uniformly at random from the set of bits in which x^j and $y^j = f(x^j)$ differ, we have $Z_j = 1$ with probability p_j, where p_j is the fraction of the bits i in which x^j and $y^j = f(x^j)$ differ for which $x_i^j \neq a_i$. Now the key observation is that since $d(y^j, a) \leq d(x^j, a)$, among the bits where x^j and y^j differ, there are at least as many bits in which a agrees with y^j as there are bits in which a agrees with x^j. Hence $p_j \geq 1/2$. Note that, given p_j, the random variable Z_j is independent of all previous $Z_{j'}$, $j' < j$.

To complete the analysis, we introduce a related and well-understood process P. In this process, a sequence of integers in the range $[-m, m]$ is constructed, starting with integer 0, and moving from each integer to the next by either incrementing or decrementing the integer by 1, where both possibilities occur with probability $1/2$, and the decision at each point in time is independent of earlier decisions. The process ends when one of the two endpoints $-m$ or m is reached. The following is well-known.

Lemma 3.19 *The expected number of steps taken by process P is m^2.*

Proof. Let Y_i be $+1$ or -1 depending on whether P increments or decrements the number at step i, and $Y_i = 0$ otherwise. Then $|\sum_i Y_i| = m$. Furthermore, $E(Y_i Y_j) = 0$ for $i < j$, because Y_j is equally likely to have value $+1$ or -1, given Y_i. Therefore

$$m^2 = E\left(\left(\sum_i Y_i\right)^2\right) = E\left(\sum_i Y_i^2\right) = E\left(\sum_i |Y_i|\right).$$

The sum on the right counts the number of steps taken. □

If we consider the integers in absolute value, then we have a sequence of integers in the range $[0, m]$, the integer 0 is always followed by 1, and the remaining integers are still either incremented or decremented with probability $1/2$; the expected time until m is reached is still m^2.

Consider now the process Q consisting of the sequence of integers $d_j = m - d(x^j, a)$ from the above algorithm. These integers also move in the range $[0, m]$, increasing or decreasing by one at each step, with $d_{j+1} = 1$ if $d_j = 0$ and $d_{j+1} = d_j + 1$ with probability at least $1/2$. The sequence d_j generated by process Q can be viewed as a subsequence of the sequence generated by process P. Whenever P increases its integer, so does Q; this can be ensured since P increases its integer with probability $1/2$, while Q increases its integer d_j with probability at least $1/2$. On the other hand, it can happen that Q increases its integer d_j, while P decreases its integer. In that case, we extend the sequence P until its integer coincides with d_j again; the intermediate values on P are all smaller than d_j. As a result, when P reaches m, Q must also reach m, and the expected number of steps is at most m^2, since the sequence from Q is a subsequence of the sequence from P. If $d_j = m$, then x^j is a fixed point of f. Therefore the expected number of steps until a fixed point is reached is at most m^2. Markov's inequality states that a nonnegative random variable will exceed its expected value by a factor of λ with probability at most $1/\lambda$. Therefore the probability that no fixed point will be reached within $3m^2$ steps, if a fixed point exists, is at most $1/3$. It follows that the probability of not reaching a fixed point within $3cm^2$ steps is at most $(1/3)^c$.

Theorem 3.20 *There is a Monte Carlo algorithm for finding a fixed point of a nonexpansive mapping on the m-cube that runs in $O(m^2)$ time, with high constant probability.*

The same algorithm can be applied to finding a simultaneous fixed point of many nonexpansive mappings f^j, by choosing at each step a j such that $y = f^j(x) \neq x$ and a random i such that $y_i \neq x_i$. Noting that a 2SAT clause $(x_i = a) \vee (x_j = b)$ describes the fixed point set of a nonexpansive mapping given by a single X_{ij}^{ab} gate (see section 3.2.3), this gives a randomized 2SAT algorithm found in Papadimitriou [90].

We shall present two improvements. The first improvement does not decrease the $O(m^2)$ running time, but reduces the number of queries to f to just $O(m^{4/3})$. This is mainly of interest from the point of view of parallel computation, since all bit changes from one query to the next can be performed in parallel, and also as a comparison with the lower bounds of the next section. The second improvement will be given in section 3.4.3, and consists of a method for transforming Monte Carlo fixed point algorithms into Las Vegas algorithms without affecting the complexity by more than a constant factor. This means that the algorithms can also prove that a fixed point does not exist, and the expected running time is still $O(m^2)$.

The overall strategy for reducing the number of queries is the following. First, we observe that in the above randomized algorithm, as long as the current point x satisfies $d(x, f(x)) \geq 2d$ for some given d, one can choose d random coordinates where x and $f(x)$ differ, and the progress in the algorithm is then essentially the same as in the case of d coordinates chosen one by one at random. This reduces the number of queries to $O(m^2/d)$, with $O(d)$ time per query, and provides eventually a point x with $d(x, f(x)) \leq 2d$. Next, we observe that the first algorithm from the previous section, which performs $O(m^2)$ queries in $O(m)$ time per query, can be modified when such a point x is known, so that it only requires $O(dm)$ queries and $O(d)$ time per query to find a fixed point. By means of randomization, the expected number of queries for this second phase can be reduced to $O(\sqrt{dm})$. The total number of queries is therefore $O(m^2/d + \sqrt{dm})$, with $O(d)$ time per query. Letting $d = m^{2/3}$ gives the stated bound.

The bound for the first phase is obtained, as before, by using a process P' that corresponds to the worst-case values of the probabilities involved in the algorithm. This time a sequence of integers is constructed in the interval $[-m+d, m-d]$, starting from 0. The boundary m is replaced by $m - d$ because once x is within distance d from a fixed point, it is within distance $2d$ of $f(x)$, and the first phase can be terminated. We assume that a subset of d elements from a set consisting of k 1s and k -1s is chosen uniformly at random, where $k \geq d$, and that these d elements are added to the current integer to obtain a new integer. The d elements correspond to the choice of d coordinates among those where x and $f(x)$ differ, the assumption that the number of 1s and -1s is the same provides a worst-case bound as before, and the assumption $2k \geq 2d$ corresponds to the fact that $d(x, f(x)) \geq 2d$. The process terminates when the integer obtained is outside the range $[-m+d, m-d]$.

Lemma 3.21 *The expected number of steps taken by process P' is at most $2m^2/d$.*

Proof. Let Y_i be the ith element chosen by P', either a $+1$ or a -1. Each such Y_i belongs to a set of d consecutive Y_j chosen simultaneously. If the current sum leaves the range $[-m+d, m-d]$, then all subsequent Y_i have value 0. Since only d of the Y_i are chosen simultaneously, and the sum of these d Y_i is in the range $[-d, d]$, we can infer that $|\sum_i Y_i| \leq m$. Furthermore, $E(Y_i Y_j) = 0$ as long as the elements $i < j$ do not belong to the same set of d elements chosen at once. If the two elements belong to the same set of d elements, and $Y_i = 1$, then Y_j is 1 with probability $(k-1)/(2k-1)$ and -1 with probability $k/(2k-1)$, with a symmetric case if $Y_i = -1$. Therefore $E(Y_i Y_j) = -1/(2k-1)$, and

$$m^2 \geq E((\sum_i Y_i)^2) = E(\sum_i Y_i (\sum_j Y_j)) = E(\sum_i |Y_i|(1 - \frac{d-1}{2k-1})) \geq \frac{1}{2} E(\sum_i |Y_i|).$$

The sum on the right counts the number of steps taken times d, proving the lemma. □

The proof that P' provides a bound to the expected number of steps taken by the algorithm is similar to the argument for P in the previous algorithm, and we omit the details. This completes our discussion of the $O(m^2/d)$ bound on the number of queries in the first phase. Each query changes only d coordinates, and therefore takes $O(d)$ time. We now turn to the second phase.

The second phase is based on the first algorithm from the last section. Suppose that f has a fixed point and that a point x with $d(x, f(x)) = d$ is known. We show how a point x' with $d(x', f(x')) < d$ can be found with $O(m)$ queries. As a result, a fixed point can be obtained with $O(dm)$ queries. The algorithm first computes the points $x^j = f^{(j)}(x)$ for $j = 1, 2, \ldots$. If $d(x^j, x^{j+1}) < d$, then $x' = x^j$ is the required point. As long as $d(x^j, x^{j+1}) = d$, each x^j differs from the next one in d coordinates. If we iterate f more than m/d times, then the total number of coordinate changes for successive x^j is greater than m, so there must exist a coordinate i that changes twice, say $x_i^j \neq x_i^{j+1} = x_i^l \neq x_i^{l+1}$. Since f has a fixed point, there must exist a value a (either 0 or 1) such that $f_i(a) = a$, say $a = x_i^j$ (the case $a = x_i^l$ is similar). As before, we iterate the network $(N, \{i\})$, starting with $x^j = az^0$ and using a as input, until the output value $y^{k+1} = a$ is obtained, for some $k \leq m+1$. The first special case of the basic equation from section 3.1.2 gives then

$$d(z^k, z^{k+1}) + 1 \leq d(z^0, z^1),$$

so $x' = az^k$ must indeed satisfy $d(x', f(x')) < d$. The number of queries to decrease $d(x, f(x))$ is $O(m)$, and the time per query is $O(d)$ since z^i and z^{i+1} differ in at most d coordinates.

We now show how the total number $O(dm)$ of queries can be reduced to $O(\sqrt{d}m)$ expected queries using randomization. We again try the two possible values for a, namely $a = x_i^j$ and $a = x_i^l$, and iterate the network $(N, \{i\})$ starting with x^j and with x^l, in parallel, until the corresponding value a appears at the output of *one* of the two computations. We then let x' be one of the two resulting $y^k z^k$, chosen from the two such points obtained by the two parallel computations at random. If this random choice is the one that produced the value a at the output, then we have indeed decreased $d(x, f(x))$ as before. Otherwise, we know at least that $d(x, f(x))$ has not increased. The expected number of such computations for $d(x, f(x)) = d$ to decrease down to 0 is therefore at most $2d$. It remains to be shown that if the number of such computations is some given quantity d, then the expected number of queries is $O(\sqrt{d}m)$.

Suppose that there is a fixed point $a\hat{z}$ with $a = x_i^j$ (the case $a = x_i^l$ is similar). If the network $(N, \{i\})$ is iterated starting with $az^0 = x^j$ using $x_i^j = a$ as input, so that a does not appear at the output except possibly at the last step, then the second case of the basic equation from section 3.1.2 gives

$$d(z^k, \hat{z}) + \sum_{0 < i \leq k} d(y^i, a) \leq d(z^0, \hat{z}),$$

which can be rewritten as

$$d(z^k, \hat{z}) + d(y^k, a) + k - 1 \leq d(z^0, \hat{z}) + d(x_i^j, a).$$

If the network is iterated starting with $\overline{a}z^0 = x^l$ using $x_i^l = \overline{a}$ as input, so that $\overline{a}$ does not appear at the output except possibly at the last step, then the second case of the basic equation gives

$$d(z^k, \hat{z}) + \sum_{0 < i \leq k} d(y^i, a) \leq d(z^0, \hat{z}) + \sum_{0 \leq i < k} d(\overline{a}, a),$$

which can be rewritten as

$$d(z^k, \hat{z}) + d(y^k, a) \leq d(z^0, \hat{z}) + k - 1 + d(x_i^l, a).$$

Therefore, if k queries are made, then the distance to the fixed point $a\hat{z}$ does not increase by more than $k - 1$, and in fact decreases by at least $k - 1$ with probability at least $1/2$, depending on whether the chosen $y^k z^k$ is the one obtained from the computation for x^j or for x^l. The worst-case scenario in terms of the distance to the fixed point is then one in which, if $k + 1$ queries are made, then the distance to the fixed point increases or decreases by k, and each alternative occurs with probability $1/2$. To model this process, we consider a sequence of d random variables (one for each of the d computations) $Y_1, Y_2, \ldots, Y_d$, such that given Y_i and $|Y_j|$ with $j > i$, the variable Y_j takes value $|Y_j|$ or $-|Y_j|$ with equal probability (so that $E(Y_iY_j) = 0$). These variables represent the change k in the distance to the fixed point as a result of each computation. We again make the problem symmetric in positive and negative values, and assume that the sum of the Y_j remains in the interval $[-m, m]$. The total number of queries is then $2\sum_j(|Y_j| + 1)$ (i.e., $k + 1$ queries for each of the two parallel computations). The $O(\sqrt{d}m)$ bound on the number of queries follows from the following lemma. The $O(d)$ time bound per query follows from the fact that each query differs from the previous one in at most d coordinates.

Lemma 3.22 *Suppose that the d random variables $Y_1, Y_2, \ldots, Y_d$ satisfy $|\sum Y_j| \leq m$ and $E(Y_iY_j) = 0$ for $i \neq j$. Then $E(\sum_j |Y_j|) \leq \sqrt{d}m$.*

Proof. We have

$$\begin{aligned} m^2 &\geq E\left(\left(\sum_j Y_j\right)^2\right) = E\left(\sum_j Y_j^2\right) = dE\left(\frac{1}{d}\sum_j |Y_j|^2\right) \\ &\geq dE\left(\left(\frac{1}{d}\sum_j |Y_j|\right)^2\right) \geq \frac{1}{d}\left(E\left(\sum_j |Y_j|\right)\right)^2, \end{aligned}$$

where the last two inequalities follow from the fact that the average of the squares is at least the square of the average (i.e., $E(Z^2) - (E(Z))^2 = E((Z - E(Z))^2) \geq 0$). □

Combining the two phases with $d = m^{2/3}$ gives the following:

Theorem 3.23 *There is a Monte Carlo algorithm for finding a fixed point of a nonexpansive mapping on the m-cube that runs in $O(m^2)$ time and makes $O(m^{4/3})$ queries, with high constant probability.*

3.1.7 Lower Bounds

In section 3.1.5, the stability question for nonexpansive bases was reduced to the evaluation question for nonexpansive bases, using $\mathcal{NC}^1$ reductions. This reduction implies, in particular, that if the evaluation question could be solved efficiently in parallel, say within the class $\mathcal{NC}$, then the stability question could also be solved efficiently in parallel. The parallel complexity of the evaluation question for nonexpansive bases has remained open for some time. The problem is not believed to be complete for $\mathcal{P}$ (which would be evidence of hardness from the point of view of parallel computation), mainly because of the lack of fanout; the problem does not appear to be in the class $\mathcal{NC}$, mainly because certain seemingly sequential problems, such as the lexicographically first maximal matching problem, are equivalent to the evaluation question for comparators. This has led to the conjecture that the evaluation and stability questions over nonexpansive bases define new complexity classes within $\mathcal{P}$ that are incomparable to the class $\mathcal{NC}$ [78, 109].

In this section we prove explicit lower bounds on the parallel complexity of the evaluation and stability questions over nonexpansive bases. In order to obtain explicit parallel lower bounds, we shall consider circuits of large nonexpansive (or scatter-free) gates that can be accessed as oracles in the nonexpansive (resp. scatter-free) model. Although the evaluation question for scatter-free circuits seems to lie outside $\mathcal{NC}$, it has been shown that a considerable amount of parallelism can still be achieved. Mayr and Subramanian showed that a scatter-free circuit of size m can be evaluated in $\sqrt{m}$ parallel time, up to polylogarithmic factors, with a polynomial number of processors; Anderson reduced the processor count to $\sqrt{m}$ [78]. The basic idea of the algorithm is the following. By the definition of scatter-free gates, no gate has more outputs than inputs. If the circuit has more than $\sqrt{m}$ inputs, then we can use these input values in parallel to simplify the gates that the inputs feed into, reducing the size of the circuit by $\sqrt{m}$. Therefore only $\sqrt{m}$ such simplifications will be performed. If the circuit has $\sqrt{m}$ or fewer inputs, then the number of inputs to the circuit can be reduced by 1 as follows. For each gate with the same number of inputs and outputs, and each possible assignment of a value to one of its inputs, determine which output gets assigned a value, and determine this value as well. This information can be represented with a directed edge from the input of a gate (with its assigned value) to the corresponding output of the gate (with the resulting value). A chosen input of the circuit (with its assigned value) determines then a path through the circuit that assigns values to the links along the path and ends in an output of the circuit or in a gate with fewer outputs than inputs; this path can be found efficiently in parallel. Now these links with assigned values can be removed from the circuit, reducing the number of inputs to the circuit while maintaining the property that no gate has more outputs than inputs. After $\sqrt{m}$ such simplifications, the circuit will have no inputs and therefore no gates, so the values of all coordinates are determined.

We shall prove a lower bound on the parallel complexity of evaluating monotone scatter-free circuits. The main idea is to devise circuits for which propagations (the paths through the circuit evaluated in the algorithm described above) will fail to assign a value to some specific output. The circuit will resemble a sorting network, in that inputs of value 0 tend to give value 0 to the leftmost outputs under propagations, and inputs of value 1 tend to give value 1 to the rightmost outputs under propagations, leaving intermediate outputs unassigned throughout most of the execution of any given algorithm.

For monotone networks, the nonexpansive and the scatter-free gates coincide (see Lemma 2.7). In fact, we can adopt the nonexpansive model and only consider oracle gates that answer queries about complete assignments to the inputs of the gates; the queries for partial assignments can be simulated as in the proof of Lemma 2.7.

The circuits will use gates g^{ab} with $I(g^{ab}) = O(g^{ab}) = \{1, 2, \ldots, n\}$, n even; both a and b are n-bit assignments with half the bits equal to 0 and half the bits equal to 1. The gate $g = g^{ab}$ satisfies $g(a) = b$, but has $g(x) = \text{SORT}(x)$ for inputs x that are 'far' from a. The precise definition is as follows. Let $d_0(a, x)$ be the Hamming distance between a and x in the coordinates i where $a_i = 0$, and define $d_1(a, x)$ similarly for the coordinates i where $a_i = 1$. Thus $d(a, x) = d_0(a, x) + d_1(a, x)$. If $d_0(a, x) = k_0$ and $d_1(a, x) = k_1$, then we first change the k_0 largest coordinates that equal 0 in b into 1s to obtain an assignment z, then change the k_1 smallest coordinates that equal 1 in z into 0s to obtain $y = g(x)$. The gate g can be defined with a comparator circuit, although we shall assume that the circuit for g is not known to the algorithm, which can only use g as an oracle to evaluate specific input assignments (or equivalently, partial input assignments). The comparator circuit that computes g is the following. We start with the assignment $w = b$, and repeatedly modify w. For each coordinate i where $a_i = 0$, we compare x_i with each w_j in turn, with j decreasing from n down to 1. After each comparison, the 'max' (OR) output of the comparison gives the value of w'_j, and the 'min' (AND) output of the comparison gives a value to be compared with the next w_{j-1}. Therefore $w' = w$ if $x_i = 0$, while w' differs from w only in the largest coordinate of w equal to 0 if $x_i = 1$. It follows that the k_0 largest coordinates of b equal to 0 will be replaced by 1s as claimed, giving the word z. The second stage is similar, starts with $w = z$, and compares each coordinate x_i such that $a_i = 1$ to each w_j in turn, for j increasing from 1 to n, leaving the 'min' in w'_j and using the 'max' for the next comparison with w_{j+1}.

If $x = a$, then $k_0 = k_1 = 0$, so in the definition of $g = g^{ab}$ we start from b and make no changes, so that $g(a) = b$ as claimed. The following lemma characterizes the words that are sorted by g.

Lemma 3.24 *If $d_0(a, x) = k_0$ and $d_1(a, x) = k_1$, let i_0 be the $(k_0 + 1)$th largest coordinate in b among the coordinates i with $b_i = 0$ (let $i_0 = 0$ if no such coordinate exists), and let i_1 be the $(k_1 + 1)$th smallest coordinate in b among the coordinates i with $b_i = 1$ (let $i_1 = n + 1$ if no such coordinate exists). Then $g^{ab}(x) = \text{SORT}(x)$ if and only if $i_0 < i_1$.*

Proof. Suppose that $i_1 \leq i_0$. Then after the changes in the definition of $g = g^{ab}$ are made, starting from b, the resulting output $y = g(x)$ will have $y_i = 0$ for $i = i_0$, and $y_i = 1$ for $i = i_1$, so y is not sorted. Suppose that we have instead $i_0 < i_1$. Then after the first phase of the definition of g, the coordinates with $i > i_0$ are $y_i = 1$, and the coordinates with $i \leq i_0$ are $y_i = b_i$. Since $i_1 > i_0$, after the second phase we have $y_i = 0$ for all $i \leq i_0$, and some initial segment of the values $i > i_0$ is changed to $y_i = 0$ as well, so y is sorted. □

The circuits that we shall use will be the composition of k gates $g^j = g^{a^{j-1}, a^j}$ with $1 \leq j \leq k$, using a^0 as the input assignment. Since $g^j(a^{j-1}) = a^j$, and a^0 is the input to the first gate g^1 in the circuit, the circuit must output $g^k(g^{k-1}(\ldots g^2(g^1(a^0))\ldots)) = a^k$. The assignments a^j will be chosen at random from a particular distribution, so that a deterministic lower bound on the probabilistic behavior of any deterministic algorithm on this distribution yields a randomized lower bound as well by a result of Yao [118]. An algorithm can query gates g^j, but does not have direct access to the words a^{j-1}, a^j used in defining g^j. The algorithms that we consider can make up to p queries to each gate g^j in one step, and must obtain the correct output within t steps. We view p and t respectively as lower bounds on the number of processors and the parallel time complexity.

Theorem 3.25 *The parallel time complexity of the evaluation question for monotone scatter-free circuits of size m, with $p \geq m$ processors, is $\Omega((m/\log p)^{1/3})$.*

Proof. The size of the circuits described above is $m = kn$. Let $l = \lceil \sqrt{n(\log p + c \log m)} \rceil$, with c a constant to be chosen later. Define a^j by setting the first lj coordinates to 0, the last lj coordinates to 1, and choosing the remaining $n_j = n - 2lj$ coordinates uniformly at random from the set of all assignments of n_j bits that have the same number of 0s and 1s (and independently for all j). We shall choose $n = 4lk$, so that the $2lj$ bits fixed to 0 or 1 do not account for more than half of the bits of a^j. We shall prove that any deterministic algorithm that takes fewer than k steps will give an incorrect value for the output bit $a^k_{n/2}$ with probability close to $1/2$ over the distribution on the a^j. By the choice of the parameters, we have $k^3 = mn/(4l)^2 \approx m/(16(\log p + c \log m))$, giving an $\Omega((m/\log p)^{1/3})$ lower bound on the number of steps.

Given any deterministic algorithm A that seeks the value of the output bit $a^k_{n/2}$, we define a simulating algorithm A' as follows. The algorithm A' mimicks A, but for $1 \leq j \leq k$ it replaces all the input queries to g^j performed by A within the first $j - 1$ steps with a query to SORT. At step j, after the queries to g^j for step j have been made but before they are answered, algorithm A' chooses a^j from the appropriate distribution (with a^0 chosen before the first step). It then checks whether the input queries x to g^j that were performed during the first $j - 1$ steps were such that $g^j(x) = \text{SORT}(x)$, i.e., whether the answers that it gave using SORT instead of g^j were correct. If this equality fails to hold for some such x, then A' fails. Otherwise, algorithm A' uses g^j (and not SORT) for the calls to g^j performed at step j and subsequent steps. If algorithm A' (mimicking A) terminates in fewer than j steps, then A' chooses a^j when it terminates, does the appropriate checking for g^j at that point, and if it does not fail it returns the output value indicated by the simulation.

If A' does not fail, then all its queries give the same values as the corresponding queries in A, so the executions of the two algorithms are undistinguishable, and in particular A' returns the same output value that A gives. Since a^k is not chosen by A' until step k, i.e., until after the output value has been returned, and since the two values 0 and 1 for $a^k_{n/2}$ are equally likely because of the symmetry between 0 and 1 in the random choice of the assignment a^k, the probability that A' either fails or gives an incorrect value for $a^k_{n/2}$ is at least $1/2$. We shall show that the probability of A' failing is very small, so A', and therefore A, must give an incorrect value with probability close to $1/2$ or greater.

Since a^{j-1} is only chosen by A' after the queries for the first $j - 1$ steps have been made, the values of a^{j-1} and a^j are independent from the queries x to g^j made during the first $j - 1$ steps. The number of such queries is at most $(j - 1)p < kp$. We show that given kp arbitrary such query points x in the n-cube, the probability that $g^j(x) \neq \text{SORT}(x)$ for even one of these kp points x is very small, for a random choice of a^{j-1} and a^j. Therefore the probability that A' fails at step j is very small.

Using the notation from the lemma, we show that the probability that $i_1 \leq i_0$ for a given point x is very small, for a random choice of $a = a^{j-1}$ and $b = a^j$. We first assume that the middle bits of a^{j-1} and of a^j are chosen uniformly at random from all n_{j-1}-bit and n_j-bit assignments respectively, and then condition on the event E that the same number of 0s and 1s is chosen for both a^{j-1} and a^j to obtain the correct distribution. We seek a bound on the probability $\Pr(i_1 \leq i_0|E) = \Pr((i_1 \leq i_0) \wedge E)/\Pr(E)$. To bound the numerator, let U be the number of bits among the middle n_{j-1} bits of a in which x differs from a. This random variable has a binomial distribution $U \sim B(n_{j-1}, 1/2)$, and using the notation from the lemma, we have $k_0 + k_1 \geq U$. The middle n_j bits of b can be selected as follows. First choose bits from right to left, calling a bit *bad* if it has value 0, and stopping when either k_0 bad bits have been chosen or all n_j bits have been chosen. If we reach k_0 bad bits first, then the remaining bits are chosen from left to right, calling a bit *bad* if it has value 1. Note that the k_0th bad bit i from right to left has $i > i_0$, and that we can only have $i_1 \leq i_0$ if at least k_1 bad bits are obtained from left to right, i.e., if the total number of bad bits V satisfies $V \geq k_0 + k_1$. Since each bit is bad with probability $1/2$, the number V of bad bits also has a binomial distribution $V \sim B(n_j, 1/2)$. Furthermore, by the preceding inequalities, we can only have $i_1 \leq i_0$ if $V \geq U$, or equivalently if $W \geq n_{j-1}$, where $W = V + (n_{j-1} - U) \sim B(n_{j-1} + n_j, 1/2)$. Let $\mu = (n_{j-1} + n_j)/2$ be the mean of this distribution, and recall that $l = (n_{j-1} - n_j)/2$. By the Chernoff bound (see Chernoff [19], Anglin and Valiant [3]),

$$\Pr(W \geq n_{j-1} = \mu + l) \leq e^{-l^2/\mu} \leq e^{-(\log p + c \log m)} = 1/(pm^c).$$

This gives an upper bound on the numerator $\Pr((i_1 \leq i_0) \wedge E)$. The probability $\Pr(E)$ that exactly half of the n_{j-1} and n_j middle bits in each of a and b are 0s is asymptotically $(1/\sqrt{\pi n_{j-1}/2})(1/\sqrt{\pi n_j/2}) \geq 1/(\pi n/2)$, so $\Pr(i_1 \leq i_0|E) \leq c'n/(pm^c)$ is an upper bound on the probability that $g^j(x) \neq \text{SORT}(x)$, for some constant c'.

Therefore the probability that even one of the kp query points x to g^j does not satisfy $g^j(x) = \text{SORT}(x)$ is at most $c'kn/m^c$, and this gives an upper bound on the probability that A' fails because of a query to g^j. Over the k possible values of j, the probability that A' fails is at most $c'k^2n/m^c \leq c'/m^{c-2}$. Since A and A' give the same output value when A' does not fail, and A' fails or gives an incorrect output value with probability at least $1/2$, the probability that A gives an incorrect output value is at least $1/2 - c'/m^{c-2}$, and this probability approaches $1/2$ as $m \to \infty$ for any choice of $c > 2$. □

The preceding lower bound relies on the fact that the gates g^{ab} in the circuit have many inputs and outputs, so that the function they compute (expressed by the values of a and b) cannot be determined with a reasonable number of queries to g^{ab} alone. The lower bound does not apply if all the gates in the circuit are small. On the other hand, since g^{ab} can be computed by a comparator circuit, we can view the result as a parallel lower bound on the comparator circuit evaluation question when certain portions of the circuit have been hidden inside 'black boxes' g^{ab}. Of course, if the description of the black box as a comparator circuit is given explicitly as above, then we can easily determine a and b. On the other hand, if p is super-polynomial in n, then the gates g^j used are $(\frac{n}{2} - \omega(\sqrt{n \log n}))$-sorters, and it is not known whether sorting networks computing such functions can be distinguished from sorters in polynomial time (see section 2.4.2). This suggests that it may be possible to encode the gates g^{ab} in such a way that they cannot reveal the value of b in polynomial time unless $\mathcal{P} = \mathcal{NP}$, giving then some evidence that the evaluation question may be hard to parallelize in the case of comparator circuits as well.

The stability question in the scatter-free case can be solved in linear time using the algorithm of Mayr and Subramanian [78]. We do not know whether the stability question has a parallel algorithm with a polynomial number of processors whose time complexity improves this $O(m)$ upper bound by more than a polylogarithmic factor, other than by means of interior point methods for linear programming as described in section 3.4.6 for the case of gates of constant size; the best known lower bound is the $\Omega((m/\log m)^{1/3})$ bound derived above for circuits. The proof of Theorem 3.18 reduces the stability question for a network of size m to the evaluation question for m different circuits, each of size $O(m^2)$. Therefore any improvement over the $\sqrt{m}$ parallel time complexity (up to polylogarithmic factors) of evaluating scatter-free circuits leads to sublinear algorithms for the stability question; for this reason, the gap between the square-root parallel upper bound and the cube-root parallel lower bound deserves further study.

An interesting model for the study of the stability question in the monotone scatter-free case is the following essentially equivalent question. Let f be a monotone scatter-free gate with the same input and output sets, with $f(0) = 0$ and $f(1) = 1$. If we start with the all-zero input assignment and set coordinate 1 to value 1, then some output i_1 gets value 1; we then set input i_1 to 1 as well, so that some new output i_2 gets value 1; and so on, until $i_k = 1$ and a cycle of coordinates has been set to 1. The question, in the black-box model, is whether coordinate 2 is among the coordinates set to 1 by this process started at coordinate 1. In the case of comparators, this question becomes a simple process on a digraph G where each vertex has in-degree 2 and out-degree 2; the two edges out of each vertex are labelled 'first' and 'second'. Starting with some specified edge, traverse a forward path, following the 'first' edge out of a vertex the first time the vertex is visited, and following the 'second' edge the second time, until the starting edge is reached. We then ask whether some other chosen edge was visited.

For the stability question on general nonexpansive networks of size m, an $O(m^{4/3})$ randomized sequential upper bound on the number of queries was given in section 3.1.6, and an $O(m)$ time deterministic parallel algorithm with a polynomial number of processors follows from the proof of Theorem 3.18; in the case of gates of constant size, an upper bound close to $\sqrt{m}$ is given in section 3.4.6. This upper bound does not apply when the gates are not of constant size, e.g., when the X_k gate is used, only an upper bound close to $\sqrt{km}$ is obtained. Naor [83] has obtained a nonexpansive lower bound which is stronger than the above parallel scatter-free lower bound. The basic idea is to consider the question of finding a fixed point for the mapping $g^a(x) = x_1 \ldots x_{i-1} a_i x_{i+1} \ldots x_m$, where a is an m-bit assignment and i is the smallest coordinate in which a and x differ, with $g^a(a) = a$. Thus a is the only fixed point, and it can be shown that if a is chosen uniformly at random from all m-bit assignments, then p queries only reveal about $\log p$ bits of a (from left to right), giving an $\Omega(m/\log p)$ lower bound for the stability question. Note that this lower bound is very close to the upper bounds just mentioned. The problem of finding a can also be viewed as an evaluation problem by considering the circuit of size $m' = m^2$ that computes $g^{a(m)}(0) = a$, giving an $\Omega(\sqrt{m'}/\log p)$ lower bound for the evaluation question. Although the mappings g_a turn out to be scatter-free, the lower bounds do not apply in this case to the scatter-free model, where queries consisting of partial input assignments are allowed. In fact, one can determine a directy by testing all the partial assignments to at most two inputs of g^a simultaneously, and then performing one additional call to g^a. (If $y = g^a(x)$, then for all $1 < i < m$, setting $x_i x_{i+1} = a_i \overline{a_{i+1}}$ gives output values $y_0 y_i = a_0 a_i$, while any input assignment of the form $x_i x_{i+1} = \overline{a_i} b_{i+1}$ would leave y_i unassigned, so the $4(n-2)$ partial input queries of the form $x_i x_{i+1} = b_i b_{i+1}$ with $1 < i < m$ are sufficient to determine all a_i with $i < m$, and a single additional query $x = a_1 a_2 \ldots a_{m-1} b_m$ yields $y = a$.) Therefore the problem of finding the fixed point a is actually easy from the point of view of parallel complexity in the stronger (with respect to lower bounds) scatter-free model.

3.2 Fixed Points and Retracts

In the last section we gave algorithms for finding a fixed point of a nonexpansive mapping on the hypercube. In this section we turn our attention to the set of all fixed points. An special operation, the median operation, plays a crucial role in this study.

3.2.1 Fixed Points and Medians

A *median graph* is a graph G with the following property: For all triples of vertices x, y, z in G, there is a unique vertex t in G such that t lies on a shortest path from x to y, on a shortest path from x to z, and on a shortest path from y to z. The unique vertex t is called the *median* of x, y and z, denoted $\text{med}(x, y, z)$. Median graphs have been studied by Avann [7] and by several other authors. See Bandelt and Hedlíková [10] for a survey up to 1980. The following is known (see Fig. 3.4):

Lemma 3.26 *The hypercube is a median graph. In the hypercube, if $t = \text{med}(x, y, z)$, then $t_i = \text{maj}(x_i, y_i, z_i)$, where* maj *is the majority function.*

Proof. Every set of three vertices in the hypercube has two that agree in coordinate i, for each i. If t lies on a shortest path from x to y, and $x_i = y_i$, then we must have $t_i = x_i = y_i$. Therefore any median point t must satisfy $t_i = \text{maj}(x_i, y_i, z_i)$.

On the other hand, if t is defined by $t_i = \text{maj}(x_i, y_i, z_i)$, then given a pair of vertices in the triple x, y, z, say the pair x, y, the vertex t agrees with x and y in every bit where x and y agree. Therefore $d(x, t) + d(t, y) = \sum_i d(x_i, t_i) + d(t_i, y_i) = \sum_i d(x_i, y_i) = d(x, y)$, and t lies on a shortest path from x to y. □

The importance of medians for the study of fixed points of nonexpansive mappings is given by the following observation:

Lemma 3.27 *Let f be a nonexpansive mapping on the hypercube, and suppose that x, y and z are fixed points of f. Then $\text{med}(x, y, z)$ is also a fixed point of f.*

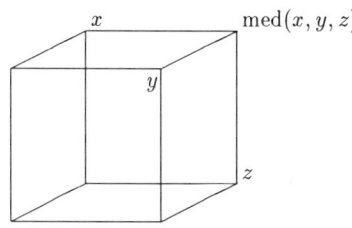

Figure 3.4: The median function.

Proof. Since $t = \text{med}(x, y, z)$ lies on a shortest path from x to y, we have $d(x, y) = d(x, t) + d(t, y)$. Therefore $d(x, y) \leq d(x, f(t)) + d(f(t), y) \leq d(x, t) + d(t, y) = d(x, y)$, and equality holds throughout. It follows that that $f(t)$ lies on a shortest path from x to y, and similarly also on a shortest path from x to z and on a shortest path from y to z. Therefore $f(t)$ is the median of x, y and z, i.e., $f(t) = t$. □

Given two assignments x, y with $S(x) = S(y)$, let $x \wedge y$ denote the assignment z such that $z_i = x_i \wedge y_i$ for all i, and let $x \vee y$ denote the word z such that $z_i = x_i \vee y_i$ for all i. Note that, for every triple x, y, z, we have $\text{med}(x, y, z) = (x \wedge y) \vee (x \wedge z) \vee (y \wedge z) = (x \vee y) \wedge (x \vee z) \wedge (y \vee z)$.

Corollary 3.28 *Let f be a monotone nonexpansive mapping on the hypercube, and suppose that x and y are fixed points of f. Then $x \wedge y$ and $x \vee y$ are also fixed points of f.*

Proof. Since f is monotone, it has a zero-most fixed point a^0 and a one-most fixed point a^1 (see section 2.3.3). Therefore $a^0 \leq x, y \leq a^1$. This implies that $\text{maj}(a_i^0, x_i, y_i) = x_i \wedge y_i$ and $\text{maj}(a_i^1, x_i, y_i) = x_i \vee y_i$, so $\text{med}(a^0, x, y) = x \wedge y$ and $\text{med}(a^1, x, y) = x \vee y$. □

Lemma 3.26 says that the set of fixed points U of a nonexpansive mapping is closed under medians, that is, if $x, y, z \in U$ then $\text{med}(x, y, z) \in U$. Such a set is called a *median set*. If, in addition, the mapping is monotone, then the set U is closed under $\wedge$ and $\vee$ and therefore forms a sublattice of the *distributive lattice* defined by the hypercube (hence a distributive lattice as well). Both distributive lattices and median sets have been studied extensively from an algebraic point of view, in terms of the axiomatic properties of the operations $\vee$ and $\wedge$, called *meet* and *join* (in the case of distributive lattices), and the properties of the median operation (in the case of median sets). A *median algebra* is a set with a ternary operation 'med' defined on it, satisfying (1) $\text{med}(x, x, y) = x$, (2) $\text{med}(x, y, z) = \text{med}(y, x, z) = \text{med}(y, z, x)$, and (3) $\text{med}(\text{med}(x, y, z), u, v) = \text{med}(x, \text{med}(y, u, v), \text{med}(z, u, v))$. Every finite median algebra can be represented as a median set in a hypercube, or as a median graph, with the median operation coinciding with the graph median. Median algebras can be viewed as a special kind of semilattice, called a *median semilattice*, by choosing a single element a and defining $x \leq_a y$ when $\text{med}(a, x, y) = x$, with $x \wedge y = \text{med}(a, x, y)$. If there are two elements a^0, a^1 such that $\text{med}(a^0, x, a^1) = x$ for all x, then the median algebras coincide with the distributive lattices, with $x \wedge y = \text{med}(a^0, x, y)$ and $x \vee y = \text{med}(a^1, x, y)$. See Birkhoff [15] for an introduction to the theory of lattices, and Bandelt and Hedlíková [10] for an introduction to median algebras.

A *retraction* on a graph G is a nonexpansive mapping f on G such that $f(x)$ is a fixed point of f for all vertices x in G. Therefore f is a retraction if its range coincides with its fixed point set, or equivalently, if $f^{(2)} = f$. A *retract* of G is a set U of vertices in G such that U is the range of some retraction on G. Retractions and retracts play an important role in the study of periodic points of nonexpansive mappings, because of the following observation.

Lemma 3.29 *If f is a nonexpansive mapping on a graph G with M vertices, then $f^{(M!)}$ is a retraction, and the corresponding retract is the set of periodic points of f.*

Proof. Let $g = f^{(M!)}$. If x is a periodic point of f, then its period is at most M and must divide $M!$. Therefore $g(x) = x$. If x a vertex of G, then $f^{(k)}$ is a periodic point of f for all $k \geq M - 1$, so $g(x)$ is a periodic point of f. Therefore the set of periodic points of f, the range of g, and the fixed point set of g coincide. □

A set of vertices in a graph G is *connected* if it induces a connected subgraph of G. It can be shown that the median graphs are precisely the graphs that can be characterized as vertex-induced subgraphs of a connected median set in a hypercube, with the median operation in the graph coinciding with the median operation in the hypercube (see Mulder [80]).

The range of a nonexpansive mapping f on a connected graph G induces a connected subgraph of G, because f maps paths to paths. In particular, every retract of a connected graph is connected, since it is the range of a nonexpansive mapping, namely the associated retraction. The fixed point set of a nonexpansive mapping f on a hypercube is a median set by Lemma 3.26. In particular, every retract of the hypercube is a median set, since it is the fixed point set of a retraction. This gives:

Corollary 3.30 *The vertices of a retract in the hypercube form a connected median set.*

3.2.2 Median Sets and 2SAT

We seek a characterization of median sets. An instance of 2-satisfiability on a finite set S, or 2SAT for short, is a formula in conjunctive normal form with two literals per clause, i.e., a set of clauses of the form $u \vee v$, where each of the literals u, v, is either a positive literal x_i or a negative literal $\overline{x_i}$, with $i \in S$. A clause $u \vee u$ is denoted simply by u. A *solution* for the 2SAT instance is an assignment x on S that satisfies all the clauses (makes them true). The following lemma can be found in Chung, Graham, and Saks [21], and can also be obtained from Isbell [66] and Mulder and Schrijver [81].

Lemma 3.31 *A subset of the vertices in the hypercube is a median set if and only if it is the set of solutions of a 2SAT instance.*

Proof. Let U be the set of solutions to a 2SAT instance I. We claim that U is a median set. Suppose that a, b and c are in U, and let $t = \text{med}(a,b,c)$. Let C be any clause of I, and let x_i, x_j be the two Boolean variables in the clause C. Since t_i is the majority of a_i, b_i and c_i, the value t_i can only differ from at most one of these three values. Similarly, t_j is different from at most one of a_j, b_j and c_j. Therefore $t_i t_j$ is different from at most two of the pairs $a_i a_j$, $b_i b_j$ and $c_i c_j$, and must agree with at least one of these pairs, say $t_i t_j = a_i a_j$. Since a is a solution of I, clause C holds under the assignment $x_i x_j = a_i a_j$, and hence under the assignment $x_i x_j = t_i t_j$. This shows that t satisfies every clause C in I, and is therefore in U. It follows that U is closed under medians and hence a median set.

To prove the converse, let U be a median set in the hypercube on $S = \{1, 2, \ldots, m\}$. Let I be the set of all the two-literal clauses that are satisfied by every point in U, and let V be the set of solutions of I. We want to show that $V = U$. Clearly $U \subseteq V$, since every point in S satisfies all the clauses in I. To show that $V \subseteq U$, we pick an arbitrary point a in V and show that a is in V. We will actually show that for all i,j with $1 \leq i \leq j \leq m$, there is a point a^{ij} in U such that $a^{ij}_k = a_k$ for $i \leq k \leq j$. Since $a^{1m} = a$, the fact that a is in U follows as a special case. The proof is by induction on $j - i$. The base case $j - i = 0$ is easy: Since a is in V, the 2SAT instance I does not contain the clause $x_i \neq a_i$ (i.e., the clause x_i if $a_i = 0$, the clause $\overline{x_i}$ if $a_i = 1$), so this clause is not satisfied by every point in U, i.e., some point a^{ii} in U must have $a^{ii}_i = a_i$. To carry out the inductive step, we assume the existence of the two points $a^{i(j-1)}$ and $a^{(i+1)j}$ in U, and construct the point a^{ij}. Since a is in V, the 2SAT instance I does not contain the clause $(x_i \neq a_i) \vee (x_j \neq a_j)$, so this clause is not satisfied by every point in U, i.e., some point b in U must have $b_i = a_i$ and $b_j = a_j$. Let $a^{ij} = \text{med}(b, a^{i(j-1)}, a^{(i+1)j})$. Then $a^{ij}_i = a_i$ because $b_i = a^{i(j-1)}_i = a_i$; $a^{ij}_j = a_j$ because $b_j = a^{(i+1)j}_j = a_j$; and $a^{ij}_k = a_k$ for $i < k < j$ because $a^{i(j-1)}_k = a^{(i+1)j}_k = a_k$, completing the induction. □

We summarize some known facts about 2SAT instances. Every clause $u \vee v$ in a 2SAT instance can be viewed as two implications $\overline{u} \to v$ and $\overline{v} \to u$. The *implication graph* of a 2SAT instance is a directed graph on the set of literals with an edge from u to v corresponding to each implication $u \to v$. The 2SAT instance is *acyclic*(*transitively closed*) if its implication graph is acyclic (resp. transitively closed). We say that the clause $u \to w$ *follows* from the clauses $u \to v$, $v \to w$. A 2SAT instance I can always be transformed into a transitively closed 2SAT instance by repeatedly adding to the set of clauses new clauses that follow from existing clauses. The resulting 2SAT instance is the *transitive closure* of I; the sets of solutions of a 2SAT instance and of its transitive closure are the same. The *inference graph* of a transitively closed 2SAT instance I is a directed graph whose vertices are the clauses in I, with edges from each of the two clauses $u \to v$, $v \to w$, to the clause

$u \to w$ that follows from them. If the 2SAT instance is acyclic, then the inference graph is acyclic. This can be seen by associating with each clause $u \to v$ a weight equal to the length of a longest path from u to v in the implication graph, and verifying that all edges of the inference graph go from a clause to another clause of larger weight.

If two literals u and v are reachable from each other in the implication graph of a 2SAT instance I, then every solution of I must satisfy $u = v$. Therefore if a strongly connected component (a strong component) contains both a variable x_i and its negation $\overline{x_i}$, then the 2SAT instance has no solution. If this is not the case, then every strong component can be replaced with a single new literal u, in such a way that the strong components of a variable x_i and its negation $\overline{x_i}$ are replaced by complementary literals $u, \overline{u}$. The resulting 2SAT instance I' is acyclic, and there is a one-to-one correspondence between the solutions of I' and the solutions of I, under the mapping that assigns to each literal in I the value of the literal in I' corresponding to its strong component. The instance I' is called the *acyclic representation* of I.

An acyclic 2SAT instance always has solutions. Given its acyclic implication graph, one can find a linear extension of this graph, and assign the value 1 to u if and only if $\overline{u}$ precedes u in this linear extension. This assignment satisfies all the clauses: If a clause $u \vee v$ is violated, then u precedes $\overline{u}$ and v precedes $\overline{v}$ in the linear extension because $u = v = 0$, and furthermore $\overline{u}$ precedes v and $\overline{v}$ precedes u because of the implications associated with the clause $u \vee v$, contradicting acyclicity. This argument also establishes the known fact that 2SAT instances can be solved in time linear in the size of the instance (see, e.g., Even, Itai, and Shamir [29]).

A variable in a 2SAT instance is *trivial* if it has the same value in all solutions. If x_i is reachable from $\overline{x_i}$ (or $\overline{x_i}$ is reachable from x_i) in the implication graph, then the variable x_i is trivial and has value 1 (resp. 0) in all solutions. If neither of x_i and $\overline{x_i}$ is reachable from the other in a solvable 2SAT instance, then in the acyclic representation there are linear extensions in which x_i precedes $\overline{x_i}$ and vice versa, so x_i is not trivial. Two nontrivial variables x_i and x_j in a 2SAT instance are *equivalent* if either $x_i = x_j$ in all solutions, or $x_i = \overline{x_j}$ in all solutions. It can be shown that two nontrivial variables x_i, x_j are equivalent if and only if x_i is in the same strong component as x_j or $\overline{x_j}$ in the implication graph.

Cook and Luby [25] showed that a 2SAT instance can be solved in $\mathcal{NC}^2$. We show that the complexity can be reduced if the 2SAT instance is transitively closed.

Lemma 3.32 *The problem of solving transitively closed* 2SAT *instances is in logspace uniform* $\mathcal{NC}^1$.

Proof. The 2SAT algorithm just described requires finding a linear extension of an acyclic implication graph and determining for each variable x whether x occurs before or after $\overline{x}$ in the linear extension. If the implication graph is transitively closed, then a linear extension can be defined by using the outdegree of each vertex, breaking ties arbitrarily: If u precedes v in the implication graph, then u will have outdegree larger than the outdegree of v. The task of comparing outdegrees of two vertices can be performed in $\mathcal{NC}^1$. To remove the acyclicity assumption, note that strong components are easily identified in a transitively closed implication graph, because two vertices are in the same strong component when there are edges joining them, in both directions. We can select one literal from every strong component, say the lowest-numbered literal, also in $\mathcal{NC}^1$. The selected literals induce then an acyclic implication graph. □

The *compatibility graph* of a 2SAT instance is an undirected graph on the set of literals with an edge (u, v) for each clause $u \vee v$; we always include in the compatibility graph the edges $(x_i, \overline{x_i})$ corresponding to clauses $x_i \vee \overline{x_i}$. A 2SAT instance is *monotone* if all its clauses are of the form $x_i \vee x_j$, with no negated variables. A 2SAT instance is *bipartite* if its compatibility graph is bipartite. Note that if A and B are the two vertex sets of the bipartite graph, then one of $x_i, \overline{x_i}$ is in A and the other one in B, for each variable x_i. By a simple renaming that replaces some variables with negated variables, we can assume that $\overline{x_i}$ is in A and x_i is in B for all variables x_i. The bipartite 2SAT instance is then in *normal form*. Note that a bipartite instance in normal form is one in which all clauses are of the form $\overline{x_i} \vee x_j$. Such clauses can be viewed as implications $x_i \to x_j$ on unnegated variables. A 2SAT instance I is bipartite and in normal form if and only if the all-zeros and all-ones assignments are solutions of I. This can be used in conjunction with Lemma 3.31 to show:

Corollary 3.33 *A set of vertices in the hypercube is a distributive lattice if and only if the nontrivial variables of the associated* 2SAT *instance form a bipartite* 2SAT *instance.*

The distinction between connected and arbitrary median sets is characterized in terms of the associated 2SAT instance by the following:

Corollary 3.34 *A set of vertices in the hypercube is a connected median set if and only if it is the set of solutions of a* 2SAT *instance with no equivalent variables.*

Proof. Let U be a median set, and let I be the corresponding 2SAT instance. Suppose that I has two equivalent variables x_i and x_j, say $x_i = x_j$ for every point x in U (the case $x_i = \overline{x_j}$ is identical). Since x_i and x_j are nontrivial variables, there must exist two points a and b in U such that $a_i = a_j = 0$ and $b_i = b_j = 1$. However no point c in U has $c_i \neq c_j$. It follows that there is no path contained in U from a to b, because adjacent vertices along a path differ in only one coordinate, and so the path would necessarily go through a point c with $c_i \neq c_j$. Therefore the set U is disconnected.

To prove the converse, let I be a 2SAT instance and let U be the set of solutions to I. The set U is a median set. Suppose that U is disconnected. Choose two points a and b in different components of the subgraph induced by U, for which $d(a, b)$ is minimum. Clearly $d(a, b) \geq 2$, for otherwise a and b are in the same component. Let i and j be two distinct coordinates such that $a_i \neq b_i$ and $a_j \neq b_j$. Suppose that $a_i = a_j = 0$ and $b_i = b_j = 1$ (the other cases are identical). We claim that every point in U satisfies $x_i = x_j$. Suppose that some point c in U has $c_i \neq c_j$. Then $v = \text{med}(a, b, c)$ has $v_i = c_i$ and $v_j = c_j$. Furthermore,

v is in U since U is a median set; $v \neq a, b$ since $v_i \neq v_j$; and v lies on a shortest path from a to b by the definition of median, so that $d(a, v), d(v, b) < d(a, b)$. By minimality of $d(a, b)$, the point v must be in the same component as a and in the same component as b, contradicting the fact that a and b are in different components. Therefore no such c exists, and all the points x in U satisfy $x_i = x_j$, implying that I has two equivalent variables x_i and x_j. □

We give two illustrations of these results.

Corollary 3.35 *Let f be a nonexpansive mapping on the hypercube, and suppose that for all coordinate pairs i, j and all binary values a, b, there exists a fixed point x for f with $x_i = a$ and $x_j = b$. Then all the vertices of the hypercube are fixed points of f.*

Proof. The 2SAT instance characterizing the set of fixed points of f has no clauses, since a clause $(x_i \neq a) \lor (x_j \neq b)$ would exclude the point x with $x_i = a$ and $x_j = b$. Therefore all vertices of the hypercube are solutions of the 2SAT instance and hence fixed points. □

We have seen (in Theorem 3.18) that the stability question for nonexpansive networks, as a decision problem, can be reduced to the evaluation question for nonexpansive circuits. The following result extends that result to the search problem.

Theorem 3.36 *The stability question for networks over a nonexpansive basis Ω, as a search problem, reduces to the evaluation question for circuits over Ω. In fact, finding a transitively closed 2SAT instance characaterizing the stable configurations for networks over a nonexpansive basis Ω reduces to the evaluation question for circuits over Ω.*

Proof. The stable configurations of a network, given an input assignment, are the fixed points of its internal function f. The set of fixed points can be characterized by a 2SAT instance I consisting of all the clauses $(x_i \neq a) \lor (x_j \neq b)$ that are satisfied by all fixed points. By Theorem 3.9, $f_{\{i,j\}}$ has a fixed point $x_i x_j = ab$ precisely when f has a fixed point x with $x_i x_j = ab$, i.e., when the clause $(x_i \neq a) \lor (x_j \neq b)$ is in I. Therefore, in order to find the clauses in I, it is sufficient to find the fixed points of $f_{\{i,j\}}$ for all i, j, and this can be done by computing each f_{ij} on each of the four possible input values (two values if $i = j$). Each of these computations can be reduced to the evaluation question by Lemma 3.12. Therefore the problem of finding I reduces to the evaluation question. By Lemma 3.32, a particular stable configuration, i.e., a solution to I, can also be found in $\mathcal{NC}^1$. □

Note that the proof constructs $O(m^2)$ circuits over Ω, at most four for each pair $\{i, j\}$. Each of these circuit has depth $O(m)$ and size $O(m^2)$ by the construction from Lemma 3.12.

3.2.3 Representation by Circuits

Section 3.2.1 showed that fixed points and retracts are related to median sets and connected median sets respectively. We use here the representation of median sets as 2SAT instances from section 3.2.2 to show that this link provides a complete characterization of fixed point sets and retracts. The equivalence between connected median sets and retracts is due to Bandelt [9]; the following representation can also be found in Chung, Graham, and Saks [21].

Lemma 3.37 *Every connected median set in the hypercube is the retract defined by a retraction f, where f is computed by a circuit over $\{X\}$. If the connected median set is a distributive lattice, then f is computed by a circuit over $\{C\}$.*

Proof. Given a connected median set U in the hypercube, let I be the corresponding transitively closed acyclic 2SAT instance. Assume that I has no trivial variables. Associate with clause $(x_i = a) \vee (x_j = b)$ in I a gate $g = X_{ij}^{ab}$ on the hypercube. This gate computes $g(x) = y$ as follows. For $i' \neq i, j$, $y_{i'} = x_{i'}$; furthermore, $y_i y_j \oplus ab = X(x_i x_j \oplus ab)$. Gate g can be described using an X gate and negations, which can in turn be described using an X gate as well. Since U is connected, the inference graph of I is acyclic, so the clauses can be listed in an order L such that every clause precedes in L the clauses that it follows from. Build a circuit that computes the function f on the hypercube defined by applying to the input x the functions X_{ij}^{ab} in the order in which the corresponding clauses $(x_i = a) \vee (x_j = b)$ appear in L.

If x is in U, then x satisfies all the clauses $(x_i = a) \vee (x_j = b)$, so $X_{ij}^{ab}(x) = x$ and therefore $f(x) = x$. We now show that if L' is a prefix of L, then the range of the function $f_{L'}$ defined by applying to the input x the functions associated with the clauses in L' in the proper order satisfies all the clauses in L'. The proof is by induction on the length of L'. If L' is empty, then no clauses need to be satisfied and the claim is vacuously true. We assume that the claim is true for $y = f_{L'}(x)$ and prove it for $z = X_{ij}^{ab}(y)$, where $(x_i = a) \vee (x_j = b)$ is the clause following L' in L. So y satisfies all the clauses in L', and z satisfies the clause $(x_i = a) \vee (x_j = b)$ by the definition of X_{ij}^{ab}. Let C be a clause in L'. If C does not involve x_i or x_j, then C must hold for z because it holds for y. Say C involves x_i but not x_j. Suppose that C is of the form $(x_i = a) \vee (x_k = c)$ with $k \neq j$. Then either $y_i = a$, in which case $z_i = a$ and C holds for z, or $y_k = c$, in which case $z_k = c$ and C holds for z. The other case is where C is of the form $(x_i = \overline{a}) \vee (x_k = c)$ with $k \neq j$. In this case, the clause $(x_j = b) \vee (x_k = c)$ follows from the clause C and the clause following L' in L, and must then be in L'. Therefore y satisfies the clause $(x_j = b) \vee (x_k = c)$, and either $y_j = b$, in which case $z_j = b$ and z satisfies C, or $y_k = c$ and $z_k = c$ satisfies C. The remaining possibility, namely C involving both i and j, cannot happen. For if C is of the form $(x_i = a) \vee (x_j \neq b)$ (or $(x_i \neq a) \vee x_j = b)$ then all solutions must satisfy $x_i = a$ (resp. $x_j = b$) and then x_i (resp. x_j) is a trivial variable. If C is of the form $(x_i \neq a) \vee (x_j \neq b)$, then all solutions satisfy $x_i = a$ if and only if $x_j \neq b$, so x_i and x_j are equivalent variables. Therefore z satisfies all the clauses in L' as well, completing the induction. Since $f = f_L$, it follows that $f(x)$

satisfies all the clauses in L, and hence all the clauses in I, so $f(x) \in U$. This shows that f is a retraction and that the corresponding retract is the set U.

In the case of a distributive lattice, we have a bipartite 2SAT instance in normal form with clauses of the form $(x_i = 0) \vee (x_j = 1)$, and the gates X_{ij}^{01} are comparators. Trivial variables such as $x_i = a$ can be handled separately by setting the ith output of f to the constant value a independently of the inputs (using a constant gate). □

This establishes the converse of the results in section 3.2.1, proving Bandelt's theorem (as well as a monotone version of this theorem).

Theorem 3.38 *A set of vertices in the hypercube is a retract if and only if it is a connected median set. A set of vertices in the hypercube is a retract that can be defined by a monotone retraction if and only if it is a connected distributive lattice.*

One disadvantage of the preceding construction is that it uses a transitively closed 2SAT instance, so the number of gates can be quadratic in the number of clauses of a given 2SAT instance that characterizes the connected median set. If we are only interested in achieving a median set as the set of fixed points of an arbitrary nonexpansive mapping, not necessarily a retraction, then a construction with only a linear number of gates can be given, and the connectedness assumption can be removed.

Lemma 3.39 *Every median set in the hypercube is the fixed point set of a mapping f, where f is computed by a circuit over $\{X\}$. If the median set is a distributive lattice, then f is computed by a circuit over $\{C\}$.*

Proof. Every median set U in the m-cube can be characterized by a 2SAT instance I. We first consider the case where I is acyclic. Construct a circuit that applies to the input x the functions X_{ij}^{ab} corresponding to clauses $(x_i = a) \vee (x_j = b)$ in I, in arbitrary order. (These functions were defined in the proof of the previous lemma.) Let f be the mapping computed by this circuit. If $x \in U$, then x satisfies all the clauses in I, so $X_{ij}^{ab}(x) = x$ and $f(x) = x$. On the other hand, if x violates some clause $(x_i = a) \vee (x_j = b)$, then $X_{ij}^{ab}(x) \neq x$. This means that some coordinate i is such that if we consider the value of the ith coordinate y_i as it traverses the gates in the circuit, starting from the input value $y_i = x_i$, this value changes at some gate. More specifically, if the gate X_{ij}^{ab} produces an output different from its input, then its input satisfies $y_i y_j = \overline{ab}$ and its output satisfies $y_i y_j = ab$. Let T be the nonempty set of clauses corresponding to all such gates.

Suppose that $f(x) = x$. Then every y_i that changes its value as it traverses the circuit must change at least twice (in fact an even number of times), once from 0 to 1, and once from 1 to 0. Therefore for every clause $(x_i = a) \vee (x_j = b)$ in T, the gate X_{ij}^{ab} changes $y_i = \overline{a}$ to $y_i = a$, so there must be a clause $(x_i = \overline{a}) \vee (x_k = c)$ in T for which the corresponding gate $X_{ik}^{\overline{a}c}$ changes $y_i = a$ to $y_i = \overline{a}$. If we view the clauses in T as implications in the implication graph, then for every implication $(x_j = \overline{b}) \to (x_i = a)$ in T there is an implication $(x_i = a) \to (x_k = c)$ in T. Thus every implication can be matched with another implication, along an infinite path in the subgraph of the implication graph defined by T. This path must eventually cycle, contradicting the fact that I is acyclic. This shows that we cannot have $f(x) = x$ unless T is empty, i.e., unless x satisfies all the clauses in I. Therefore the fixed points of f are the points in U.

To remove the acyclicity assumption, recall that every 2SAT instance I has an acyclic representation I' in which the strong components shrink to single literals u_i and $\overline{u_i}$, so that the solutions to I can be obtained from the solutions to I' by giving to the variables in each strong component the value of the corresponding u_i or $\overline{u_i}$. In other words, the solutions to I differ from the solutions to I' only in the presence of multiple copies, some of them negated, of each coordinate from I'. We must therefore be able to obtain, given a mapping f' on assignments $z' = z'_1 z'_2 \ldots z'_n$ a mapping f on assignments $z = z_0 z_1 z_2 \ldots z_n$ such that z is a fixed point of f if and only if $z_0 = z_1$ and $z_1 z_2 \ldots z_n$ is a fixed point of f' (in other words, coordinate 0 is a copy of coordinate 1, for all fixed points). The mapping f can be obtained from f' by padding, defining $y = f(x)$ by letting $y_1 = x_0$ and $y_0 y_2 \ldots y_n = f'(x_1 x_2 \ldots x_n)$. The fixed points of f are then the assignments $z = z_0 z_1 \ldots z_n$ such that $z_1 = z_0$ and $z_0 z_2 \ldots z_n = f'(z_1 z_2 \ldots z_n)$, as needed. A similar construction, using negations, gives a negated copy $z_0 = \overline{z_1}$ of z_1 (just define $y = f(x)$ by letting $y_1 = \overline{x_0}$ and $\overline{y_0} y_2 \ldots y_n = f'(x_1 x_2 \ldots x_n)$). Therefore every median set is the set of fixed points of a mapping defined by a circuit over $\{X\}$.

In the case of a distributive lattice, we again have a bipartite 2SAT instance in normal form with clauses of the form $(x_i = 0) \vee (x_j = 1)$, and the gates X_{ij}^{01} are comparators. Furthermore, cycles in the implication graph involve only positive or only negative literals, so no negations are introduced in the reduction to the acyclic case. □

We have therefore shown:

Theorem 3.40 *A set of vertices in the hypercube is the fixed point set of a nonexpansive mapping if and only if it is a median set; it is the fixed point set of a monotone nonexpansive mapping if and only if it is a distributive lattice.*

The problem of counting solutions to a bipartite 2SAT instance in normal form was shown to be #$\mathcal{P}$-complete by Provan and Ball [95]. We can construct a circuit over $\{C\}$ that computes a function f whose fixed points are the points of the distributive lattice defined by a given bipartite 2SAT instance in normal form. Since the stable configurations of $N = \{f\}$ correspond to fixed points of f, we obtain the following:

Corollary 3.41 *The counting problem for networks over $\{C\}$ is #$\mathcal{P}$-complete.*

The question of whether an acyclic bipartite 2SAT instance has a solution with two particular variables set to 0 and to 1 respectively, is complete for nondeterministic logspace. See Jones, Lien, and Laaser [71], and also Immerman [61]. If we represent the 2SAT instance by a comparator network as before, then it follows that the question of whether a comparator network has a stable configuration with two particular links set to 0 and to 1 respectively. If we break these two links, and assign them these two values as inputs, then the existence of a solution must imply that the same two values will appear at the two corresponding outputs after the evaluation of the convergent network; conversely, if the correct values appear at the outputs, then we can obtain a stable configuration for the original network from a stable configuration for the network with the two links broken, and such a configuration always exists by monotonicity. We are therefore left with the problem of evaluating a convergent network, which in turn reduces to the evaluation of a comparator circuit.

Corollary 3.42 *The evaluation question for circuits over $\{C\}$ is hard for nondeterministic logspace.*

3.3 Periodic Points and Isomorphisms

We have seen that the fixed point set and the periodic set of nonexpansive mappings have a simple characterization in terms of 2SAT instances. We shall show that a nonexpansive mapping, when restricted to its periodic set, also admits a simple characterization. This characterization is then used to give an efficient algorithm for finding the 2SAT instance associated with the fixed point set.

3.3.1 Distances and Periodic Points

If an edge of the hypercube is given an orientation, it becomes an *ordered* pair (x, y). We can then define an equivalence relation ρ between edges: two edges are related by ρ if they traverse the same dimension in the same direction. Formally, $(x, y)\rho(x', y')$ if, for some coordinate i, $y = x \oplus e^i$, $y' = x' \oplus e^i$, and $x_i = x'_i$. The equivalence class of an oriented edge $(x, x \oplus e^i)$ under ρ can be named by the pair (i, x_i); for convenience, we introduce *signed coordinates* $i = (i, 0)$ and $\bar{i} = (i, 1)$, with $\bar{\bar{i}} = i$ by convention. The equivalence relation ρ can be characterized in terms of distances.

Lemma 3.43 *Two oriented edges (x, y) and (x', y') in the hypercube are related by ρ if and only if $d(x, x') + 1 = d(y, y') + 1 = d(x, y') = d(y, x')$.*

Proof. If the two edges are related by ρ, and we let $d = d(x, x')$, then clearly $d(y, y') = d$ and $d(x, y') = d(y, x') = d + 1$. Conversely, if the distance properties hold, x, y differ in coordinate i, and x', y' differ in coordinate j, then $x'_j = x_j$ since $d(x, y') = d(x, x') + 1$, and $y'_j = y_j$ since $d(y, x') = d(y, y') + 1$, so x and y differ in bit j and therefore $i = j$, implying that the two edges are related by ρ. $\square$

A *signed permutation* on $\{1, 2, \ldots, m\}$ is a permutation σ on the signed coordinates $S = \{1, 2, \ldots, m, \bar{1}, \bar{2}, \ldots, \bar{m}\}$ such that $\sigma(i) = j$ if and only if $\sigma(\bar{i}) = \bar{j}$, for all $i, j \in S$. For convenience, we introduce the notation $x_{\bar{i}} = \overline{x_i}$. Given a signed permutation σ, we can now define a mapping $g = g_\sigma$ on the m-cube by letting $g(x_1 \ldots x_m) = x_{\sigma(1)} \ldots x_{\sigma(m)}$. Thus g

permutes and perhaps negates some coordinates as indicated by σ. The mappings g_σ are all isomorphisms of the hypercube. A strong converse to this statement is the following.

Lemma 3.44 *Let g be a mapping on a connected set U in the hypercube with the property that $d(g(x), g(y)) = d(x, y)$ for all $x, y \in U$. Then there exists a signed permutation σ such that the isomorphism g_σ on the hypercube satisfies $g(x) = g_\sigma(x)$ for all $x \in U$.*

Proof. Let G be the connected subgraph induced by U, and let G' be the connected subgraph induced by the images of the points in U under g. Then the equivalence classes $j, \bar{j}$ of edges $\{x, y\}$ in G under ρ correspond to equivalence classes $i, \bar{i}$ of edges $\{g(x), g(y)\}$ in G' under ρ, since ρ can be defined in terms of distances and g preserves distances. The partial mapping from i to j is one-to-one, and can be extended to a permutation μ on the set of coordinates $\{1, 2, \ldots, m\}$. Given two points x and y in U, there is a path from x to y in G because G is connected, and a corresponding path from $g(x)$ to $g(y)$ in G' under the mapping g. Each oriented edge along the path from x to y traverses a dimension $\mu(i)$, and the corresponding edge along the path from $g(x)$ to $g(y)$ traverses dimension i. Therefore, if $y = z \oplus x$, then $g(y) = (z_{\mu(1)} \ldots z_{\mu(m)}) \oplus g(x)$. If we fix a particular $x \in U$, then $g(y) = (y_{\mu(1)} \ldots y_{\mu(m)}) \oplus u$ for all $y \in U$, where $u = (x_{\mu(1)} \ldots x_{\mu(m)}) \oplus g(x)$. If we let σ be the signed permutation defined by $\sigma(i) = \mu(i)$ if $u_i = 0$, $\sigma(i) = \overline{\mu(i)}$ if $u_i = 1$, and $\sigma(\bar{i}) = \overline{\sigma(i)}$, then $g(y) = g_\sigma(y)$. $\square$

The connectedness assumption is crucial. As an example, no nonexpansive mapping on the 4-cube can satisfy $f(1000) = 0000$, $f(0100) = 1100$, $f(0010) = 1010$ and $f(0001) = 0110$, because $f(0000)$ would have to be adjacent to each of 0000, 1100, 1010, 0110; yet the distances between the four antecedents and the distances between their images are all equal to 2.

A gate is *trivial* if some of its outputs are obtained by permuting, and possibly negating, some subset of the inputs, while the remaining outputs are constants (either 0 or 1, for all values of the inputs). A gate g in a circuit K is *useful* if there does not exist a trivial gate g' such that all stable configurations of K (determined by the possible choices of input assignments) are also stable configurations of the circuit obtained by replacing g with g' in K. If g is monotone, then g' is also required to be monotone.

Theorem 3.45 *A nonexpansive circuit with m inputs has at most $m2^{2m-1}$ useful gates.*

Proof. Let $g_1, \ldots, g_r$ be the gates of the circuit, listed in a linear order consistent with the partial order defined by the underlying acyclic graph of the circuit, from inputs to outputs. We can assume that the gate g_i operates on all the outputs of g_{i-1}: the outputs of g_{i-1} that are not used as inputs by g_i are simply returned unchanged as outputs of g_i. We can also add some extra inputs to g_i that do not affect the outputs of g_i, and add some extra constant outputs to g_i, so that all the gates g_i now have the same number of inputs and outputs $n \geq m$. The added $n - m$ inputs of g_1 that do not affect the outputs of g_1 are assigned constant values in some arbitrary manner. Therefore all added links have constant values in all stable configurations of the circuit. Given an input assignment x on $\{1, 2, \ldots, m\}$ for the circuit, let $h_i(x)$ be the assignment on $\{1, 2, \ldots, n\}$ that describes the values assigned to the n inputs of g_i in the stable configuration corresponding to the input assignment x (or the n outputs of the circuit if $i = r + 1$). Therefore $h_{i+1}(x) = g_i(h_i(x))$, where $h_1(x)$ is the n-input assignment consisting of the m-input assignment x extended with the $n - m$ constant values for the added inputs.

Define a potential $\Phi_i = \sum d(h_i(x), h_i(y))$, where the sum is over all 2^{2m} pairs x, y of assignments on $\{1, 2, \ldots, m\}$. Note that $\Phi_1 = (m/2)2^{2m}$, since the expected value of $d(x, y)$ for two assignments x, y chosen at random is $m/2$. Furthermore,

$$\Phi_{i+1} = \sum d(g_i(h_i(x)), g_i(h_i(y))) \leq \sum d(h_i(x), h_i(y)) = \Phi_i.$$

Suppose that $\Phi_{i+1} = \Phi_i$ for some i. This can only happen if $d(g_i(z), g_i(t)) = d(z, t)$ for all z, t such that $z = h_i(x)$ and $t = h_i(y)$. Let U be the set of all points z in the n-cube such that $z = h_i(x)$ for some x in the m-cube. The set U is connected because a path from x to y in the m-cube maps to a path from $z = h_i(x)$ to $t = h_i(y)$ in U under the nonexpansive mapping h_i. Therefore the mapping g_i and the set U satisfy the conditions of Lemma 3.44, and so there exists a trivial gate g'_i such that $g'_i(z) = g_i(z)$ for all $z \in U$. Since all stable configurations (corresponding to input assignments x) assign a value $z \in U$ to the inputs of g_i, we can infer that g_i is not useful. In fact g_i is still not useful even if we remove the links that were added to make the number of inputs and outputs match for each gate, since these links have the same constant value in all stable configurations, Therefore all useful gates g_i satisfy $\Phi_{i+1} < \Phi_i$. Given the value of Φ_1 computed above, and the obvious lower bound $\Phi_i \geq 0$, the potential Φ_i can only decrease $(m/2)2^{2m}$ times as i increases, so the circuit has at most $(m/2)2^{2m}$ useful gates.

It can be verified that for monotone circuits, the trivial gates found can be taken to be monotone (since g_i is monotone). □

For comparator circuits, Graham [45] established a bound on the number of useful gates which is quadratic in m. The proof extends to other scatter-free gates such as the X gate, and in fact we shall use the idea of Graham's proof in section 3.5.1 to obtain a result for arbitrary scatter-free gates. It seems then plausible that the bound in the preceding proof can be reduced to a bound polynomial in m, perhaps quadratic in m, at least in the scatter-free case, but we do not know at present whether this is feasible.

The theorem gives an algorithm for deciding whether a given gate with m inputs can be represented by a circuit over a given nonexpansive basis. The complexity of the algorithm is doubly exponential in m.

Corollary 3.46 *The question of whether a gate g with m inputs can be computed by a circuit over a fixed nonexpansive basis Ω can be answered by a nondeterministic algorithm in time exponential in m.*

CHAPTER 3. STABILITY IN NONEXPANSIVE NETWORKS

Proof. The nondeterministic algorithm guesses a circuit containing at most $m2^{2m-1}$ gates from Ω, plus negations to simulate gates that are not useful if Ω is non-monotone, then verifies that it computes g by testing the 2^m possible input assignments. The algorithm can be made deterministic, with a further exponential increase in its time complexity. □

For our purposes, the most useful application of Lemma 3.44 is in the study of the periodic behavior of nonexpansive mappings.

Theorem 3.47 *Let f be a nonexpansive mapping on the hypercube, and let U be the set of periodic points of f. Then there exists an isomorphism g_σ on the hypercube such that $f(x) = g_\sigma(x)$ for all $x \in U$.*

Proof. The periodic points of f induce a retract of the m-cube and hence a connected set. For periodic points x, y, we know that $d(f(x), f(y)) = d(x, y)$. Hence the conditions of the lemma are satisfied and the result follows. □

3.3.2 Efficient Representation of Stable Configurations

We now use the structural properties obtained so far to represent the stable configurations of a network efficiently (under a given input assignment). Suppose that we are given a network and an input assignment, and let f be the corresponding internal function; suppose that we are also given a fixed point x of f, say the all-zero assignment $x = 0$. The idea of the algorithm is then to (a) find the signed permutation σ associated with the isomorphism g promised by Theorem 3.47 (recall that $g(x_1, \ldots, x_m) = x_{\sigma(1)} \ldots x_{\sigma(m)}$), and then (b) use σ to find the 2SAT characterization for the fixed point set guaranteed by Lemmas 3.27 and 3.31. This representation can then be used, for example, to enumerate the stable configurations efficiently.

Since $f(0) = 0$, the signed permutation σ can be restricted to a permutation on $\{1, \ldots, m\}$. The disjoint cycles of the permutation σ will give cycles in the underlying acyclic graph of the network when we associate the coordinate x_i of a configuration x with the link i in the network. These cycles in the network are found as follows. Starting with an arbitrary coordinate (i.e., a link in the network), we build a path of coordinates (links) in the network. The next coordinate on the path is always found using the *successor* function; each new coordinate j on the path is colored *green*, and its predecessor on the path is remembered in pred[j]. If the path dies off because *successor* fails, then all coordinates on the path are re-colored *black* and a new path is started somewhere else. Otherwise, the path must eventually cycle, in which case all coordinates on the cycle are re-colored *red*; for these red coordinates j, the value of $\sigma(j)$ is then given by pred[j]. The green path is then continued from the last green edge on the path. At the end of the algorithm, we only have black coordinates and cycles of red coordinates; if we now set $\sigma(j) = j$ for all black coordinates, we have obtained the permutation σ and hence the isomorphism g.

Algorithm *cycles*;
pick a coordinate i and color i green;

repeat
    execute $successor(i)$;
    if $successor$ returns a coordinate j
    then if j is uncolored
        then $\{\text{pred}[j] \leftarrow i;\ i \leftarrow j;\ \text{color } i \text{ green};\}$
        else {**if** pred$[j]$ is undefined
            then {pred$[j] \leftarrow i$; pick a new uncolored i and color i green
            (if such an i exists);}
            else interchange pred$[j]$ and i;
            follow the pred pointers from j and color all coordinates reached red;}
    else follow the pred pointers from i, color all
        coordinates j reached black (including i) and set pred$[j] \leftarrow j$,
        pick a new uncolored i and color i green (if such an i exists);
until all coordinates are either red or black.

The function *successor*, on input i^0, iteratively changes different coordinates i in x until the corresponding coordinate j that changes in $y = f(x)$ is not red, then returns this coordinate j provided it is not a black coordinate. Therefore either j is uncolored and the path can be extended, or j is green and a cycle has been closed.

Function $successor(i^0)$;
set all coordinates unmarked;
$x \leftarrow 0;\ y \leftarrow f(x);\ i \leftarrow i^0$;
loop
    $x \leftarrow x \oplus e^i;\ y' \leftarrow y;\ y \leftarrow f(x)$;
    if $y = y' \oplus e^j$ for some unmarked non-black coordinate j
    then if j is red
        then $i \leftarrow \text{pred}[j]$; mark j;
        else return j
    else return *"fail"*;
endloop;

Let $\sigma(j)$ be the final value of pred$[j]$ produced by the algorithm.

Lemma 3.48 *Let 0 be a fixed point of the nonexpansive mapping f. If x is a periodic point of f, then $f(x) = g(x)$, where g is the isomorphism of the m-cube given by $g(x_1 \ldots x_m) = x_{\sigma(1)} \ldots x_{\sigma(m)}$ in the algorithm above.*

Proof. We shall show that three invariants are maintained by the algorithm, for all periodic points x and $y = f(x)$: (1) $x_i = 0$ if i is black, (2) $x_i \leq y_j$ if j is green and pred$[j] = i$; and (3) $x_i = y_j$ if j is red and pred$[j] = i$. At the end of the algorithm, $\sigma(i) = i$ if i is black because pred$[i] = i$, and $\sigma(j) = i$ if i is red and pred$[j] = i$; so $x_{\sigma(j)} = y_j$ for all j, proving that $f(x_1 \ldots x_m) = x_{\sigma(1)} \ldots x_{\sigma(m)}$.

The invariants hold vacuously at the beginning of the algorithm, since all coordinates are uncolored. Suppose that at some stage $successor(i^0)$ is called, with i^0 green, and also suppose that there is a periodic point $\hat{x}$ with $\hat{x}_{i^0} = 1$. We want to show that if $\hat{y} = f(\hat{x})$, then $successor$ will return a green or uncolored coordinate j with $\hat{y}_j = 1$. Note that $d(f(0), \hat{y}) = d(0, \hat{x})$ by Lemma 3.2, so if x moves from 0 to $\hat{x}$ along a shortest path, then $y = f(x)$ must move from $f(0)$ to $\hat{y}$ along a shortest path. This is precisely what happens when $successor(i^0)$ is called. The initial assignment $x \leftarrow x \oplus e^i$ with $x = 0$ and $i = i^0$ clearly moves x from 0 towards $\hat{x}$. Suppose that up to right after some execution of $x \leftarrow x \oplus e^i$, the points x and y have moved along shortest paths from 0 to $\hat{x}$ and from $f(0)$ to $\hat{y}$ respectively. Since the last assignment $x \leftarrow x \oplus e^i$ moved x towards $\hat{x}$, we have $d(x, \hat{x}) < d(y', \hat{y})$, so by Lemma 3.1, the assignment $y \leftarrow f(x)$ must move y towards $\hat{y}$ along some dimension j. Then $y = y' \oplus e^j$ for some coordinate j, this coordinate j is non-black by invariant (1) since $\hat{y}_j = 1$, and j is unmarked because each dimension is traversed at most once in a shortest path. If j is either green or uncolored, then $successor$ returns j with $\hat{y}_j = 1$ as claimed. If j is red and $\text{pred}[j] = i'$, then $\hat{x}_{i'} = \hat{y}_j = 1$ by invariant (3). Furthermore, coordinate i' is different from all previously encountered i because i' is red and the only red j such that $\text{pred}[j] = i'$ was unmarked up to now. So the next execution of $x \leftarrow x \oplus e^i$ with $i = i'$ sets $x_{i'} = 1$ and moves x towards $\hat{x}$. Hence our shortest path assumption for x and y holds through one more iteration. Since there is only a finite supply of unmarked red j, eventually some j which is either green or uncolored and satisfies $\hat{y}_j = 1$ is obtained and returned, as claimed.

This fact about $successor$ is all we need to know about this function, and we use it to show that the three invariants are preserved. The coordinates are partitioned at all times into three types: black coordinates, cycles of red coordinates, where $\text{pred}[j]$ always indicates the red coordinate preceding j in the cycle, and a single path of green coordinates, where i is the last coordinate on the path, $\text{pred}[j]$ always indicates the coordinate preceding j on the path, and $\text{pred}[j]$ is undefined for the first coordinate on the path. The input i to $successor$ is always green. If the output j is uncolored, then setting $\text{pred}[j] = i$ and coloring j green preserves invariant (2), because $successor$ guarantees that if $x_i = 1$ then $y_j = 1$ for x and $y = f(x)$ periodic. If the output j is green, then again setting $\text{pred}[j] = i$ preserves invariant (2) for the same reason, and also completes a cycle of green coordinates. Interchanging $\text{pred}[j]$ and i in fact separates this cycle from the path, leaving a shorter path whose last coordinate i is the coordinate that used to precede j on the path before the cycle was completed. (In the special case where j was the first coordinate on the path, the completed cycle covers the entire path, and a new path is started somewhere else by picking some uncolored i and coloring i green.) We must show that coloring all coordinates in the cycle red preserves invariant (3). Let i be a coordinate on this cycle, let x be a periodic point, let i^l be the coordinate that is obtained by moving forward in the cycle (following the pred pointers backwards) l times, starting at i, and let $x^l = f^{(l)}(x)$. Then $x_{i^l}^l \leq x_{i^{l+1}}^{l+1}$ by invariant (2), because $\text{pred}[i^{l+1}] = i^l$. But $x^{M!} = x^0$ and $i^{M!} = i^0$ (since periods and cycle lengths are at most $M = 2^m$), so we must in fact have $x_{i^l}^l = x_{i^{l+1}}^{l+1}$ for all $0 \leq l < M!$. In particular, if $j = i^1$ and $y = x^1$, then invariant (3) holds for i, j, x and y. We must finally

CHAPTER 3. STABILITY IN NONEXPANSIVE NETWORKS   74

show that invariant (1) is preserved when all coordinates in the green path are colored black. Invariant (1) holds for the last coordinate i on the path, because *successor* failed for this last coordinate and we know that *successor* does not fail if there exists a periodic point x with $x_i = 1$. If invariant (1) holds for some coordinate j on the path, then it must also hold for $i = \text{pred}[j]$, because all periodic points x and $y = f(x)$ satisfy $y_j = 0$ by invariant (1) and $x_i \le y_j$ by invariant (2), so that $x_i = 0$. Therefore invariant (1) holds for all coordinates on the green path when the path is re-colored black. □

The algorithm also gives the 2SAT clauses that we need to characterize the fixed points of f. Since every periodic point satisfies $f(x) = g(x)$, we can break the fixed point condition $f(x) = x$ into two separate conditions $f(x) = g(x)$ and $g(x) = x$. The first condition is guaranteed by clauses on the black and on the red coordinates. If i is a black coordinate, then we include a clause that forces $x_i = 0$. Let k and k' be red coordinates. We say that k' is *implied* by k if the variable i takes the value k' during some call to $successor(k)$ in the execution of the algorithm. We include a clause $\overline{x_k} \vee x_{k'}$ whenever k' is implied by k, and a clause $\overline{x_k} \vee \overline{x_{k'}}$ whenever the m-bit vector x with a 1 in each coordinate implied by k or by k' (or by both) and a 0 everywhere else does not satisfy $f(x) = g(x)$. The remaining condition $g(x) = x$ is guaranteed by clauses that force $x_{\sigma(j)} = x_j$ for all red coordinates j.

Lemma 3.49 *The 2SAT clauses mentioned above characterize the set of all fixed points of f.*

Proof. The set U of all points x that satisfy $f(x) = g(x)$ is a median set, because U is the set of fixed points of $g^{-1} \circ f$. By Lemma 3.16, there is a 2SAT instance I that characterizes the set U. We first claim that if $\hat{x}$ is in U and i is black, then $\hat{x}_i = 0$. If this were not the case, consider the black coordinate i^0 such that $\hat{x}_{i^0} = 1$ that was colored black earliest during the algorithm. Consider the last call to $successor(i^0)$ during the execution of the algorithm. If $\hat{y} = f(\hat{x}) = g(\hat{x})$, then $d(0, \hat{x}) = d(f(0), \hat{y})$, because g preserves distances. So if x moves from 0 to $\hat{x}$ along a shortest path, then $y = f(x)$ will move from $f(0)$ to $\hat{y}$ along a shortest path. This is what happens during the execution of *successor*, because x_{i^0} is set to 1 the first time x is changed, moving x closer to $\hat{x}$, and whenever y is moved closer to $\hat{y}$ by setting $y_j = 1$ for some j such that $\hat{y}_j = 1$, the following assignment to x sets $x_i = 1$ for $i = \sigma(j)$, and since $\hat{x}_{\sigma(j)} = \hat{y}_j = 1$, this assignment moves x in turn closer to $\hat{x}$. Therefore the call to *successor* can only fail if y_j is set to 1 for some black j. But then $\hat{x}_j = \hat{x}_{\sigma(j)} = \hat{y}_j = 1$ and j was colored black before i^0 was, a contradiction. On the other hand, if *successor* does not fail and outputs some j with $\hat{y}_j = 1$, then i^0 will be eventually colored black by the main algorithm only if j is colored black before i^0 is colored black, and we have as before $\hat{x}_j = \hat{x}_{\sigma(j)} = \hat{y}_j = 1$ with j colored black before i^0, a contradiction.

We can therefore include clauses $\hat{x}_i = 0$ in I for all black coordinates i, and then simplify I so that the remaining clauses only mention variables that correspond to red coordinates. There can only be in I two types of clauses involving two variables x_k and $x_{k'}$ with k and k' red, namely clauses of the form $\overline{x_k} \vee x_{k'}$ and clauses of the form $\overline{x_k} \vee \overline{x_{k'}}$; no clause can be of the form $x_k \vee x_{k'}$ since the all-zero assignment is in U. We find all such clauses. Suppose that k' is implied by k, and that there is a point $\hat{x}$ in U with $\hat{x}_k = 1$. As we argued in the previous paragraph, a call to *successor*(k) will only set x_i to 1 if this assignment moves x closer to $\hat{x}$, i.e., if $\hat{x}_i = 1$. Since k' is implied by k, some call to *successor*(k) sets $x_{k'}$ to 1, and so $\hat{x}_{k'} = 1$. This shows that $\hat{x}_k \leq \hat{x}_{k'}$ and justifies all the clauses of the form $\overline{x_k} \vee x_{k'}$ where k' is implied by k.

Note that if k' is implied by k, then in fact $x_{k'}$ is set to 1 during the *last* call to *successor*(k). This is so because every execution of *successor*(k) is an extension of the previous execution of *successor*(k): Once a coordinate j becomes red or black, it will remain the same color until the end of the algorithm and with pred[j] unchanged. Therefore the point x^k with a 1 in each coordinate implied by k and a 0 everywhere else is precisely the *last* value taken by x during the *last* call to *successor*(k) for this k. The corresponding bits set to 1 in $y = f(x)$ during this execution of *successor* are precisely the bits j for which bit $i = \sigma(j)$ is set to 1 in x (in particular, the last such j will have $\sigma(j) = k$ once the red cycle containing k is closed). Therefore the final values of x and $y = f(x)$ after the last call to *successor* satisfy $y_j = x_{\sigma(j)}$ for all j, or $y = g(x)$. Thus $f(x^k) = g(x^k)$ and x^k is in U. So there can not be clauses in I of the form $\overline{x_k} \vee x_{k'}$ where k' is not implied by k: the point x^k is in U but would violate such a clause. This shows that we already have *all* the clauses of the form $\overline{x_k} \vee x_{k'}$ with k and k' red.

Finally, if some point x in U has $x_k = x_{k'} = 1$, then x must necessarily have $x_i = 1$ if i is implied by k or by k'. The median $x^{kk'} = \text{med}(x, x^k, x^{k'})$ is then in U, and $x_i^{kk'} = 1$ if and only if i is implied by k or by k' (just take bitwise majority). So we can decide whether we can include in I a clause $\overline{x_k} \vee \overline{x_{k'}}$, i.e., whether some point x in U has $x_k = x_{k'} = 1$, by testing just one such x, namely the point $x^{kk'}$ with $x_i^{kk'} = 1$ if and only if i is implied by k or by k', for membership in U.

This completes the characterization of U, i.e., of the points x such that $f(x) = g(x)$. The points x such that $g(x) = x$ are precisely the points with $x_{\sigma(j)} = x_j$, by definition, and we need only consider j red because $\sigma(j) = j$ for j black. The condition is enforced by clauses $\overline{x_{\sigma(j)}} \vee x_j$ (the converse $\overline{x_j} \vee x_{\sigma(j)}$ is redundant because $\sigma^{-1} = \sigma^{M!-1}$ is a power of σ). □

Recall that the gatewidth of a network N is $\max_{g \in N} \min(|I(g)|, |O(g)|)$.

Theorem 3.50 *In a nonexpansive network of size m and gatewidth c, there is a 2SAT instance with $O(cm)$ clauses that characterizes all the stable configurations and can be found in $O(c^2 m)$ time. All the clauses are local, involving either two inputs or an input and an output to the same gate.*

Proof. The key observation to bound both the running time and the number of clauses is that if pred[j] = i, then i and j are an input and an output, respectively, to the *same* gate h (ignore the special case where we set pred[j] = j when j is colored black). This invariant is preserved inductively because *successor*(i^0) operates entirely within the gate h one of whose inputs is i^0, i.e., during the execution of *successor* the coordinate j can only be an output to h and the coordinate i can only be an input to h. So *successor* returns an output j of h, and setting $\mathrm{pred}[j] = i^0$ preserves the invariant. Therefore each call to *successor* runs in time $O(c)$ (either i or j can only take c different values). After each call to *successor*, some i is either colored or re-colored. The number of such colorations is $2m$, because each coordinate is colored twice, green the first time, and either red or black the second time. The running time is thus $O(cm)$.

All the clauses but those of the form $\overline{x_k} \vee \overline{x_{k'}}$ with k, k' red are found directly during the execution of the algorithm in $O(cm)$ time, and so there are at most $O(cm)$ of them. The clauses $\overline{x_{\sigma(j)}} \vee x_j$ involve an input and an output to the same gate, and the clauses $\overline{x_k} \vee x_{k'}$ where k' is implied by k involve two inputs to the same gate.

The clauses $\overline{x_k} \vee \overline{x_{k'}}$ must also involve two red inputs to the same gate. If k and k' are inputs to two different gates h and h', then the point x^k with a 1 in each coordinate implied by k and 0 elsewhere only has 1s in inputs to h, and the corresponding $x^{k'}$ only has 1s in inputs to h'. The corresponding $f(x^k) = g(x^k)$ and $f(x^{k'}) = g(x^{k'})$ will have 1s at the outputs of h and h', and if $x^{kk'}$ has a 1 where either x^k has a 1 or $x^{k'}$ has a 1, then $f(x^{kk'})$ has a 1 where either $f(x^k)$ or $f(x^{k'})$ has a 1, so that $f(x^{kk'}) = g(x^{kk'})$ and the clause $\overline{x_k} \vee \overline{x_{k'}}$ is not included. So such a clause can only be included if k, k' are red inputs to the same gate. For every choice of a red k, the number of red inputs k' to the same gate that k is an input to is at most c (since pred pairs-up red outputs and red inputs to the same gate), so the number of clauses of this type is also $O(cm)$. The time to test each of them is at most $O(c)$, because only the $2c$ coordinates implied by k or by k' must be set to 1, bounding the total running time by $O(c^2 m)$. □

Knuth [74] found a simple characterization of the behaviour of the algorithm outside the periodic set, for points obtained using red coordinates alone in the calls to *successor*. We state this characterization but omit the proof. Given a red coordinate i, let L_i be the list of coordinates implied by i, in the order that they appear during a call to *successor*. Then the lists must satisfy the property that if j is in L_i, then the elements in L_j, excluding those that occur in L_i before j and excluding j itself, must constitute a prefix of L_i without the elements up to j. Furthermore, any collection of lists satisfying this property can be the set of lists for some appropriate f; in fact such an f can be described by means of a comparator circuit.

It can be shown that the output of the algorithm is independent of the choices made by the algorithm, i.e., the isomorphism g and the 2SAT clauses are the same regardless of how the starting fixed point 0 is chosen and of the order in which uncolored coordinates are picked by the algorithm. This is not a priori obvious since there can be more than one isomorphism g that coincides with f in the set of periodic points (although two such g can only differ in the coordinates where all the periodic points of f agree). We do not know whether the specific isomorphism g obtained by the algorithm has a natural combinatorial interpretation.

The 2SAT instance for the set of points that satisfy $f(x) = g(x)$ is acyclic and transitively closed. The same cannot be said about the 2SAT instance for the fixed points, whose strong components correspond to the cycles of σ, and where taking the transitive closure could increase the number of clauses from $O(m)$ to $\Omega(m^2)$. The situation is in a sense worse for periodic points. A periodic point x must not only satisfy $f(x) = g(x)$, but also $f(x^r) = g(x^r)$ for $x^r = f^{(r)}(x) = g^{(r)}(x)$ and $r \geq 0$; conversely, if $f(x^r) = g^{(r+1)}(x)$ for all $r \geq 0$ then x is periodic. So we must replace the clauses listed above involving two variables x_k and $x_{k'}$ by corresponding clauses involving $x_{\sigma^{(r)}(k)}$ and $x_{\sigma^{(r)}(k')}$ with $0 \leq r \leq m^2$. This can increase the number of clauses quadratically, even if we do not take the transitive closure (an example of this phenomenon is Subramanian's function in section 3.5.1). It is therefore a fortunate fact that, at least when the transitive closure is not taken, the fixed point set only requires $O(m)$ clauses. Much of the material in the next two chapters concerns algorithms that gain efficiency by not computing this transitive closure explicitly.

3.3.3 Efficient Representation in Scatter-Free Networks

In the scatter-free case, Mayr and Subramanian [78] showed that a single stable configuration can be found in $O(m)$ time. The algorithm that we just described can then be used to find a succint representation of all the stable configurations, in $O(c^2 m)$ time. In practice, it is convenient to have a combined algorithm that obtains the representation for the stable configurations directly. We give here such an algorithm. One of the differences between this algorithm and that of Mayr and Subramanian, is that their algorithm chooses boolean values for certain coordinates by testing the two possible values in parallel and choosing the one that is recognized as valid first. Since we are equally interested in all the stable configurations, we need to considuer all the possible values for the links of the network.

The first step of the algorithm is a simplification like the one performed in [78]. We start by eliminating all inputs and outputs of the network as in section 3.1.5. Now the following simplification is repeatedly performed. If a gate has more outputs than inputs, then at least one of its outputs must be a constant b, by the definition of scatter-free gates. The corresponding link is then removed from the network: the gate that this link feeds into is simplified by assigning to the corresponding input the constant value b, and discarding that input. The process is repeated until all the gates have at most as many outputs as inputs. Since the total number of inputs and the total number of outputs of the gates in the network are both equal to the number of links, it follows then that all the gates must have the same number of inputs and outputs. A partial assignment of values to k of the inputs of a gate must now assign constant values to k of the outputs of the gate, by the definition of scatter-free gates.

The rest of the algorithm is similar to the nonexpansive algorithm from the last section. The main difference is that colors (green, red and black) and red cycles are defined for *signed coordinates* rather than just coordinates. The two complementary signed coordinates i, $\bar{i}$ are thus handled separately by the algorithm.

Algorithm *cycles*;

pick a signed coordinate i and color i green;
repeat
   execute *successor*(i);
   if *successor* returns a signed coordinate j
   then if j is uncolored
      then {pred$[j] \leftarrow i$; $i \leftarrow j$; color i green;}
      else {**if** pred$[j]$ is undefined
         then {pred$[j] \leftarrow i$; pick an uncolored i and color i green
           (if such an i exists);}
         else interchange pred$[j]$ and i;
           follow pred pointers from j and color all signed coordinates reached red};
   else {follow the pred pointers from i, color all
      signed coordinates j reached black (including i), set pred$[j] \leftarrow j$,
      and output the clause "$\overline{x_j}$";
      pick a new uncolored i and color i green (if an such i exists);}
until all signed coordinates are either red or black.

Function *successor*(i^0);
make all variables x_i in the partial assignment x unassigned;
$i \leftarrow i^0$;
loop
   assign $x_i = 1$ in the partial assignment x;
   let x_j be a literal that gets assigned $x_j = 1$ in the partial assignment $f(x)$;
   if j is not black then
   then if j is red
      then {$i \leftarrow$ pred$[j]$;
           output the clause "$x_{i^0} \rightarrow x_i$";
           if x has $x_i = 0$ **then** return *"fail"*;}
      else return j;
   else return *"fail"*;
endloop;

It can be shown that if $i, \bar{i}$ are both red and $j = \text{pred}[i]$, then $\bar{j} = \text{pred}[\bar{i}]$.

If we now add to the clauses output by the algorithm the clauses "$x_i = x_{\text{pred}[i]}$", then one can show, as in the algorithm for the nonexpansive case, that the stable configurations must satisfy all the clauses, and that any configuration that satisfies all the clauses is stable.

Theorem 3.51 *In a scatter-free network of size m and gatewidth c, there is a 2SAT instance with $O(cm)$ clauses that characterizes all the stable configurations and can be found in $O(cm)$ time. All the clauses are local, involving either two inputs or an input and an output to the same gate.*

Note that a factor of c is saved in the complexity of the algorithm with respect to the more general nonexpansive case.

3.4 Unstable Networks and Fixed Cubes

The algorithms of the last section enabled us to find a characterization of the behavior of nonexpansive mappings in the set of periodic points, under the assumption that at least one stable configuration exists. It is natural to ask whether similar algorithms can be developed in the absence of stable configurations. It turns out that the proper substitute for a fixed point in nonexpansive mappings without fixed point is a fixed cube.

A *subcube* of the hypercube is a subgraph induced by the vertices x that satisfy some set of constraints of the form $x_i = a_i$. Such a partial assignment leaves some set of coordinates x_i unassigned: we say that these coordinates are *contained* in the subcube. Given a nonexpansive mapping f on the hypercube, a *fixed cube* is a subcube Q that is invariant under f, that is, f restricted to Q is an isomorphism on Q. For complete assignments, the corresponding subcubes are simply vertices of the hypercube, and the fixed cubes are fixed points.

Bandelt and Vel [11] showed that every nonexpansive mapping on the hypercube has a fixed cube. Their proof is implicitly based on the study of the distance center of median graphs. A characterization of the distance center of a median graph was obtained independently by Nieminen [87].

The *distance center* of a connected graph G is the set U of vertices $x \in V(G)$ that minimize $\sum_{y \in V(G)} d(y, x)$. We now state Nieminen's result, and give a simple proof based on the connection between median graphs and retracts.

Lemma 3.52 *The distance center of a connected median set in the hypercube is a subcube.*

Proof. A connected median set U in the hypercube is a retract of the hypercube; let g be a retraction whose range is U. Let V be the set of vertices x in the hypercube that minimize $\sum_{y \in U} d(y, x)$. By Lemma 3.1, we have $\sum_{y \in U} d(y, g(x)) \leq \sum_{y \in U} d(y, x)$. Furthermore, if x is not in U, then $y = g(x)$ is a vertex in U for which $d(y, g(x)) = 0 < d(y, x)$, so the preceding inequality is in fact strict. It follows that the set of minimizing vertices V is contained in U.

On the other hand, $\sum_{y \in U} d(y, x) = \sum_i (\sum_{y \in U} d(y_i, x_i))$. Let V_i be the set of x_i that minimize $\sum_{y \in U} d(y_i, x_i)$. Then $x \in V$ if and only if $x_i \in V_i$ for all i. So V is specified by conditions on the individual coordinates x_i, and therefore V is a subcube. □

This lemma gives Bandelt and Vel's theorem:

Theorem 3.53 *Every nonexpansive mapping on a hypercube has a fixed cube.*

Proof. The set U of periodic points of a nonexpansive mapping f is a connected median set, and f restricted to U is an isomorphism of the graph induced by U. The distance center of a graph is invariant under isomorphisms, since distances are preserved by isomorphisms.

Therefore f restricted to the distance center V of U is an isomorphism of V. By the lemma, V is a subcube and hence a fixed cube of f. $\square$

Bandelt and Vel stated this theorem for nonexpansive mappings on an arbitrary median graph, not necessarily the hypercube. Since a median graph is a connected median set in a hypercube, and thus a retract of the hypercube defined by a retraction g, any nonexpansive mapping on a median graph can be extended to a nonexpansive mapping on the hypercube by pre-composing it with the corresponding retraction g. Thus the fixed cube theorem for median graphs follows from the fixed cube theorem in the hypercube.

While distance centers provide a simple proof of the theorem, they give little insight into how fixed cubes can be found. In fact, we have the following.

Lemma 3.54 *Given the 2SAT description of a connected median set U in the hypercube, the problem of finding the distance center of U is #$\mathcal{P}$-hard.*

Proof. Let U and V be connected median sets in the m-cube and the m'-cube respectively. Assume, by renaming if necessary, that U contains the all-zero point 0_m and V contains the all-one point $1_{m'}$. The *union* of U and V is a median set with $|U| + |V|$ vertices in the $(m + 1 + m')$-cube, namely the vertices of the form $x11_{m'}$ with $x \in U$ and the vertices of the form $0_m 0y$ with $y \in V$. These two pieces of the union are connected by the pair of adjacent vertices $0_m 11_{m'}, 0_m 01_{m'}$. In fact, the union is itself a connected median set: the 2SAT clauses that characterize it are the clauses on the first m bits that characterize U, the clauses on the last m' bits that characterize V, and the additional clauses of the form $x_i \to x_j$ for the values i, j such that either i is one of the first m coordinates and $j = m+1$, or $i = m+1$ and j is one of the last m' coordinates.

This construction enables us to do two things. First, given an m-bit integer K, we can construct a 2SAT instance characterizing a median set V on an m'-cube with $m' = O(m)$ such that $|V| = K$. This is achieved by representing K as a sum of powers of the form 2^n; an n-cube (trivially a connected median set) has 2^n vertices, and the sum needed to obtain a connected median set with K vertices is obtained by taking unions. Second, given two connected median sets U and V, we can decide whether $|U| < |V|$, $|U| > |V|$ or $|U| = |V|$ by finding the distance center of the union W of U and V. This is achieved by using the fact, obtained in the proof of Lemma 3.52, that the possible values of x_i for x in the distance center are those that minimize $\sum_{y \in W} d(y_i, x_i)$. If $i = m+1$, then the sum equals $|U|$ if $x_i = 0$, and equals $|V|$ if $x_i = 1$. Therefore, in order to compare $|U|$ to $|V|$, it is sufficient to determine whether the distance center of W contains points x with $x_i = 0$, points x with $x_i = 1$, or both.

We can therefore use distance centers to compare $|U|$ to an arbitrary m-bit integer K. A binary search on K enables us then to determine $|U|$ exactly. On the other hand, we know that evaluating $|U|$, the number of solutions to a 2SAT instance, is #$\mathcal{P}$-complete, and so finding the distance centers that enable us to do this evaluation is a #$\mathcal{P}$-hard problem. $\square$

The preceding result holds even if U is a distributive lattice, since the counting problem for 2SAT instances is #$\mathcal{P}$-complete even in the bipartite case.

This result tells us that if we want to find fixed cubes efficiently for a nonexpansive mapping f, then we must look for a fixed cube other than the one defined by the distance center of the periodic set. There are at least two ways of achieving this. One of them, and perhaps the most useful for us, will be to look for fixed cubes that depend on both the periodic set and on the specific isomorphism of this set defined by f. A second way, interesting mainly from a theoretical point of view, enables us to find a fixed cube that depends on the periodic set alone, just like in the case of the distance center. The construction is as follows.

Given a connected median set U described by a transitively closed, acyclic 2SAT instance, we define the *heart* of U by the following procedure. Consider all the clauses $u \vee v$, and find all the variables x_i such that precisely one of the two complementary literals $x_i, \overline{x_i}$ occurs as a literal in these clauses. If it is x_i that occurs, set $x_i = 1$ and remove all the clauses involving x_i, otherwise set $x_i = 0$ and remove all the clauses involving $\overline{x_i}$. Note that the removed clauses are satisfied by the assignment to x_i. This results in a 2SAT instance with fewer clauses, and the elimination process can be repeated. If we reach a point where no clauses can be eliminated, we stop. The resulting assignment of boolean values to a subset of the coordinates defines then a subcube Q, the heart of U.

We must first show that the heart of U is indeed contained in U. Equivalently, all the clauses that characterize U must be satisfied by the partial assignment obtained with the above procedure, or in other words, the set of clauses remaining when the elimination process terminates must be empty. To prove this, we consider the implication graph of the 2SAT instance, which we know is acyclic. If there remains a clause, corresponding to a directed edge in the implication graph, then there must remain an edge of positive indegree and zero out-degree in the implication graph; the consequent of this implication is then a literal that can be eliminated by the algorithm, so the elimination process has not terminated.

The procedure for finding the heart produces the same result if U is transformed by an isomorphism, because an isomorphism of U simply permutes, and possibly negates, the variables x_i. This gives:

Theorem 3.55 *The heart of a connected median set U is a fixed cube of every isomorphism of U. The heart of the periodic set of a nonexpansive mapping f is a fixed cube of f.*

3.4.1 Minimal Fixed Cubes

We shall be mainly interested in the *minimal fixed cubes* of a nonexpansive mapping f, i.e., the fixed cubes that do not contain a smaller fixed cube. Given an isomorphism $g = g_\sigma$ of the hypercube that coincides with f in the set of periodic points of f, we say that a cycle C of σ is *stable* if it contains at most one of $i, \bar{\imath}$ for each coordinate i. Otherwise the cycle is *unstable*. Note that an unstable cycle C must in fact contain $\bar{j}$ for each j in C, because if $j = \sigma^{(k)}(i)$, then $\bar{j} = \sigma^{(k)}(\bar{i})$.

The following result can be seen in the scatter-free case from Lemma 7.8 and Theorem 7.24 in Subramanian [109].

Theorem 3.56 *For every minimal fixed cube Q of a nonexpansive mapping f, the coordinates contained in Q are those that belong to unstable cycles of the associated isomorphism g_σ.*

Proof. If i is contained in the fixed cube Q, then Q has two vertices x, y such that $x_i \neq y_i$. If j belongs to the same cycle C of σ as i, say $i = \sigma^{(k)}(j)$, then $f^{(k)}(x)$ and $f^{(k)}(y)$ differ in coordinate j, so j is also contained in Q. It follows that the coordinates contained in Q consist of entire cycles of σ. Suppose that C is not contained in Q, and choose some $i \in C$. Then the value of x_i is the same for all vertices x in Q. It follows that $\bar{i}$ is not in C; for otherwise we would have $\bar{i} = \sigma^{(l)}(i)$ for some l, so that x and $f^{(l)}(x) = g_\sigma^{(l)}(x)$ would disagree in coordinate i, for all x in Q. This shows that C is a stable cycle. Therefore all unstable cycles are contained in Q.

In the other direction, suppose that C is a stable cycle and that C is contained in Q. Let Q' be the subcube of Q obtained by setting $x_i = 1$ for each i in C. This assignment is not contradictory, i.e., does not set both $x_i = 1$ and $x_{\bar{i}} = \overline{x_i} = 1$, because C contains at most one of $i, \bar{i}$. Now, if x is in Q' and $y = f(x) = g_\sigma(x)$, then for each i in C we have $y_i = x_{\sigma(i)} = 1$, so y is in Q'. Therefore the subcube Q' is also a fixed cube. It follows that Q cannot be a minimal subcube unless all the cycles it contains are unstable cycles. □

Several useful corollaries follow from this theorem.

Corollary 3.57 *Let f be a nonexpansive mapping on a hypercube and let g_σ be an isomorphism that coincides with f in the periodic set of f. Then the minimum of $d(x, f(x))$ over all x in the hypercube equals the number of unstable cycles of σ.*

Proof. Suppose that x minimizes $d(x, f(x))$, and let $y = f(x)$. Since $d(f(x), f(y)) \leq d(x, y)$, we have $d(y, f(y)) \leq d(x, f(x))$, so y also achieves the minimum. We can therefore repeatedly replace x by its image under f without increasing $d(x, f(x))$. Eventually, we obtain a minimizing point x that is also a periodic point, so that $f(x) = g_\sigma(x)$ and $d(x, f(x)) = d(x, g_\sigma(x))$. Let $y = g_\sigma(x)$. If C is a cycle of σ and $y_i = x_{\sigma(i)} = x_i$ for all i in C, then C must be a stable cycle; otherwise, choosing l so that $\sigma^{(l)}(i) = \bar{i}$, we could show $x_{\bar{i}} = x_i$, a contradiction. It follows that for every unstable cycle C, there must exist an i in C such that $y_i \neq x_i$, and therefore $d(x, y)$ is at least as large as the number of unstable cycles of σ.

On the other hand, we can construct a point x such that $d(x, f(x))$ is at most the number of unstable cycles of σ. Let Q be a minimal fixed cube, and define a point x in Q by assigning a value to the coordinates of x contained in Q, as follows. Note that an unstable cycle has even length $2l$, since it consists of pairs $i, \bar{i}$. Furthermore, each such pair satisfies $\bar{i} = \sigma^{(l)}(i)$, i.e., the two elements i and $\bar{i}$ are opposite in the cycle. (Otherwise, if the two elements were closer in the cycle, say $\bar{i} = \sigma^{(k)}(i)$ with $k < l$, then $i = \sigma^{(k)}(\bar{i})$ by the definition of a signed permutation, giving $i = \sigma^{(2k)}(i)$ with $2k < 2l$, a contradiction.) For each cycle C contained in Q, necessarily an unstable cycle, pick an i in C, and set $x_j = 1$ for all $j = \sigma^{(k)}(i)$ with $0 \leq k < l$. Note that this considers precisely one element of each pair $j', \overline{j'}$, so the assignment is consistent and assigns values to all coordinates. If $y = f(x) = g_\sigma(x)$, then $y_j = x_{\sigma(j)} = 1 = x_j$ for all j such that x_j was set to 1 above, with the only exception of $j = l - 1$. (For the remaining j, we have $0 \leq k < l - 1$, and therefore $0 \leq k + 1 < l$, so $x_{\sigma(j)} = 1$.) Therefore x and y differ in only one coordinate per unstable cycle, proving that $d(x, y)$ does not exceed the number of unstable cycles. □

In particular, since a fixed point is a point x such that $d(x, f(x)) = 0$, we have:

Corollary 3.58 *Let f be a nonexpansive mapping on a hypercube and let g_σ be an isomorphism that coincides with f in the periodic set of f. Then f has a fixed point if and only if σ has no unstable cycles.*

By composing the mapping f with a retraction, we also obtain:

Corollary 3.59 *Let f be a nonexpansive mapping on the hypercube and let U be a connected median set such that f maps U into U. Then U contains a minimal fixed cube of f, and f has a fixed point in the hypercube if and only if f has a fixed point in U.*

Proof. Let g be a retraction of the hypercube whose range and set of fixed points is U, and let h be the composition $h(x) = f(g(x))$. The fixed cubes of h are contained in U since the range of h is contained in U, so these fixed cubes are fixed cubes of f as well. Furthermore, the fixed cubes of f contained in U are fixed cubes of h as well. Therefore the fixed cubes of h are the fixed cubes of f that are contained in U. Since h has a fixed cube, it follows that f has a fixed cube contained in U, and therefore has a minimal fixed cube contained in U as well. If f has a fixed point in U, then f has a fixed point in the hypercube. Conversely, if f has a fixed point in the hypercube, then this fixed point is a minimal fixed cube of f that contains no coordinates, and since all minimal fixed cubes contain the same set of coordinates by the theorem, a minimal fixed cube of f in U contains no coordinates either and must therefore be a fixed point of f in U. □

Theorem 3.56 can also be used to give an alternative proof of the fact that a nonexpansive mapping f on a hypercube has a fixed point if and only if each projection f_i has a fixed point, which played a crucial role in the development of fixed point algorithms in section 3.1. The interesting direction is the 'if' direction. Let f be a nonexpansive mapping on a hypercube with coordinate set S. If f_i has a fixed point a, and we let Q be a fixed cube of the restriction $h = f_{a, U \setminus \{i\}}$, then every x in Q satisfies $f(ax) = f_i(a)h(x) = ah(x)$, so Q yields a fixed cube of f that does not contain coordinate i. Since all minimal fixed cubes contain the same set of coordinates by Theorem 3.56, it follows that no minimal fixed cube contains coordinate i. Therefore, if each f_i has a fixed point, we can conclude that the minimal fixed cubes of f contain no coordinate i, so the minimal fixed cubes are in fact fixed points.

3.4.2 Algorithms for Fixed Cubes

The $O(m^3)$ fixed point algorithm of section 3.1.5 can be adapted so that it always finds a minimal fixed cube (in particular a fixed point if one exists). If f has a fixed cube Q that does not contain coordinate i, with $x_i = a$ for all x in Q, then $f_i(a) = a$. Conversely, as we observed in the last section, if $f_i(a) = a$, then f has a fixed cube Q that does not contain coordinate i, with $x_i = a$ for all x in Q. As a result, we can consider the different coordinates i in turn, test whether $f_i(a) = a$ for some a, and if this test succeeds, restrict the mapping f to the restriction $h = f_{a, S \setminus \{i\}}$ as in section 3.1.5. In the end, we are left with a mapping f' on a smaller hypercube Q such that none of the projections f'_i contains a fixed point, so all coordinates i are contained in a minimal fixed cube, and the minimal fixed cube of this mapping is just Q itself.

It can also be shown that the algorithm of section 3.3.2, with no modifications, will find the isomorphism g_σ coinciding with f in the periodic set when started at a point x minimizing $d(x, f(x))$ and belonging to a fixed cube. In fact, by removing from the 2SAT instance produced by the algorithm all occurrences of coordinates S contained in a minimal fixed cube, we obtain a 2SAT instance on the remaining coordinates, whose solutions specify subcubes containing precisely the coordinates in S: these subcubes, it turns out, are precisely the minimal fixed cubes.

3.4.3 From Monte Carlo to Las Vegas

Minimal fixed cubes provide a simple way of transforming Monte Carlo fixed point algorithms with one-sided error like the ones in section 3.1.6, into Las Vegas algorithms, without affecting their time bound. (A Monte Carlo algorithm with one-sided error can only give the wrong answer in one of the two possible answers 'yes' or 'no', gives the wrong answer with only a small probability, and runs within the stated time bound; in our case, the algorithms err in asserting that there is no fixed point when a fixed point exists only with small probability. A Las Vegas algorithm always gives the correct answer, with the stated time bound holding only in expectation.)

Suppose that f has no fixed point, and suppose that we know a coordinate i contained in a minimal fixed cube. Then f_i has no fixed point, because as we saw in section 3.4.1, if f_i has a fixed point then there exists a fixed cube that does not contain i, so that all minimal fixed cubes exclude coordinate i. To prove that f_i has no fixed point, it is sufficient to compute $f_i(0) = 1$ and $f_i(1) = 0$, and this can be done with $O(m)$ queries in $O(m^2)$ time by Lemma 3.10. Once these two values are known, we have shown that f_i has no fixed point, and therefore f has no fixed point. The complexity in terms of the running time and number of queries is no larger (in fact smaller, for the number of queries) than the complexity of the Monte Carlo algorithms of section 3.1.6.

The main goal in our approach will the be to find the appropriate coordinate i contained in a minimal fixed cube, without testing all m possible values for i. The proofs of correctness and the analysis for the algorithms of section 3.1.6 are based on the existence of some unknown fixed point a. It turns out that these algorithms have the property that if we replace the fixed point a by a fixed cube Q, the proofs of correctness remain virtually unchanged. For the time bound, the algorithms can be made to provide a coordinate i at each step, so that if we charge the time taken at each step to the corresponding coordinate, then the time charged to coordinates not contained in the fixed cube Q (the *active* time) admits the same time bound analysis. As a result, if the algorithm is expected to find a fixed point within time T with probability at least p, yet the mapping has no fixed point, then the fact that no fixed point has been found after $2T$ steps can be used to infer that with probability at least p, the active time was only T, and therefore at least half of the steps were charged to a coordinate contained in Q. Choosing one such step at random, we obtained with probability at least $1/2$ a coordinated contained in Q. Combining both probabilities gives a probability of at least $p/2$ that a coordinate contained in Q will be obtained, and hence that a proof that there is no fixed point can be obtained as explained above, in an additional $O(m^2)$ time with $O(m)$ queries. We now have a modified algorithm that can, with constant probability, either find a fixed point if one exists, or prove that none exists. The probability that neither alternative occurs decreases exponentially if the algorithm is executed many times, so by repeating the execution until one of the two conclusive alternatives takes place, we obtain an algorithm running in expected $O(T)$ time.

We illustrate this idea only for the first (and simplest) Monte Carlo algorithm in section 3.1.6, although the same idea works for the more complicated version, with the appropriate choice of coordinate to which each step is charged. In the simple algorithm, we charge each step to the corresponding coordinate i that is changed in x. In the original analysis of the algorithm, the key observation was the fact that this change in coordinate i would change $d(x, a)$, where a is a fixed point, by '+1' or by '−1', and that the '−1' case had probability at least $1/2$. If we replace a by a fixed cube Q, then the same argument holds for $d(x, Q)$, provided that i is chosen from coordinates that are not contained in Q; if i is contained in Q, then $d(x, Q)$ remains unchanged. As a result, the analysis of the active time (time spent in coordinates not contained in Q) is the same as the analysis of the running time of the original algorithm, meeting the conditions required to transform the Monte Carlo algorithm into a Las Vegas algorithm.

3.4.4 A Small Certificate for Mappings without Fixed Point

We have seen that it is possible to determine whether a nonexpansive mapping f on the m-cube has a fixed point in time polynomial in m. This shows that if f has no fixed point, then it is not necessary to know the images of all 2^m points to prove it: a number of points polynomial in m is sufficient. We shall prove that in fact, four points suffice to prove that a mapping has no fixed point.

A *certificate* for a nonexpansive mapping f without fixed point is a set of points in the hypercube such that every mapping f' that coincides with f on these points has no fixed point. For some mappings f, the size of the smallest certificate is 4. For example, if f is the mapping on the 2-cube defined by $f(x_1 x_2) = \overline{x_1}\,\overline{x_2}$, then f has no fixed points, but if a nonexpansive mapping f' agrees with f in three of the four points, it can still have the fourth one as a fixed point. The proof of the existence of a four-point certificate for all nonexpansive mappings without fixed point is based on the following lemma, which says that gates satisfying a certain properties can be decomposed into smaller gates. Given an assignment x, define $y = \overline{x}$ by letting $y_i = \overline{x_i}$.

Lemma 3.60 *Given a nonexpansive mapping f on the hypercube, suppose that there exist points x, y and a partition of the coordinate set into two disjoint sets S and T, such that $f(x_S x_T) = y_S y_T$, $f(x_S \overline{x_T}) = y_S \overline{y_T}$, $f(\overline{x_S} x_T) = \overline{y_S} y_T$, and $f(\overline{x_S}\, \overline{x_T}) = \overline{y_S}\, \overline{y_T}$. Then $f(z) = f_S(z_S) f_T(z_T)$ for all points z.*

Proof. Let $m = |S \cup T|$ be the total number of coordinates. Then $d(x_S x_T, \overline{x_S}\, \overline{x_T}) = m = d(y_S y_T, \overline{y_S}\, \overline{y_T})$, and $d(\overline{x_S} x_T, x_S \overline{x_T}) = m = d(\overline{y_S} y_T, y_S \overline{y_T})$. Given $z = z_S z_T$, let $w = z_S x_T$, and let $z' = f(z)$, $w' = f(w)$. We want to show that $z'_S = w'_S$. Since $d(x_S, \overline{x_S}) = d(x_S, z_S) + d(z_S, \overline{x_S})$ and $d(x_T, \overline{x_T}) = d(x_T, z_T) + d(z_T, \overline{x_T})$, we have $d(x_S x_T, \overline{x_S}\, \overline{x_T}) = d(x_S x_T, w) + d(w, z) + d(z, \overline{x_S}\, \overline{x_T})$. Taking images under f, and using the nonexpansiveness of f, we obtain $d(y_S y_T, \overline{y_S}\, \overline{y_T}) = d(y_S y_T, w') + d(w', z') + d(z', \overline{y_S}\, \overline{y_T})$. Separating the S and T parts, and using the triangle inequality on the two parts, we get $d(y_S, \overline{y_S}) = d(y_S, w'_S) + d(w'_S, z'_S) + d(z'_S, \overline{y_S})$. Similarly, replacing x_S and y_S with $\overline{x_S}$ and $\overline{y_S}$ respectively throughout this argument, we obtain $d(\overline{y_S}, y_S) = d(\overline{y_S}, w'_S) + d(w'_S, z'_S) + d(z'_S, y_S)$. Adding these two equalities gives

$$2d(y_S, \overline{y_S}) = d(y_S, w'_S) + d(w'_S, \overline{y_S}) + 2d(w'_S, z'_S) + d(y_S, z'_S) + d(z'_S, \overline{y_S})$$
$$\geq 2d(y_S, \overline{y_S}) + 2d(w'_S, z'_S)$$

by the triangle inequality, so $d(w'_S, z'_S) = 0$ and therefore $z'_S = w'_S$. It follows that $z'_S = [f(z)]_S$ depends on z_S alone, and not on z_T, and therefore $z'_S = f_S(z_S)$ for all values of z_T. The same argument, with the roles of S and T reversed, shows that $z'_T = f_T(z_T)$ for all values of z_S. □

We are now ready to show that every nonexpansive mapping without fixed point has a four-point certificate.

Theorem 3.61 *Let f be a nonexpansive mapping on the m-cube without fixed point. Then there exist four points x^1, x^2, x^3, x^4, such that if f' is a nonexpansive mapping on the m-cube satisfying $f'(x^i) = f(x^i)$ for $1 \leq i \leq 4$, then f' has no fixed point.*

Proof. Let Q be a minimal fixed cube of f, and let $g = g_\sigma$ be an isomorphism that coincides with f in the periodic set of f. Let R be the set of coordinates contained in Q, and let x^1 and x^2 be two points in Q such that $x_i^2 \neq x_i^1$ for all $i \in R$. Pick an odd number of unstable cycles of σ, and let S be the set of coordinates that belong to these cycles. By Theorem 3.56, we have $S \subseteq R$. Let $T = R \setminus S$. Let x^3 and x^4 be the points in Q such that $x_i^3 \neq x_i^1$ if and only if $i \in S$, and $x_i^4 \neq x_i^3$ for all $i \in R$. We shall show that x^1, x^2, x^3, x^4 constitute a four-point certificate.

Suppose that $f'(x^i) = f(x^i)$ for $1 \leq i \leq 4$, and that f' has a fixed point. If z is any point in Q, then z is on a shortest path from x^1 to x^2, and since $d(f'(x^1), f'(x^2)) = d(x^1, x^2)$, the point $f'(z)$ must be on a shortest path from $f'(x^1)$ to $f'(x^2)$; these two points are in Q, so $f'(z)$ is also in Q. This shows that f' maps Q to itself, so by Corollary 3.59, f' must have a fixed point z in Q.

Restrict the mapping f' and the four points x^1, x^2, x^3, x^4 to Q (i.e., to the coordinates in R). If we write the point $x = x^1$ as $x = x_S\, x_T$, then the other three points are $x^2 = \overline{x_S}\, \overline{x_T}$, $x^3 = \overline{x_S}\, x_T$, and $x^4 = x_S\, \overline{x_T}$. Furthermore, observe that if we complement the values of the coordinates in some cycle of σ for some point z, then the same coordinates get complemented in $g_\sigma(z)$. As a result, the points $y^i = f'(x^i) = f(x^i) = g_\sigma(x^i)$ can be expressed with respect to $y = y^1$ by $y = y_S\, y_T$, $y^2 = \overline{y_S}\, \overline{y_T}$, $y^3 = \overline{y_S}\, y_T$, and $y^4 = y_S\, \overline{y_T}$. The conditions of Lemma 3.60 are then met, and we can write f' as $f'(z) = f'_S(z) f'_T(z)$. To prove that f' has no fixed point, it suffices to prove that f'_S has no fixed point. We know that $f'_S(x_S) = y_S$, and $f'_S(\overline{x_S}) = \overline{y_S}$. Therefore every point z in the domain of f'_S lies on a shortest path from x_S to $\overline{x_S}$, its image $f'_S(z)$ lies on a shortest path from y_S to $\overline{y_S}$, and $d(x_S, z) = d(y_S, f'_S(z))$. Furthermore, we claim that $d(x_S, y_S)$ is odd. To show this, it is sufficient to show that the distance in coordinates belonging to each of the unstable cycles that form S is odd, since S consists of an odd number of such cycles. If C is such a cyle, of length $2l$, and i is in C, then the elements $\sigma^{(k)}(i)$ for $0 \leq k < l$ contain precisely one of each pair $j, \bar{j}$ in C, and $\sigma^{(l)}(i) = \bar{i}$ (see the proof of Corollary 3.57). The distance in coordinates belonging to C is then

$$\sum_{0 \leq k < l} d(x_{\sigma^{(k)}(i)}, y_{\sigma^{(k)}(i)}) = \sum_{0 \leq k < l} d(x_{\sigma^{(k)}(i)}, x_{\sigma^{(k+1)}(i)}) \equiv d(x_i, x_{\sigma^{(l)}(i)}) = d(x_i, x_{\bar{i}}) = 1 \pmod{2},$$

so the distance is indeed odd. Therefore

$$d(z, f'_S(z)) \equiv d(z, x_S) + d(x_S, y_S) + d(y_S, f'_S(z)) = 2d(x_S, z) + d(x_S, y_S) \pmod{2}$$

is also odd, proving that f'_S cannot have a fixed point. It follows that f' has no fixed point in Q, and therefore no fixed point anywhere. □

Note that the distance between periodic points and their respective images under f is in fact odd for coordinates contained in unstable cycles, and even for coordinates contained in stable cycles, so this distance is odd if and only if the number of unstable cycles is odd. In that case, we can let $S = R$ and and $T = \emptyset$ in the preceding construction, obtaining a two-point certificate, which is clearly optimal. When the distance between periodic points and their respective images is even, we only seem to be able to construct four-point certificates. We shall show below that this parity distinction is inherent, by proving that in the even case the number of points needed for a certificate is at least 3. Along the way, we examine certain cases of the general (and very difficult) question of characterizing the partial mappings on the hypercube that can be extended to complete nonexpansive mappings.

Lemma 3.62 *Suppose that the points $x^1, x^2, x^3, y^1, y^2, y^3$ in the hypercube satisfy $d(y^i, y^j) \leq d(x^i, x^j)$ for all i, j. Then there exists a nonexpansive mapping f on the hypercube such that $f(x^i) = y^i$ for all i.*

Proof. Let $t = \operatorname{med}(x^1, x^2, x^3)$ and $u = \operatorname{med}(y^1, y^2, y^3)$. Let T be the tree consisting of the union of three shortest paths joining t to x^1, x^2 and x^3 respectively, and let U be the analogous tree joining u to y^1, y^2 and y^3. Both trees are connected median subsets of the hypercube, and hence retracts of the hypercube. Let h be a retraction whose range is T. If we can construct a nonexpansive mapping f from T to U that satisfies the required constraints, then by pre-composing f with the retraction h we shall obtain a mapping defined in the entire hypercube. To define f on T, it is sufficient to give a point $t' = f(t)$ in U such that $d(t', y^i) \leq d(t, x^i)$ for all i, because then the path joining t to x^i in T can be mapped to the path joining t' to y^i in U. Note that we cannot have $d(t, x^i) < d(u, y^i)$ for two distinct values of i, say $i = 1, 2$, because then $d(x^1, x^2) < d(y^1, y^2)$. Therefore $d(u, y^i) \leq d(t, x^i)$ for at least two values of i, say $i = 2, 3$. If $d(u, y^1) \leq d(t, x^1)$ as well, then $t' = u$ is an appropriate choice. If $d(t, x^1) < d(u, y^1)$, then we let t' be the point at distance $d(t, x^1)$ from y^1 along the path joining y^1 and u in U. Then $d(t', y^1) = d(t, x^1)$ and $d(t', y^i) = d(y^1, y^i) - d(t', y^1) \leq d(x^1, x^i) - d(t, x^1) = d(t, x^i)$ for $i = 2, 3$, giving again an appropriate t'. □

The analogous characterization question for four points and their images seems more delicate and is still open. Pairwise distances as in the preceding lemma are not enough to describe how the points are arranged, and do not lead to an appropriate characterization, as illustrated by the example following Lemma 3.44.

Corollary 3.63 *A nonexpansive mapping f on the hypercube without fixed point has a two-point certificate if and only if the periodic points of f are at odd distance from their images.*

Proof. We already know that if the distance between periodic points and their images is odd, then a two-point certificate exists. In the other direction, suppose that a two-point certificate exists, say $f(x^i) = y^i$ for $i = 1, 2$ with $d(y^1, y^2) \leq d(x^1, x^2)$. Then no nonexpansive mapping f' such that $f'(x^i) = y^i$ for $i = 1, 2$ can have a point z such that $f'(z) = z$. By the lemma, this implies that no point z satisfies $d(y^i, z) \leq d(x^i, z)$ for both values $i = 1, 2$. Let Q be the smallest cube containing both y^1 and y^2, and let $n_i = d(x^i, Q)$. Note that the two y^i disagree in the coordinates contained in Q. We partition the coordinates contained in Q into four sets: The set S_{ij} contains the coordinates where x^1 agrees with y^i and x^2 agrees with y^j, for $i, j = 1, 2$. Let $k_{ij} = |S_{ij}|$. Consider a point z in Q that agrees with y^2 for all coordinates in S_{11}, agrees with y^1 for all coordinates in S_{22}, and agrees with y^1 in l out of the k_{21} coordinates in S_{21}, for some $0 \leq l \leq k_{21}$. Then $d(x^1, z) - d(y^1, z) = n_1 + k_{22} + (2l - k_{21})$ and $d(x^2, z) - d(y^2, z) = n_2 + k_{11} + (k_{21} - 2l)$. By assumption, these two quantities cannot be both nonnegative, i.e., we cannot have $k_{21} - k_{22} - n_1 \leq 2l \leq k_{21} + k_{11} + n_2$ for any integer choice of l between 0 and k_{21}.

Since the choice $l = k_{21}/2$ satisfies this constraint, we infer that k_{21} must be odd. For k_{21} odd, the choice $l = (k_{21} - 1)/2$ will satisfy the constraint unless $k_{22} = n_1 = 0$, and the choice $l = (k_{21} + 1)/2$ will satisfy the constraint unless $k_{11} = n_2 = 0$. We have therefore shown that k_{21} is odd and $k_{11} = k_{22} = n_1 = n_2 = 0$. It follows that x^1 and x^2 are both in Q and disagree in all coordinates contained in Q so that $d(x^1, x^2) = d(y^1, y^2)$. Therefore every point u in Q lies on a shortest path from x^1 to x^2, so $f(u)$ lies on a shortest path from y^1 to y^2, and belongs also to Q. Since f maps Q to Q, there must exist a periodic point of f in Q. Observe now that $d(x^i, y^i) = k_{21}$ is odd for $i = 1, 2$. Because of the shortest path condition, $d(f(u), y^i) = d(u, x^i)$ for all u in Q, so $d(f(u), u)$ is odd. In particular, $d(f(u), u)$ is odd for some periodic point u of f, and therefore odd for all periodic points of f. □

We do not know whether there are nonexpansive mappings without fixed points where periodic points are at even distance from their respective images, and which admit a three-point certificate; we do know that such a certificate could not consist of periodic points alone. A full characterization of the distinction between mappings with three-point and four-point certificates would probably require a better understanding of partial mappings on four points.

3.4.5 The Convergent Network Representation is Decidable

Given a nonexpansive network N that converges to a gate g and an input assignment x, we can use the algorithm of section 3.1.3 to evaluate $g(x)$. By trying all possible input assignments, we can completely determine the gate g. On the other hand, it is not clear whether, given a basis Ω and a gate g, it is possible to determine whether there exists a network over Ω that converges to gate g. The main difficulty is that there is no obvious bound on how large such a network might be (compare to Theorem 3.45 in the case of circuits). We shall give here a bound on the size of the smallest network over a basis Ω that converges to a gate g, therefore obtaining an algorithm for the problem of representation of gates by convergent networks.

Consider for a moment the special case where, for all input assignments, the network has a stable configuration consistent with the input that gives the corresponding output assignment. One can verify that a configuration is stable by testing the inputs and outputs of each gate separately, so the test for stability reduces to local tests at each gate. Since the gates are chosen from a given basis Ω, their size is bounded, and we can hope to be able to use this boundedness to obtain an algorithm for the representation question. By contrast, in order to verify a periodic point, when no fixed point exists, we need some sort of global examination of the network, whose size is not a priori bounded.

To overcome this obstacle, in the absence of a fixed point, we shall look for its natural generalization, namely a fixed cube. Since fixed cubes consist of periodic points, they can be used in the definition of convergent networks. A subcube is a partial assignment to links of the network; to verify that a subcube is a fixed cube, it suffices to check at each gate of the network that its partial input assignment produces its partial output assignment, and that the remaining unassigned output links are simply a permutation (up to negations) of the unassigned input links. This gives the required local test at each gate. If this test succeeds, we say that the gate is *consistent* with the partial input and output assignments.

If the gate g to which the network must converge has n inputs, then we will choose 2^n fixed cubes $Q(x)$, one for each of the 2^n possible input-output assignments x and $g(x)$ for the network. For a choice of 2^n fixed cubes $Q(x)$, one for each x, each gate g' in the network can be viewed as a mapping from the 2^n partial assignments to its inputs in $(x, g(x), Q(x))$ to the 2^n partial assignments to its outputs in $(x, g(x), Q(x))$ defined by the 2^n possible choices of x. A gate in the network can be replaced by any other gate that is consistent with these partial input and output assignments without affecting the validity of the fixed cubes and therefore without affecting the gate g to which the network converges. Given the possible choices for a single triple $(x, g(x), Q(x))$, there are three alternatives for each link in the network: the link can be either 0, 1, or unassigned. This gives $A = 3^{2^n}$ possible choices for each link over all possible choices of 2^n triples $(x, g(x), Q(x))$. We represent each such choice by a vector with A components: the ith possible choice is represented by a vector V_i with the ith component equal to 1 and the remaining components equal to 0. If a gate g' in the network has k inputs, we can choose one of the A possible choices for each of its inputs (corresponding to choices of 2^n triples $(x, g(x), Q(x))$) and check whether the gate is consistent with the resulting mapping from 2^n partial input assignments to 2^n partial output assignments. If it is, then we say that the gate *generates* the vector $V(g')$ with A components obtained by adding the vectors V_i chosen for each of its inputs and subtracting the resulting vectors V_i for each of its outputs. Note that a gate in Ω with k inputs can generate up to A^k vectors, one for each choice of 2^n values (0, 1, or unassigned) for each of its k inputs. On the other hand, the gate g to which the network must converge generates a single vector U with A components that describes the mapping from the 2^n possible inputs to the resulting outputs.

If we have a network that converges to g, and we choose a fixed cube for each of the 2^n possible input assignments, then the sum $\sum_{g'} V(g')$ of the vectors generated by each of the gates under the choice of fixed cubes equals the vector U generated by the gate g, because the vectors V_i corresponding to internal links of the network are added once (as inputs to a gate) and subtracted once (as outputs of a gate), leaving in the sum only the inputs of the network, which are added once, and the outputs of the network, which are subtracted once. Conversely, if we can represent the vector U generated by the gate g as a sum of vectors $V(g')$ generated by gates g' in Ω, then we can construct a network by selecting, for each vector $V(g')$ in the sum, the gate g' that generates it, and matching, as far as possible, inputs of gates that give a '+1' in one of the A components to outputs of gates that give a '−1' in the same component. The unmatched inputs and outputs still add up to the vector U generated by g, but now no unmatched input and output correspond to the same component of this vector, so there are no cancellations in the sum and these inputs and outputs must then coincide with inputs and outputs of g. If g has any additional inputs and outputs, then their vector sum is zero, which means that the additional outputs are simply copies of the additional inputs and can be described by additional ID gates from inputs to outputs of the network. This gives a network over Ω that converges to g.

We have therefore reduced the problem of deciding whether g can be represented by a network over Ω to the problem of deciding whether the vector with A components generated by g is a nonnegative integer linear combination of the vectors generated by the gates of Ω, where a gate in Ω with k inputs generates up to A^k vectors. This is an integer program, which admits a nondeterministic polynomial-time algorithm (see Borosh and Treybig [18] and Karp [72]).

Theorem 3.64 *There is a nondeterministic algorithm that decides whether there is a network over a given nonexpansive basis Ω that converges to a given gate g; the time complexity of the nondeterministic algorithm is doubly-exponential in the number of inputs of g.*

In fact, the nondeterministic algorithm has time complexity $2^{O(k2^n)}|\Omega|$ if g has n inputs and the gates in Ω have at most k inputs. The key to the decidability of the problem is therefore the fact that if an integer program has a solution, it has a relatively small solution, with numbers whose description is no larger than a polynomial in the size of the input integer program [18]. In our case, this size is doubly-exponential in n, and so the numbers themselves, which give the total number of links used in the network, may be triply-exponential in n. This provides a bound, perhaps excessively large, on the number of gates needed for a network that converges to a given n-input gate.

There is a weaker notion of convergence which is easier to handle. Let $x^{[r]}$ denote the assignment that consists of r copies of the assignment x (after creating r different coordinates $(i,1),\ldots,(i,r)$ for each coordinate $i \in S(x)$). A network *converges weakly* to a gate g if it converges to a gate g' such that $g'(x^{[r]}) = (g(x))^{[r]}$ for some $r \geq 1$. If we remove the integrality constraint in the integer program defined above, we obtain a linear program that can be solved *deterministically* in polynomial time. The solutions will be rational numbers; if we multiply them by their least common denominator r, then we obtain an integer linear combination of vectors generated by gates in Ω whose sum is the vector generated by g multiplied by r, so the corresponding network converges to a gate g' mapping r copies of x to r copies of $g(x)$. We therefore have a deterministic algorithm for weak convergence whose complexity is comparable to that of the nondeterministic algorithm for (strong) convergence given above.

A seemingly even weaker notion of convergence is that of asymptotic convergence. A family of networks *converges asymptotically* to a gate g if the ith network in the family converges to a gate g^i such that $g^i(x^{[r_i]}) = (g(x))^{[s_i]}$ for some $r_i, s_i \geq 1$, and $\lim_{i\to\infty} r_i/s_i = 1$. This notion was used by Mayr and Subramanian [78] to represent convergent networks N by circuits, by means of the composition circuit (N, k) of section 2.1.1 (see Fig. 2.3). By the definition of convergent networks, if we fix an arbitrary internal assignment z, then for every input assignment x to N, the corresponding input assignment $(z,0)(x,0)(x,1)\ldots(x,k-1)$ will be mapped by the circuit to an output assignment $(y^1,1)(y^2,2)\ldots(y^k,k)(z',k)$ such that, for some integer a, independent of k, we have $y^j = g(x)$ for all $j > a$. Then $\lim_{k\to\infty} k/(k-a) = 1$, and we have a family of circuits that converges asymptotically to g.

Suppose now that we have a family of circuits or networks over Ω that converges asymptotically to a gate g. The vector generated by g is the difference $U - V$ of the vectors corresponding to the inputs and to the outputs of g. By asymptotic convergence, the vectors $r_i U - s_i V$ are nonnegative linear combinations of vectors generated by gates in Ω. Therefore $(U - V) + (1 - s_i/r_i)V$ is also a nonnegative linear combination, and using the fact that $s_i/r_i \to 1$ we can conclude that $U - V$ is a nonnegative linear combination as well, because the space of nonnegative linear combinations contains all its limit points. This shows that there is a network over Ω that converges weakly to g as well. These observations are summarized in the following theorem.

Theorem 3.65 *Given a finite nonexpansive basis Ω and a gate g, there is family of circuits (or a family of networks) over Ω that converges asymptotically to g if and only if there is a network over Ω that converges weakly to g. Such a network can be found deterministically in time doubly-exponential in the number of inputs of g.*

Therefore the deterministic complexity of the representation by weakly convergent networks is at present comparable to the complexity of the representation by circuits from section 3.3.1, despite the discrepancy (from exponential to triply-exponential) in the known bounds on the size of the resulting circuits and networks.

In the definition of weak convergence, it is sufficient to consider fixed points, instead of fixed cubes as in the case of strict convergence. The reason for this is that weak convergence implies asymptotic convergence using circuits; for asymptotic convergence using circuits, only fixed points are needed, so only vectors $V(g')$ that assign values $0, 1$ to all inputs and outputs of g' are used; and asymptotic convergence implies in turn weak convergence by a limiting process that can be carried out using the same vectors $V(g')$.

Using an implementation of linear programming, we have executed the algorithm presented here to prove that the 4-input, 1-output monotone nonexpansive gate $g(x_1 x_2 x_3 x_4) = (x_1 \wedge x_2) \vee (x_2 \wedge x_3) \vee (x_3 \wedge x_4)$ cannot be obtained from a weakly convergent network over $\{C\}$. This in turn implies that the 4-input, 4-output monotone nonexpansive gate $h(x_1 x_2 x_3 x_4) = y_1 y_2 y_3 y_4$ with $y_1 = x_1 \wedge x_2 \wedge x_3 \wedge x_4$, $y_2 = (x_1 \wedge x_2) \vee (x_2 \wedge x_3) \vee (x_3 \wedge x_4)$, $y_3 = (x_1 \wedge x_3) \vee (x_1 \wedge x_4) \vee (x_2 \wedge x_4)$, $y_4 = x_1 \vee x_2 \vee x_3 \vee x_4$, cannot be obtained from a weakly convergent network over $\{C\}$. It then follows, by the theorem below, that this gate cannot be obtained from a weakly convergent network over $\{X\}$ either, since $\{X\}$ is essentially the same as $\{NOT, C\}$.

A monotone nonexpansive gate is *complete* if it has the same number of inputs and outputs, and none of its outputs is constant. Subramanian [109] showed that all monotone nonexpansive gates with no constant outputs can be extended to complete gates by adding zero or more additional outputs. A complete gate g satisfies $g(0) = 0$, $g(1) = 1$, and more generally, if $g(x) = y$, then x and y have the same number of 0s and the same number of 1s. Given a monotone gate g, its *dual* is the monotone gate g' defined by $g'(x) = y$ if and only if $g(\overline{x}) = \overline{y}$. A monotone basis Ω is *self-dual* if the dual of every gate in Ω is also in Ω.

CHAPTER 3. STABILITY IN NONEXPANSIVE NETWORKS   93

Note that the basis $\{C\}$ is self-dual [109]. The following theorem says that negations are not needed in the representation of complete monotone gates.

Theorem 3.66 *Suppose that there is a network over $\Omega \cup \{\text{NOT}\}$ that converges weakly to g, where Ω is a self-dual monotone nonexpansive basis, and g is a complete monotone nonexpansive gate. Then there is a network over Ω that converges weakly to g.*

Proof. Let N be a network over $\Omega \cup \{\text{NOT}\}$ that converges weakly to g. We can assume, as we said before, that convergence is by means of fixed points (stable configurations) for every choice of input assignments. To eliminate negations, we make two copies i_0, i_1 of each coordinate i in the network, and set up the gates in the new network N' so that the two copies i_0, i_1 can have as value the original value of coordinate i and its negation, respectively, in N'. This is accomplished for a NOT gate with input i and output j by two ID gates with input i_0 and output j_1, and with input i_1 and output j_0, respectively. For a gate $g \in \Omega$, given an input assignment x and its negation $\overline{x}$, we can obtain the output assignment $g(x)$ and its negation $\overline{g(x)}$ using the gate g and its dual g'. The resulting network N' over Ω converges weakly to a monotone gate h such that $h(x\overline{x}) = y\overline{y}$ for $y = g(x)$. Our aim is to show that it is possible to partition the network into two disconnected pieces, a piece N_0 that has the inputs x and the outputs y, and a piece N_1 that has the inputs $\overline{x}$ and the outputs $\overline{y}$. Then the piece N_0 converges weakly to g.

Note that all the gates of N' must be complete gates; for none of them can have more outputs than inputs, since all monotone nonexpansive gates are scatter-free; and if one of them actually has fewer outputs than inputs, then the network will have fewer outputs than inputs, and this is impossible because h has the same number of inputs and outputs. We can assume that gates obtained from gates g' in Ω by projection are also in Ω; this assumption is justified by the substitutivity property (see section 3.1.2), which makes it possible to replace every occurrence of such a projection with a small convergent network consisting of a g' gate with the appropriate links.

We are interested in the values $z_i(x)$ taken by the coordinates i of N' in stable configurations, over the possible assignments to x, the input to g. We say that a coordinate i is *minimal* if there is no coordinate j such that $z_j(x) \leq z_i(x)$ for all x, and $z_j(x) < z_i(x)$ for some x. We claim that one can assume that no coordinate has $z_i(0) = z_i(1) = 0$. For if this holds for some coordinate i, it must hold for some minimal coordinate i. One can verify directly that this property does not hold for inputs or outputs of the network, because $h(0) = 0$ and $h(1) = 1$. Therefore this minimal coordinate i is an input to a gate g' in N'; setting i to 0 sets some output j of g' to 0 since g' is scatter free, so $z_j(x) \leq z_i(x)$ for all x, and by minimality $z_j(x) = z_i(x)$. We can then simplify N', identifying output j with input i, and remove input i and output j from gate g' by projection (by means of the network consisting of g' alone with output j linked to input i). Since this simplification makes the network smaller, it must eventually terminate, at which point no coordinate with $z_i(0) = z_i(1) = 0$ exists. A similar argument makes it possible to assume that there is no coordinate with $z_i(0) = z_i(1) = 1$.

We can therefore assume that there are only two types of coordinates, namely $z_i(0)z_i(1) = 01$ and $z_i(0)z_i(1) = 10$. We claim further that we can assume that these two types of coordinates participate in *different* gates. A complete monotone nonexpansive gate has the same number of inputs set to 0 (to 1) as it has outputs set to 0 (to 1). Therefore the number of inputs to a gate g' of type 01 (type 10) is the same as the number of ouputs of type 01 (type 10). If we let S and T be the sets corresponding to the two types of input coordinates in g', and we use the same sets for the output coordinates of the same types under some appropriate renaming, then $g'(0_S 0_T) = 0_S 0_T$, $g'(0_S 1_T) = 0_S 1_T$, $g'(1_S 0_T) = 1_S 0_T$, $g'(1_S 1_T) = 1_S 1_T$, where the first and last equation hold because g' is a complete monotone gate, and the two middle equations hold by considering the input assignments $x = 0$ and $x = 1$ and using the definition of the two types. By Lemma 3.60 we must have $g'(x_S x_T) = g'_S(x_S) g'_T(x_T)$, i.e., the gate g' can be decomposed into two gates, one for coordinates of type 01 and one for coordinates of type 10. This in turn decomposes the network into two parts. If $h(x\bar{x}) = y\bar{y}$ with $y = g(x)$, then the inputs that give x and the outputs that give y are in the part of type 01, while the inputs that give $\bar{x}$ and the outputs that give $\bar{y}$ are in the part of type 10.

□

The preceding proof can also be carried out for the standard notion of convergence, instead of weak convergence.

3.4.6 Convergent Network Evaluation and Linear Programming

We now show that the evaluation of the gate to which a nonexpansive network converges can be viewed as a linear program. The stability question as a decision problem is simply the evaluation of $O(m)$ convergent networks, since as observed in the discussion following Theorem 3.9, a network $N = \{f\}$ has a stable configuration consistent with a given input assignment if and only if $\{f_i\}$ has such a stable configuration for each link $i \in L(N)$. The stability question as a search problem requires the evaluation of $O(m^2)$ convergent networks, since as observed in the proof of Theorem 3.36, a transitively closed 2SAT characterization of the stable configurations can be obtained from the stable configurations of the $\{f_{\{i,j\}}\}$.

Let N be a nonexpansive network with transition function f. A stable configuration is a configuration x such that for each choice of a configuration a, the condition $d(f(a_{I(f)}), x_{O(f)}) \leq d(a_{I(f)}, x_{I(f)})$ holds. This condition is a linear constraint, since for example $d(a, x)$ is obtained by adding terms x_i when $a_i = 0$ and terms $1 - x_i$ when $a_i = 1$. Here the input assignment for the network may be fixed in x.

Now, this system of linear inequalities on variables $0 \leq x_i \leq 1$ has a unique solution for those variables corresponding to coordinates in $O(N)$. The reason is that given any real assignment satisfying these conditions, we may select an integer periodic configuration closest to it, and infer from these conditions that the outputs of the network take the same values for both, as in the proof of Theorem 3.4, where nonexpansiveness was used to establish convergence for networks. It remains to show that a real assignment satisfying the conditions always exists. The nonexpansive mapping f can be extended to a mapping on a continuous unit box, which is nonexpansive in the metric sense considered in Chapter 6, by defining $f(x) = \sum_a \prod_{i \in I(f)} w(x, a, i) f(a)$. Here the sum is over all integer 0-1 configurations a, and $w(x, a, i)$ equals x_i if $a_i = 1$ and equals $1 - x_i$ if $a_i = 0$. Once f has been so extended to a nonexpansive mapping on a unit box, it follows from the Brouwer fixed point theorem that a stable configuration exists, hence a real assignment satisfying the linear conditions exists. A Brouwer fixed point can be found by finding a fixed cube and returning its geometric center, with the coordinates contained in the cube set to $1/2$.

Unfortunately, this gives an exponential number of linear inequalities, since there are 2^m choices for a. However, we may consider the gates $g \in N$ separately, and require instead $d(g(b), x_{O(g)}) \leq d(b, x_{I(g)})$, where b ranges over the possible input assignments for g. When the gates are of constant size, this gives a number of constraints that is linear in the number of gates.

The polytope so defined may not have integer vertices, except for the coordinates in $O(N)$ when the coordinates in $I(N)$ are integer, which coincide with the values obtained by the evaluation of the convergent network. For monotone networks, the vertices can be forced to be integer, and therefore give the stable configurations directly. For monotone networks, the inequalities may further be decomposed as (1) $d_0(g(b), x_{O(g)}) \leq d_0(b, x_{I(g)})$ and (2) $d_1(g(b), x_{O(g)}) \leq d_1(b, x_{I(g)})$, where $d_r(b, z)$ for instance is the distance betwwen b and z in those coordinates with $b_i = r$. If outputs i whose values are forced to be 1 are identified, then $x_i = 1$ can be inferred from the inequalities (2), and similarly for outputs forced to be 0, using inequalities (1). Now, if a single b_i is set to 1, and as a result $g(b)_j$ is set to 1, then equation (2) gives $1 - x_j \leq 1 - x_i$, or $x_i \leq x_j$, corresponding to a boolean implication $x_i \to x_j$. This process can be carried further to identify red cycles of implications as in sections 3.3.2 and 3.3.3, so that the linear program is in the end transformed into an equivalent linear program involving only conditions $x_i = 0$, $x_i = 1$, and $x_i \leq x_j$. This defines a polytope whose vertices are precisely the points in the distributive lattice obtained by replacing inequalities with implications, i.e., all vertices have 0-1 coordinates and correspond to stable solutions.

We now use the linear programming formulation to give a parallel algorithm for the evaluation of the gate to which a nonexpansive network converges. When the gates have a constant number of inputs and outputs, the algorithm runs in $\sqrt{m}$ time up to logarithmic factors, with a polynomial number of processors. The following is work with Nimrod Megiddo and Serge Plotkin.

Consider a linear program of the form

$$\begin{aligned} \text{minimize} \quad & c^t x \\ \text{subject to:} \quad & Ax = b, \\ & x \geq 0. \end{aligned}$$

The dual of this linear program is

$$\text{maximize} \quad b^t \pi$$
$$\text{subject to:} \quad A^t \pi + s = c,$$
$$s \geq 0.$$

A solution (x, π, s) to this primal-dual pair is on the *central path* if $x(i)s(i)$ has the same value for all i. The *duality gap* of a solution is the difference $c^t x - b^t \pi = s^t x$. Monteiro and Adler [79] showed for a linear program on N variables that if we have an initial solution (x_0, π_0, s_0) on the central path, then for any constant $\delta > 0$, after $O(\sqrt{N} \log(s_0^t x_0))$ iterations the duality gap $s^t x$ of the current solution (x, π, s) is at most δ.

We consider linear programs of the special form minimize r, where r is a sum of terms of one of the forms $x, 1 - x$ for variables x, subject to constraints $t \leq 0$, where t is a sum of terms of one of the forms $x, 1 - x, -x, x - 1$ for variables x. All variables are constrained also by $0 \leq x \leq 1$, which can also be viewed as constraints $t \leq 0$ when written as $-x \leq 0$ and $x - 1 \leq 0$. We assume that a minimum has value $r = 0$.

Theorem 3.67 *A linear program of the special form of size m can be put into an equivalent form such that for any constant $\delta > 0$, after $O(\sqrt{m} \log m)$ iterations the duality gap of the current solution is at most δ.*

Proof. First, we ensure that each t has one more term of one of the forms $-x, x - 1$ than of one of the forms $x, 1 - x$. This difference can be decreased by one by adding a new variable x to t with $0 \leq x \leq 1$, or increased by one by adding a term $-x$ to t with $0 \leq x \leq 1$, and adding x to the objective function r.

Second, we ensure that if the coefficient of x in r is c, then the sum of the coefficients of this x in all the terms t is $1 - c$. This sum can be decreased by one by adding a constraint $-x \leq 0$, or increased by one by adding a constraint $x - 1 \leq 0$.

Third, we introduce a new variable $\alpha \geq 0$ per constraint $t \leq 0$, and replace the constraint by $t + \alpha = 0$.

Now the primal has a solution with all variables set to $1/2$ by the first condition enforced above. The dual has a solution with all $\pi(i) = -1$ and all $s(i) = 1$, by the second condition enforced above. This solution is on the central path, and satisfies the conditions of the result of Monteiro and Adler. The transformation only increased the size of the linear program by a constant factor, giving the stated bound. □

In a nonexpansive network N, the associated linear program is minimize $d(a_{I(N)}, x_{I(N)})$, where $a_{I(N)}$ is the input assignment to the network, subject to constraints $d(g(b), x_{O(g)}) - d(b, x_{I(g)}) \leq 0$, with $0 \leq x_i \leq 1$, for each gate g and input assignment b. The linear program is thus of the above special form, and the theorem can be applied.

CHAPTER 3. STABILITY IN NONEXPANSIVE NETWORKS

In the end, the duality gap is any constant $\delta > 0$. Let a be a periodic configuration consistent with the input assignment $a_{I(N)}$ which is closest to x. Thus $d(f(a_{I(f)})_{L(N)}, x_{L(N)}) \geq d(a_{L(N)}, x_{L(N)})$. On the other hand, when the linear program is put in the equivalent form, we obtain $d(f(a_{I(f)}), x_{O(f)}) - \sum y_i \leq d(a_{I(f)}, x_{I(f)})$, where the terms $-y_i$ were added to the objective function. This can be rewritten as $d(a_{O(N)}, x_{O(N)}) + d(f(a_{I(f)})_{L(N)}, x_{L(N)}) \leq d(a_{L(N)}, x_{L(N)}) + d(a_{I(N)}, x_{I(N)}) + \sum y_i$. But $d(a_{I(N)}, x_{I(N)}) + \sum y_i \leq \delta$ since all the y_i appear in the objective function, so combining the three inequalities we get $d(a_{O(N)}, x_{O(N)}) \leq \delta$. Provided that $\delta < 1/2$, this can be used to obtain the value of $a_{O(N)}$, which is the output value produced by the gate to which the nonexpansive network converges.

It is observed in Goldberg, Plotkin, Shmoys, and Tardos [40] that one step of the iteration of the interior point algorithm can be done in $O(\log^2 N)$ time using N^3 processors, with a concurrent read concurrent write algorithm.

Corollary 3.68 *The gate to which a nonexpansive network with a constant number of inputs and outputs per gate converges can be evaluated in $O(\sqrt{m} \log m)$ iterations of an interior point algorithm. The stability question for nonexpansive networks with gates of constant size can thus be solved in $\sqrt{m}$ time, up to logarithmic factors, with a polynomial number of processors.*

3.5 Non-Periodic Points and Iterates

We have examined the structure of the sets of stable and periodic configurations, as well as the behavior of the transition function on periodic configurations. We now focus on the non-periodic configurations. Outside the set of periodic points, nonexpansive mappings do not admit a succint description (see Lemma 2.8). Instead of looking for a characterization of the behavior of such mappings on non-periodic points, our goal will be to show that a relatively small number of iterations of the given nonexpansive mapping is sufficient to map an arbitrary non-periodic point to a periodic point. The bounds obtained are then used to relate the complexity of the convergence question for networks over a given nonexpansive basis to the complexity of the evaluation question over the same basis.

3.5.1 The Scatter-Free Case

Graham [45] showed that the number of useful gates in a comparator circuit is at most quadratic in the number of inputs. The basic idea of the proof seems to apply to arbitrary scatter free gates, and indeed can be used to obtain quadratic bounds for many of them. Given a scatter-free gate g with the same number of inputs and outputs, choose a coordinate $i \in I(g)$ and try the two partial input assignments $x_i = 0, 1$. This must as a result set outputs $y_j = a$ and $y_{j'} = a'$ respectively for some $j, j' \in O(g)$ and $y = g(x)$, from the definition of scatter-freedom. If $j = j'$, then either $a = a'$, in which case the output $j = j'$ is a constant a, or $a \neq a'$, in which case $y_j = x_i \oplus a$ for all inputs. These two cases indicate progress, because they either reduce the number of non-constant coordinates from inputs to outputs, in the first case, or they make it possible to subdivide the gate by splitting off input i and output j. The interesting case has $j \neq j'$ and implies that a clause $(y_j = a) \vee (y_{j'} = a')$ holds at the outputs. In many cases, this argument can be iterated, so that after each gate in the circuit, the number of clauses that hold increases, giving a quadratic bound on the number of gates by the quadratic bound on the number of clauses. We do not know whether such a construction can be carried out for arbitrary scatter-free circuits; we show that it can at least be carried out in the case where the circuit consists of the repeated application of a given scatter-free function.

Theorem 3.69 *Let f be a scatter-free mapping on the m-cube, and let x be an arbitrary point in the m-cube. Then $f^{(k)}(x)$ is a periodic point for all $k \geq \binom{m+1}{2}$.*

Proof. Let U_k be the set of all y such that $y = f^{(k)}(x)$ for some x in the m-cube, where $k \geq 0$. The set U_k induces a connected subgraph of the m-cube because f maps paths to paths, and $U_{k+1} \subseteq U_k$ for all k. We let I_k be the most restrictive 2SAT instance whose solution set contains U_k. We construct I_{k+1} by adding zero or more clauses to I_k, under the constraint that for every pair of coordinates i, j, there is never more than one clause in I_k of the form $u \vee v$, where u is x_i up to negations and v is x_j up to negations. This can be done in such a way that the clauses not included are in fact implied by existing clauses, as follows: For $i = j$, we never need the clause $x_i \vee \overline{x_i}$, and clearly only one of the two clauses $x_i \vee x_i$ and $\overline{x_i} \vee \overline{x_i}$ can hold. For $i \neq j$, we never need to include a clause if one of the two variables x_i, x_j is trivial (such a clause could be simplified to a one-literal clause), so we never have two clauses of the form $u \vee v$ and $u \vee \overline{v}$ because such clauses imply that the variable in u is trivial, and we never have two clauses $u \vee v$ and $\overline{u} \vee \overline{v}$ because they imply $u = \overline{v}$, which is impossible by connectivity unless both variables are trivial. Therefore the number of clauses cannot exceed $\binom{m+1}{2}$. Since $I_k \subseteq I_{k+1}$, this implies that we must have $I_k = I_{k+1}$ for some $k \leq \binom{m+1}{2}$. We shall prove that if $I_k = I_{k+1}$, then there is an isomorphism $g = g_\sigma$ on the m-cube such that $f(y) = g(y)$ for all $y \in U_k$. This in turn implies that $f(z) = g(z)$ for all $z = f^{(l)}(y)$ with $l \geq 0$, since $z \in U_{k+l} \subseteq U_k$. Since y is a periodic point of g, it must then be also a periodic point of f, and so U_k is contained in the periodic set of f, proving that there is a $k \leq \binom{m+1}{2}$ such that $f^{(k)}(x)$ is periodic for all x.

So suppose that $I_k = I_{k+1}$. The trivial variables in I_k and I_{k+1} are the same, so for each trivial variable x_i we set $\sigma(i) = i$ and $\sigma(\overline{i}) = \overline{i}$, and view f as a mapping on the cube of nontrivial variables. If all the pairs x, y such that $y = f(x)$ with $x \in U_k$ satisfy $y_i = x_j$ for some pair of nontrivial variables x_i, x_j, then we set $\sigma(i) = j$ and $\sigma(\overline{i}) = \overline{j}$ (here i and j are *signed* coordinates). The mapping σ has at most one value j for each i, because if it has two values j_1, j_2 then we must have $x_{j_1} = x_{j_2}$ for all $x \in U_k$, and this is impossible since U_k induces a connected graph. The mapping σ also maps at most one i to each j, because if it maps two i_1, i_2 to the same j, then we have $y_{i_1} = y_{i_2}$ for all $y \in U_{k+1}$, again contradicting connectivity.

If σ is a permutation, then it is by definition a signed permutation, and clearly $f(x) = g_\sigma(x)$ for all $x \in S_k$ as claimed. By the preceding remarks, in order to show that σ is a permutation, it is sufficient to show that for every j there exists an i such that $\sigma(i) = j$. Suppose that this is not the case. Let the *degree* of a signed coordinate i be the number of clauses of the form $x_i \vee u$ in $I_k = I_{k+1}$, where u is a literal. If x_i has degree d, then the d corresponding clauses can be represented by a single condition $x_i \vee (u_1 \wedge u_2 \wedge \ldots u_d)$. If $\sigma(i) = j$, then we claim that the degree of i is at least as large as the degree of j. To verify this, consider the clauses $x_j \vee (u_1 \wedge u_2 \wedge \ldots u_d)$ for x_j, and set the literals $\overline{x_j}, u_1, \ldots, u_d$ to 1 in x; this must as a result set literals $v_0, v_1, \ldots, v_d$ to 1 in $y = f(x)$, since f is scatter-free. If $y_i = 0$ in $y = f(x)$, then $x_j = 0$ because $\sigma(i) = j$, so $u_{j'} = 1$ in x for all $u_{j'}$ by the above clauses, so $v_{i'} = 1$ in y for all $v_{i'}$, giving clauses $y_i \vee (v_0 \wedge v_1 \wedge \ldots v_d)$. One of the $v_{i'}$ may be $\overline{y_i}$ and does not count as a clause $y_i \vee v_{i'}$, but we still have d other $v_{i'}$, showing that i has degree at least d, the degree of $j = \sigma(i)$. As a result, the elements of the range of σ of degree at least d must be images of elements of the domain of σ of degree at least d. If some signed coordinate of degree at least d is not in the domain of σ, then by the Pigeonhole Principle, some signed coordinate of degree at least d is not in the range of σ, so the largest degree of a signed coordinate not in the domain of σ is no larger than the largest degree of a signed coordinate not in the range of σ.

Pick a signed coordinate j of largest degree among those not in the range of σ. Let $x_j \vee (u_1 \wedge u_2 \wedge \ldots u_d)$ be the corresponding clauses, and let $\overline{x_j} \vee (v_1 \wedge v_2 \wedge \ldots v_{d'})$ be the clauses for $\overline{j}$. Suppose that we set the literals $x_j, v_1, v_2, \ldots, v_{d'}$ to 1 in x; this must as a result set literals $w_0, w_1, \ldots, w_{d'}$ to 1 in $y = f(x)$, since f is scatter-free. Furthermore, if the mapping σ is defined for the signed coordinate of some $w_{i'}$, then we claim that σ must map this signed coordinate to the signed coordinate of some v' in the list $v_1, \ldots, v_{d'}$. The reason for this is that if $x_j = 1$ for some $x \in U_k$, then this x has $v_{j'} = 1$ for all $v_{j'}$ in the list by the above clauses, so $y = f(x)$ has $w_{i'} = 1$, and x has $v' = 1$ by the definition of σ, so the clause $\overline{x_j} \vee v'$ holds and v' must be in the list of $v_{j'}$. By the Pigeonhole Principle, σ cannot be defined on the signed coordinates of all $w_{i'}$, since there are $d' + 1$ of them but there are only d' signed coordinates $v_{j'}$.

Pick a w_i such that σ is not defined on the corresponding signed coordinate. We just saw that if $x_j = 1$ for some $x \in U_k$, then $w_i = 1$ for $y = f(x)$. We claim that the converse is also true. Suppose that we set the literals $\overline{x_j}, u_1, u_2, \ldots, u_d$ to 1 in x; this must as a result set literals $t_0, t_1, \ldots, t_d$ to 1 in $y = f(x)$, since f is scatter-free. If $w_i = 0$ in y, then $x_j = 0$ in x, so x has $u_{j'} = 1$ for all $u_{j'}$ by the above clauses, and so y has $t_{i'} = 1$ for all i'. We therefore have clauses $w_i \vee (t_0 \wedge t_1 \wedge \ldots t_d)$ that hold for y. But we know that w_i has degree at most d, because the signed coordinate of w_i is not in the domain of σ. Therefore one of the $t_{i'}$ must be $\overline{w_i}$ (so that the clause $w_i \vee t_{i'}$ is not counted). We just observed that if $x_j = 0$ in x then $t_{i'} = 1$ in y, and so in particular $w_i = 0$ in y, proving the converse. This means that the literal w_i in y has the same value as the literal x_j in x for all $x \in U_k$ and $y = f(x)$, allowing us to map the signed coordinate of w_i to the signed coordinate of x_j under σ, and contradicting the fact that the signed coordinates of w_i and x_j are not in the domain and range, respectively, of σ. The only possibility left is that the range consists in fact of all the signed coordinates, and so σ is a signed permutation. □

There are mappings that exhibit the quadratic behaviour indicated in the theorem. The following example is due to Subramanian [107]. Let $m = 2n+1$ and define $y = f(x)$ by $y_i = x_{i-1}$ for $i \neq 1, n+1$, $y_1 y_{n+1} = \mathrm{X}(x_n x_{2n+1})$. Then the point x with $x_i = 1$ for $i = 1, n+1$ and $x_i = 0$ elsewhere is a non-periodic point adjacent to two periodic points (the two neighbors of x with a single bit set to 1). If we iterate f on x, the two bits of x set to 1 cycle through the first n and the last $n+1$ coordinates of x respectively. After $k = n(n+1) - 1$ iterations, we obtain the non-periodic point $x' = f^{(k)}(x)$ with $x'_i = 1$ for $i = n, 2n+1$ and $x'_i = 0$ elsewhere, while one more iteration yields a fixed point $f(x') = 0$. Therefore the $\binom{m+1}{2}$ bound cannot be improved below $(m^2 - 1)/4$ (for m odd).

3.5.2 The Nonexpansive Case

General nonexpansive mappings are less structured than scatter-free mappings, and it is hard to characterize their behavior on non-periodic points. Fortunately, it becomes easier to study this behavior when we consider points that are close to the periodic set, in particular, points that are adjacent to a periodic point. We shall prove a fairly tight bound for such points, and infer a weaker bound for arbitrary non-periodic points.

Theorem 3.70 *Let f be a nonexpansive mapping on the m-cube, and let x be an arbitrary point in the m-cube, at distance d from the periodic set. Then $f^{(k)}(x)$ is a periodic point for all $k \geq dm^2$.*

Proof. We prove the statement for $d = 1$. The result for general d follows by induction, because if x is at distance $d+1$ from a periodic point y, then x is at distance d from some vertex z adjacent to y, and $f^{(k)}(z)$ is periodic for $k = m^2$, reducing the distance from the periodic set to $d(f^{(k)}(x), f^{(k)}(z)) \leq d$.

Let Q be the smallest subcube containing the periodic set of f, and let $g = g_\sigma$ be the isomorphism of Q that coincides with f in the periodic set of f. Let n be the number of coordinates contained in Q. We first give a proof for the case where the point x at distance $d = 1$ from the periodic set is in Q; this means that $x \oplus e^i$ is a periodic point for some coordinate i contained in Q. Let I be the 2SAT instance that characterizes the periodic set of f within Q, consisting of all the clauses satisfied by all the periodic points of f (excluding clauses $u \vee \overline{u}$).

The set U of points x such that $f(x) = g(x)$ is a median set, because the points in S are the fixed points of $g^{-1} \circ f$. Let J be the 2SAT instance that characterizes the set U within Q, consisting of all the clauses satisfied by all the points in U. Since all the periodic points are in U, all the clauses in J are also in I.

Call a clause in I *minimal* if it does not follow from other clauses in I. It can be shown that if a clause C is minimal, then there is a point x that violates C but satisfies all other clauses in I. Futhermore, any point x that is not a solution to I must violate some minimal clause in I, by the acyclicity of the inference graph of I.

Lemma 3.71 *Given a minimal clause $x_i \vee x_j$ in I, where i,j are signed coordinates, there is a nonnegative integer k such that the clause $x_{i'} \vee x_{j'}$ is in J, where $i = \sigma^{(k)}(i')$ and $j = \sigma^{(k)}(j')$.*

Proof. Let x be a point in Q that violates only the clause $x_i \vee x_j$ in I. Then $y = g^{(k)}(x)$ violates only the clause $x_{i'} \vee x_{j'}$ in I, with i', j' as in the statement of the lemma and $k \geq 0$ arbitrary, since g is an isomorphism on the set of solutions to I which permutes coordinates and clauses as indicated by σ. If this clause is not in J, then $f(y) = g(y)$, because any other clause in J that excludes y would not be in I, contradicting containment. If this clause is not in J for any $k \geq 0$, then $f^{(k)}(x) = g^{(k)}(x)$ for all $k \geq 0$ by induction on k. Since x is a periodic point of g, it must also be a periodic point of f, contrary to the fact that x does not satisfy some clause in I. □

Lemma 3.72 *Suppose that a is a non-periodic point adjacent in Q to a periodic point $a \oplus e^i$, and that $f(a \oplus e^i) = b \oplus e^j$ with $b = f(a) \neq g(a)$. Make the coordinates i, j signed in such a way that $a_i = b_j = 1$. Then the clause $\overline{x_i} \vee x_{\sigma(j)}$ is in I.*

Proof. Let x be a periodic point such that $x_i = a_i = 1$. Then a lies on a shortest path between the periodic points $a \oplus e^i$ and x, and since $d(f(a \oplus e^i), f(x)) = d(a \oplus e^i, x)$, the point $f(a) = b$ must lie on a shortest path between the periodic points $f(a \oplus e^i) = g(a \oplus e^i) = b \oplus e^j$ and $f(x) = g(x)$. It follows that $g(a \oplus e^i)$ and $g(x)$ differ in coordinate j, so $a \oplus e^i$ and x differ in coordinate $\sigma(j)$. Since coordinate $\sigma(j)$ of $a \oplus e^i$ is the same as coordinate j of $g(a \oplus e^i) = b \oplus e^j$, we have $x_{\sigma(j)} = b_j = 1$. This shows that the implication $x_i \to x_{\sigma(j)}$ holds for all periodic points of f and gives a clause in I provided that $i \neq \sigma(j)$. But $i = \sigma(j)$ cannot hold, because then $g(a \oplus e^{\sigma(j)}) = b \oplus e^j$, implying that $g(a) = b$, contrary to one of the assumptions. □

The proof of the lemma implies in particular that if a is adjacent in Q to a periodic point $a \oplus e^i$, then $b = f(a)$ is also adjacent in Q to a periodic point $b \oplus e^j = f(a \oplus e^i)$ along some coordinate j contained in Q, since b must lie on a shortest path between the two periodic points $f(a \oplus e^i)$ and $f(x)$, for every periodic point x with $x_i = a_i$.

Let x be a non-periodic point adjacent in Q to a periodic point $x \oplus e^i$. We shall associate with each non-periodic point $x(k) = f^{(k)}(x)$ having $f^{(k)}(x \oplus e^i) = x(k) \oplus e^{i(k)}$ a minimal clause in I that $x(k)$ does not satisfy, of the form $\overline{x_{i(k)}} \vee x_{j(k)}$, where $i(k)$ and $j(k)$ are signed coordinates, and $i(k)$ is signed so that $[x(k)]_{i(k)} = 1$. Furthermore, we shall show that if the order of the two literals in these clauses is taken into account, then all these clauses are distinct. Since the periodic set is connected and the two variables in the literals $x_{i(k)}$ and $x_{j(k)}$ are nontrivial, there is at most one clause in I involving these two variables. So by distinctness the number of possible values of $k \geq 0$ with $x(k)$ non-periodic is at most $n(n-1)$, where n is the number of coordinates contained in Q. This shows that $x(n(n-1))$ is a periodic point.

The clauses are defined as follows. If $k = 0$, or if $k \geq 1$ and $x(k) \neq g(x(k-1))$, then we choose any minimal clause in I that $x(k)$ does not satisfy. Since $x(k) \oplus e^{i(k)}$ is a periodic point, it cannot be excluded by the chosen clause, so this clause must necessarily involve the literal $\overline{x_{i(k)}}$ and is therefore of the required form. If $k \geq 1$ and $x(k) = g(x(k-1))$, then we assume inductively that the chosen clause in I that $x(k-1)$ does not satisfy is $\overline{x_{i(k-1)}} \vee x_{j(k-1)}$ with $[x(k-1)]_{i(k-1)} = 1$, and let the clause for $x(k)$ be $\overline{x_{i(k)}} \vee x_{j(k)}$, where $\sigma(i(k)) = i(k-1)$ and $\sigma(j(k)) = j(k-1)$. Since g is an isomorphism on the set of solutions to I which permutes and complements coordinates as indicated by σ, this new clause is also in I, and is also minimal in I. Since g maps $x(k-1)$ to $x(k)$, this new clause exludes $x(k)$, and $f(x(k-1) \oplus e^{i(k-1)}) = x(k) \oplus e^{i(k)}$ with $\sigma(i(k)) = i(k-1)$ and $[x(k)]_{i(k)} = 1$, as required.

Partition the values of k with $x(k)$ non-periodic into intervals, with k and $k-1$ in the same interval if and only if $g(x(k-1)) = x(k)$. Since g is an isomorphism on the solutions to I which permutes and complements coordinates as indicated by σ, two literals x_i and $x_{\sigma(i)}$ must have the same outdegree in the implication graph of I. Therefore, if $k-1$ and k belong to the same interval so that $g(x(k-1)) = x(k)$, then the literals $x_{i(k-1)}$ and $x_{i(k)}$ in the clauses for $x(k-1)$ and $x(k)$ have the same outdegree. On the other hand, if $k-1$ and k are in different intervals, then $g(x(k-1)) \neq x(k)$. Letting $a = x(k-1)$ and $b = x(k)$ in Lemma 3.72 we can infer that the implication $x_{i(k-1)} \to x_{\sigma(i(k))}$ is a clause in I. Therefore the literal $x_{i(k-1)}$ has outdegree in I greater than the outdegree of the literal $x_{\sigma(i(k))}$, which is in turn is the same as the outdegree of the literal $x_{i(k)}$. Therefore the outdegree of the literals $x_{i(k)}$ remains the same within an interval but decreases from an interval to the next, showing that these literals must be different for two values of k chosen from different intervals, and so the two corresponding clauses associated with $x(k)$ must be different as well.

It remains to be shown that the clauses are different even for two values of k chosen from the same interval. If the clause associated with the first value k in the interval is $\overline{x_i} \vee x_j$, then the clause associated with the $(l+1)$th value in the interval is $\overline{x_{i'}} \vee x_{j'}$, where $i = \sigma^{(l)}(i')$ and $j = \sigma^{(l)}(j')$. The first value $l = r$ at which a repetition of the same clause occurs actually repeats the first clause $\overline{x_i} \vee x_j$. (Here r is the least common multiple of the lengths of the cycles of σ containing i and j.) The clauses then recur periodically, i.e., the $(l+r)$th clause is the same as the lth clause. By Lemma 3.71, at least one of these clauses must be in J, and by periodicity this clause in J must be a clause corresponding to one of the first r values of k in the interval. By definition of J, we have $x(k+1) \neq g(x(k))$ for this value k, so $k+1$ is not in the same interval as k, showing that the length of the interval under consideration is at most r and that no repetitions of the same clause occur within the interval.

This completes the proof that $x(n(n-1))$ is a periodic point if $x = x(0)$ is a non-periodic point adjacent to a periodic point along one of the n coordinates contained in Q. It remains to examine the case where x is a non-periodic point adjacent to a periodic point along one of the $n' = m - n$ coordinates not contained in Q. Suppose then that x is a non-periodic point adjacent to a periodic point $x \oplus e^i$ with i not contained in Q, and consider the non-periodic points $x(k) = f^{(k)}(x)$ with $f^{(k)}(x \oplus e^i) = x(k) \oplus e^{i(k)}$ and

$[x(k)]_{i(k)} = 1$ up to, but not including, the first $k = s$ such that $x(s)$ is periodic or $i(s)$ is contained in Q. Since there are only n' possible values for $i(k)$, there must exist two values $k_1 < k_2 \leq n'$ such that $i(k_1) = i(k_2)$, provided that $s \geq n'$. Let $h = f^{(L)}$, where $L = k_2 - k_1$, and let $x'(r) = x(k_1 + rL)$, $i'(r) = i(k_1 + rL)$ so that $x'(r+1) = h(x'(r))$ and $i'(0) = i'(1)$. If $i'(r-1) = i'(r)$, then $d(x'(r), x'(r+1)) \leq d(x'(r-1), x'(r)) = d(x'(r-1) \oplus e^{i'(r-1)}, x'(r) \oplus e^{i'(r)}) = d(x'(r) \oplus e^{i'(r)}, x'(r+1) \oplus e^{i'(r+1)})$, because the mapping h preserves distances on the periodic points of f. Since the two periodic points $x'(r) \oplus e^{i'(r)}$ and $x'(r+1) \oplus e^{i'(r+1)}$ agree in the coordinates $i'(r)$ and $i'(r+1)$ that are not contained in Q, this inequality can only hold (in fact with equality) if $i'(r) = i'(r+1)$. By induction, this shows that $i'(r) = i'(0)$ for all r.

Let t be the largest value for which $x'(t)$ is defined, with $i'(t) = i'(0)$. Let $y = h(x'(t))$ so that either $h(x'(t) \oplus e^{i'(t)}) = y$ or $h(x'(t) \oplus e^{i'(t)}) = y \oplus e^j$ with j contained in Q. The first case is impossible if $t \geq 1$, because we would have as before $d(x'(t), y) \leq d(x'(t-1), x'(t)) = d(x'(t-1) \oplus e^{i'(t-1)}, x'(t) \oplus e^{i'(t)}) = d(x'(t) \oplus e^{i'(t)}, y) = d(x'(t), y) - 1$, using the fact that h preserves distances on periodic points of f and that periodic points agree in $i'(t)$. In the second case, coordinate j belongs to a cycle of σ of length $l \leq 2n$. If $t \geq l$, then the periodic points $y \oplus e^j$ and $y' \oplus e^{i'(t+1-l)}$ with $y' = x'(t+1-l)$ agree in coordinate j since $y \oplus e^j = h^{(l)}(y' \oplus e^{i'(t+1-l)})$, as well as in coordinate $i'(t+1-l)$, so again $d(y', y) \leq d(x'(t-l), x'(t)) = d(x'(t-l) \oplus e^{i'(t-l)}, x'(t) \oplus e^{i'(t)}) = d(y' \oplus e_{i'(t+1-l)}, y \oplus e_j) = d(y', y) - 2$, a contradiction. This shows that $t < l$ and so $s \leq k_1 + lL \leq n' + 2nn'$.

It therefore takes at most $n' + 2nn'$ iterations of f to map x into either a periodic point or a point y adjacent in Q to a periodic point, and in the second case an additional $n(n-1)$ iterations map y to a periodic point. The theorem follows by observing that $n' + 2nn' + n(n-1) \leq (n+n')^2 = m^2$. □

The scatter-free example from the last section shows that there are mappings requiring a quadratic number of iterations, even for $d = 1$. On the other hand, we have no evidence to suggest that the bound must increase and become cubic for larger d, as the theorem indicates. By the results in the last section, if there are nonexpansive mappings that achieve a super-quadratic bound, then they can be found only outside the domain of scatter-free mappings.

It would be interesting to know whether there is a simple characterization for the mapping f restricted to the points inside the subcube Q containing the periodic set that are adjacent to the periodic set. This mapping need not be scatter free, as the following example shows. Let f be the composition of a retraction g on the 5-cube with retract U described by clauses $\overline{x_2} \vee \overline{x_3}$, $\overline{x_4} \vee \overline{x_5}$, with the retraction h on the set U with retract consisting of points in U satisfying the additional clauses $\overline{x_1} \vee (\overline{x_2} \wedge \overline{x_3} \wedge \overline{x_4} \wedge \overline{x_5})$, where $h(11010) = 01000$, $h(10110) = 00010$, $h(11001) = 00001$, $h(10101) = 00100$, $h(11000) = h(10100) = h(10010) = h(10001) = 00000$, and $h(x) = x$ for the remaining points x in U. Then the images under h of the points in U satisfying $x_1 = 1$ do not agree in any coordinate, so h cannot be extended to a scatter-free map. Contrast this with Knuth's observation (see section 3.3.2) that for a different class of points near the periodic set, the mapping is scatter-free, and can in fact be described with a comparator circuit.

3.5.3 Evaluation, Stability and Convergence

We can now show, using the results from the last section, that the three questions of evaluation, stability and convergence have the same parallel complexity for nonexpansive mappings, up to $\mathcal{NC}^1$ reductions.

Theorem 3.73 *Given a nonexpansive basis Ω, the evaluation, stability and convergence questions, both as decision and as search problems, have the same complexity, up to logspace uniform $\mathcal{NC}^1$ reductions (with the exception of the stability question as a decision problem for a monotone basis, where the answer is always 'yes').*

Proof. The evaluation question as a decision problem clearly reduces to the search problem. To reduce the search problem to the decision problem, note that each of the coordinates in a circuit can be viewed as an output of a (slightly modified) circuit, so we can decide the value of each coordinate and obtain the solution to the search problem.

The stability question, both as a decision problem and as a search problem, reduces to the evaluation question by Theorems 3.18 and 3.36. The convergence question can be reduced to the evaluation question as follows. If f is the internal function of the network for a given input assignment, and f has m inputs and m outputs, then we can construct a circuit that computes $f^{(k)}$ for $k = m^3$ from the description of f, in $\mathcal{NC}^1$. By Theorem 3.68, the circuit must give under evaluation a periodic point of f, and hence the answer to the convergence question as a search problem. The answer to the decision problem is given by whether the periodic point is a fixed point of f or not, and this is again a simple circuit evaluation.

To reduce the evaluation question to the convergence question, note that a given circuit converges to a stable configuration that gives the answer to the evaluation question. To obtain a reduction to the convergence question as a decision problem, we must construct a network that converges to a stable configuration for a given input assignment if and only if a chosen output of the given circuit produces the value 0 under evaluation. The construction requires considering several special cases, and in fact we consider more cases than we need, to avoid a similar case analysis below for the stability question. Suppose that the basis Ω is nonmonotone and contains a gate other than NOT. Then Ω must contain at least one of the two gates XOR and NAND (the composition of AND and NOT). We can then connect the chosen output of the circuit to the first input of the XOR gate, and connect the output of the XOR gate to the second input of the XOR gate. It is easy to verify that this constructed network will stabilize if and only if the circuit produces an output value 0, and that this still holds when a NAND gate is used instead of the XOR gate. If the basis is just {NOT}, then the circuit consists simply of paths from inputs to outputs. We can determine whether a path goes from a chosen input to the chosen output by connecting this output to the input, with and without an intermediate negation. If the path does not link the chosen input and output, then the added connection does not introduce a cycle, so the network stabilizes. Otherwise we have introduced a cycle, and at least one of the two constructions (with and without a negation) will not stabilize. This idea can also be used to determine which NOT gates in the circuit are on the path joining the chosen input and output. The output of the circuit is determined by whether the number of such NOT gates is even or odd, and this simple parity check can be performed in $\mathcal{NC}^1$ as well. The monotone case is handled somewhat differently, in the sense that the initial configuration of the constructed network (which was irrelevant in the preceding constructions) has some significance here. If the basis contains a gate other than ID, then it must contain at least one of the two gates OR and AND. Both cases are analogous, so we assume that the AND gate is available. We connect the output of the circuit to the first input of an AND gate, and the output of the AND gate to the second input of the AND gate via a long path of ID gates. The first two links of this path are given values 0 and 1 respectively. If the circuit outputs a 0, then all the values on this long path will stabilize to 0. Otherwise, if the path is longer than the size of the circuit, then after the output of the circuit stabilizes to 1, the 0 and the 1 that were initially at the beginning of the path will continue to cycle indefinitely, so the network does not stabilize. Finally, if only ID gates are available, then as in the case of the NOT gate, we only need to determine the input from which there is a path in the circuit ending in the chosen output. This can be done by connecting the output to a chosen input, and assigning values 0 and 1 to the first two links following the input. If we have created a cycle, this cycle will never stabilize, otherwise the network is acyclic and stabilizes.

Only the reduction from the evaluation question to the stability question remains to be considered. This reduction is essentially the same as the reduction to the convergence question. Thus, when the stability question is viewed as a search problem, a stable configuration answers the evaluation question. When the stability question is viewed as a decision problem, we need to construct a network that has a stable configuration if and only if the circuit outputs the value 0, and the constructions from the nonmonotone cases above for the convergence question go through unchanged. The monotone cases above used specific initial assignments, and so those constructions cannot be used in a reduction to the stability question as a decision problem; in fact, no such reduction is possible because this question is trivial, since monotone networks always have stable configurations. □

Of the results in this theorem, the equivalence between the evaluation question and the stability question as a decision problem was previously known in the scatter-free case, while the equivalence with the stability question as a search problem was known in the monotone case. See [78].

3.6 Discussion

Building on earlier work by Bandelt and Vel [9, 11], Graham [45], Mayr and Subramanian [78, 109], and others, we have studied the structure for nonexpansive mappings on the hypercube. The main aspects of the theory are (1) the distance structure of the mappings, which is used to show that nonexpansive networks are convergent and which leads to a polynomial-time fixed point algorithm; (2) the median structure of the fixed point set, which gives a 2SAT characterization for fixed points; (3) the existence of an isomorphism associated with the periodic set, which can be used to give an algorithm for finding the 2SAT characterization of the fixed point set; and (4) the properties of fixed cubes, which result in a better understanding of the distinction between mappings with and without fixed point. Several results and open questions come out of this study.

- In the general case, the three basic questions of evaluation, stability, and convergence have very different complexities, providing complete problems for the classes $\mathcal{P}$, $\mathcal{NP}$, and $\mathcal{PSPACE}$ respectively. For nonexpansive networks, all three problems lie within $\mathcal{P}$, and give rise to the same parallel complexity class, both as search problems and as decision problems, up to $\mathcal{NC}^1$ reductions. This class is known to be above $\mathcal{NL}$, but the relationship with the class $\mathcal{NC}$ remains open. The lower bounds seem to indicate that the class may be hard to parallelize. The one question that maintains its complexity from the general case is the counting problem, remaining complete for $\#\mathcal{P}$ even when only comparators are allowed.

- In the presence of fanout, only a few bases need to be considered. For nonexpansive networks, there is a large variety of bases, and their relative computational power is not always clear. Two operators that can be used to relate different bases are composition (simulation by circuits) and projection (simulation by convergent networks). There does not seem to be a simple set of gates that will give all the nonexpansive gates under composition, and any such set would necessarily be infinite. With the more powerful projection operator, on the other hand, it is at present unclear whether all nonexpansive gates can be obtained from a finite set of gates. At this point, we can only show that the X gate is not sufficient to build all scatter-free mappings, or even all monotone scatter-free mappings.

- Part of the difficulty in determining the gates that can be obtained by convergent networks over a given basis seems to be the fact that while we can ensure an exponential upper bound on the size of circuits needed to define a gate, we can only prove a triply exponential bound on the required size for convergent networks. In general, circuits seem easier than networks, as indicated by the difference in both sequential and parallel complexity between the evaluation and stability questions. We lack here strong enough lower bounds, however, to prove that the difference in complexity between circuits and networks is inherent.

- There are several results that seem to indicate that the nonexpansive case is harder than the scatter-free case. Algorithmically, a stable configuration in the scatter-free case can be found in linear time, but only an $O(m^2)$ bound is known in the nonexpansive case. From a structural point of view, the number of iterations required to reach the periodic set is quadratic in the scatter-free case, with a matching lower bound, but only a cubic upper bound is known in the nonexpansive case. For comparators and some other scatter-free gates, a quadratic upper bound on the number of useful gates in a circuit can be obtained; only an exponential upper bound is known in general. In each of these cases, however, the upper and lower bounds are not sufficiently close to establish a separation between scatter-free and nonexpansive bases. The class $\mathcal{CC}$ consists of problems that are reducible to comparator circuit evaluation. For many $\mathcal{CC}$-complete problems, there is a parallel algorithm with a polynomial number of processors that runs in time $\sqrt{m}$ up to logarithmic factors. This includes comparator circuit evaluation, lex-first maximal matching (Mayr and Subramanian [78]), telephone connection (Ramachandran and Wang [99]). For stable matching, which corresponds to stability for networks of comparators or X gates (see Chapter 5), and which is also $\mathcal{CC}$-complete, the same parallel time complexity can be obtained via interior point methods in linear programming, as shown in section 3.4.6.

- Given a set of points and their images in the hypercube, it is in general hard to determine whether there exists a nonexpansive mapping consistent with this partial information. We even lack a complete characterization for the case of four points and their images, which is the simplest case where the obvious distance condition for pairs of points and their images is not a strong enough constraint. We conjecture that a complete characterization in this case will show that the minimum-size certificates for nonexpansive mappings always consist of one, two or four points.

Chapter 4
Optimization and Enumeration

The set of stable configurations for a nonexpansive network, given a fixed input assignment, has a simple structure characterized by a 2SAT instance. This structure can be used to look for different stable configurations. The optimization problem asks for one that minimizes a given objective function; the enumeration problem seeks to list all of them.

Our basic approach for the optimization problem views a minimum weight solution to a bipartite 2SAT instance as a minimum cut, and then uses a max-flow algorithm in conjunction with the max-flow min-cut theorem. The nonbipartite case cannot be solved in polynomial time if $\mathcal{P} \neq \mathcal{NP}$, so we look instead for good approximations using blocking flows.

Many max-flow algorithms have been presented in the literature. We propose two new ones that are particularly efficient when the optimum flow value is small, or when the graph has small width. These two special situations will arise in the next chapter when we apply network stability to stable matching. We finally present a 2SAT enumeration algorithm that is particularly efficient for graphs with small degree, a situation that is implied by the assumption of small width and therefore arises also in stable matching.

4.1 Uncapacitated Flow

The first maximum-flow algorithm was obtained Ford and Fulkerson [36]; a polynomial time algorithm was first obtained by Edmonds and Karp [28]. More efficient algorithms have been obtained since that early work. The current best known time bound is essentially nm, up to logarithmic factors. See Goldberg, Tardos, and Tarjan [41] for a survey, and also King, Rao, and Tarjan [73] for recent results.

We study two special cases in which faster algorithms can be obtained. The model that we adopt is one with edges of infinite capacity, and integer supplies and demands at the vertices.

CHAPTER 4. OPTIMIZATION AND ENUMERATION

The first case we study is that of graphs which are not too wide. We define a measure of width and prove an $O(wm \log K)$ time bound when the width is w, where K is value of the optimal flow. The second case is that of flow problems for which the value K of the optimal flow is relatively small (not too close to m^2). Here we obtain an $O(m\sqrt{K})$ time bound. In this second case, a standard reduction yields results for the more common capacitated case (with a single source and a single sink, and infinite supply and demand), where K is now defined by adding the *finite* edge capacities alone.

We also study the problem of finding blocking flows. A result of Goldberg and Tarjan [42] gives an $O(m \log(n^2/m))$ time bound in the acyclic case, with finite capacities, improving on an $O(m \log n)$ bound of Sleator and Tarjan [105, 106]. We prove an $O(m \log(w^2/m))$ bound for the case of infinite capacities. Each of these results will have an application to 2SAT and to stable matching.

4.1.1 Flow Problems and Width

We begin by reviewing standard definitions and terminology for flow problems. A more detailed exposition can be found, e.g., in Tarjan [112]. A *capacitated flow problem* is defined by a directed graph $G = (V, E)$ with two distinguished vertices, a *source* s and a *sink* t, and a positive (possibly infinite) integer capacity $\text{cap}(u, v)$ on every edge (u, v). For convenience we define $\text{cap}(u, v) = 0$ if (u, v) is not an edge in G. A *flow* on G is an integer-valued function f on vertex pairs satisfying the following three properties: (1) $f(v, u) = -f(u, v)$. If $f(u, v) > 0$, we say that there is a flow from u to v. (2) $f(u, v) \leq \text{cap}(u, v)$. If (u, v) is an edge such that $f(u, v) = \text{cap}(u, v)$, we say that the flow *saturates* (u, v). (3) For every vertex u other than s and t, $\sum_v f(u, v) = 0$. The *value* of a flow is the net flow out of the source, $\sum_v f(s, v)$. The *maximum flow problem* is that of finding a flow of maximum value. Given a flow f, the residual graph is the graph with vertex set V, source s, sink t, and an edge (u, v) of capacity $\text{res}(u, v) = \text{cap}(u, v) - f(u, v)$ (the residual capacity) for every pair (u, v) such that $\text{res}(u, v) > 0$. A flow g in the residual graph of f can be transformed into a flow $g + f$ in the original graph (and vice versa, by subtracting f instead of adding f).

A cut $X, \overline{X}$ is a partition of the vertex set V into two parts X and $\overline{X} = V - X$ such that X contains s and $\overline{X}$ contains t. It will sometimes be convenient to view a cut as an assignment of boolean values to the vertices in V, with the vertices in X given the value 1 and the vertices in $\overline{X}$ given the value 0. The edges *across the cut* are the edges that start in X and end in $\overline{X}$. The *capacity* of a cut is the sum of the capacities of the edges across the cut. A cut of minimum capacity is a *minimum cut*. The *flow across the cut* is the sum of all $f(u, v)$ with $u \in X$ and $v \in \overline{X}$; it can be shown that the flow across any cut is equal to the flow value. By the capacity constraint, the flow across any cut cannot exceed the capacity of the cut. Therefore the value of a maximum flow is no greater than the capacity of a minimum cut. The *max-flow min-cut theorem* states that these two numbers coincide. Thus, for a minimum cut, the flow across the cut equals the capacity of the cut; this property holds for a particular cut $X, \overline{X}$ if and only if all the edges that start in X and end in $\overline{X}$ are saturated, and there is no flow along the edges that start in $\overline{X}$ and end in X.

Sometimes the flow problem has the following special structure. All directed edges have infinite capacities and join two vertices other than the source and the sink, except for finite capacity edges (s, u) joining the source to a vertex other than the sink, and finite capacity edges (v, t) joining a vertex other than the source to the sink. In that case, we shall adopt a more convenient representation. We discard the finite capacity edges, the source and the sink, and say that vertex u is a *supply vertex* with supply cap(s, u), and that vertex v is a *demand vertex* with demand cap(v, t). We can assume without loss of generality that no vertex is both a supply and a demand vertex. We apply the terms residual supply (demand) and saturated supply (demand) to a vertex u (resp. v) as they would apply to the corresponding edge (s, u) (resp. (v, t)); the flow out of u (resp. into v) in the simplified representation is just the flow along (s, u) (resp. along (v, t)).

We refer to this problem with infinite capacities alone, supplies and demands, as the *uncapacitated flow problem*. An important special case is the *bipartite matching problem*: Here the vertex set is the union of two disjoint sets $V = A \cup B$, with edges of infinite capacity directed from A to B, and where vertices in A have supply 1 and vertices in B have demand 1.

We let $n = |V|$ be the number of vertices and $m = |E|$ be the number of edges in a directed graph G. (We usually assume for convenience that the graph has no isolated vertices, so that $m \geq n/2$.) We shall also associate with G a third parameter, the *width* of the graph. We define two closely related notions of width. The *implicit width* of G is the maximum cardinality of a set of vertices U with the property that, given two distinct vertices u and v in U, there is no directed path in G from u to v. Our algorithms will actually require the notion of *explicit width*, defined as follows. A *path cover* for G is a set of vertex-disjoint directed paths in the transitive closure of G whose union contains all the vertices in G. By Dilworth's theorem [26], the minimum number of paths in a path cover for G equals the implicit width of G. We shall say that G has explicit width w' if a path cover for G consisting of w' paths is known. We shall always add the edges of the paths in the known path cover (at most $n - 1$ edges) to the graph G, in the uncapacitated flow problems; this does not affect the flow results, because these infinite capacity edges correspond to infinite capacity paths in the graph.

It is possible to find a path cover consisting of w paths for a graph of implicit width w by running a costly min-flow algorithm. For our purposes, however, even a non-optimal path cover can sometimes be useful. The lemma below implies that for our first flow algorithm, the distinction between implicit and explicit width will cost at most a logarithmic factor.

Lemma 4.1 *A greedy algorithm finds a path cover with $w' \leq w \log n$ paths for a graph of implicit width w in $O(w'm)$ time.*

Proof. First transform the graph into an acyclic graph G by merging strong components, and determine a consistent linear order, in $O(m)$ time (see Tarjan [111]). The algorithm now marks the vertices of G in some order, starting from the graph with all vertices unmarked.

At each stage, it finds in $O(m)$ time a path in G containing the largest possible number of unmarked vertices, and marks these vertices. Since a path cover with w paths exists, some path in that cover must contain at least a fraction $1/w$ of the unmarked vertices; hence each path found will mark at least this many. After $w \log n$ paths have been found, the number of unmarked vertices left is at most $n(1 - 1/w)^{w \log n} < 1$, so all vertices are marked and hence covered. □

The following result gives a simple illustration of the notion of width.

Lemma 4.2 *All the edges in the transitive closure of a directed graph which involve at least one vertex from a given path P can be obtained in $O(m)$ time. Therefore the transitive closure can be obtained in $O(wm)$ time for a graph of explicit width w.*

Proof. Let $u_1 \to u_2 \to \cdots \to u_r$ be the path P. For every vertex v, let $\phi(v)$ be the greatest i such that the edge $u_i \to v$ is in the transitive closure of the graph. If no such i exists, we let $\phi(v) = 0$. Note that an edge (u_j, v) is in the transitive closure if and only if $j \leq \phi(v)$. To obtain $\phi(v)$ for all vertices v, we consider $i = r, \ldots, 2, 1$ in turn. For each i, we find all vertices v that are reachable from u_i in the graph, set $\phi(v) = i$ since $u_i \to v$ is in the transitive closure, and remove these vertices and incident edges from the graph. After the last value $i = 0$ has been considered, we set $\phi(v) = 0$ for all vertices v that were not removed from the graph. The correctness of the algorithm follows from the fact that if $\phi(v) = i$, then a path from u_i to v cannot go through a vertex v' with $\phi(v') > i$. The time complexity is $O(m)$ since every vertex and every edge of the graph is considered and removed only once. The edges in the transitive closure that are of the form $v \to u_i$ can be obtained with the same algorithm, after reversing all directed edges in the graph. □

A capacitated flow problem can be reduced to an uncapacitated problem by using the following standard reduction. For each capacitated edge (u, v) of positive finite capacity c, introduce two new vertices x and y with demand and supply respectively, and replace (u, v) with three directed edges (u, x), (y, x) and (y, v) of infinite capacity. An amount $f \leq c$ of flow along (u, v) in the original problem can then be represented with an amount f of flow along (u, x) and (y, v), and an amount $c - f$ of flow along (y, x). The resulting uncapacitated maximum flow problem will then have a maximum flow whose value differs from that of the original capacitated flow problem by exactly the sum C of the finite capacities c. Note that if the optimal flow value for the original capacitated problem is finite, then its value is at most C. This reduction may significantly increase the number of vertices of the graph, but it does not increase the number of edges by more than a factor of 3. Therefore any time bound for the uncapacitated problem that depends only on the number of edges and the value of the optimal flow (or even on the total supply and demand) can be translated into a time bound for the capacitated problem in terms of the number of edges and the sum of the finite capacities alone.

For uncapacitated flow problems, the measures of implicit and explicit width are those associated with the underlying graph. For capacitated flow problems, we shall consider an alternative notion of width; this notion will be of interest mainly because it allows a reduction from uncapacitated networks of explicit width w to capacitated networks of width $2w$, and because we can give a fast flow algorithm for the latter. In the capacitated case, which has a source, a sink, and edges of both finite and infinite capacity, we define the *cut width* as the maximum number of edges across a *finite capacity cut*. The reduction from the uncapacitated to the capacitated case that does not increase the width by more than a factor of 2 is as follows. Given an uncapacitated problem on a graph G and a path cover of w paths for G, we construct a capacitated problem on a graph G' by adding vertices and edges to G as follows (see Fig. 4.1). For each path P in the path cover with vertices $u_1, u_2, \ldots, u_r$, and infinite capacity edges (u_i, u_{i+1}), we create a new path with new vertices $u'_1, u'_2, \ldots, u'_r, u'_{r+1} = s$ (where s is the source). We add infinite capacity edges (u'_i, u'_{i+1}) and (u'_i, u_i) for $1 \leq i \leq r$, as well as finite capacity edges (u'_{i+1}, u'_i) of capacity $C_i = \sum_{j \leq i} c_j$, where c_j is the supply of vertex u_j in the uncapacitated problem. A similar construction is carried out for the demands: We create a new path with new vertices $t = u''_0, u''_1, u''_2, \ldots, u''_r$ (where t is the sink), add infinite capacity edges (u''_{i-1}, u''_i) and (u_i, u''_i) for $1 \leq i \leq r$, as well as finite capacity edges (u''_i, u''_{i-1}) of capacity $D_i = \sum_{j \geq i} d_j$, where d_j is the demand of vertex u_j in the uncapacitated problem. This network will have cut width at most $2w$: All the finite capacity edges are on the paths of u'_i and u''_i, and the set of edges across a finite capacity cut $X, \overline{X}$ cannot contain two finite capacity edges from the same path of u'_i (or of u''_i) because if it contains two edges (u'_{i+1}, u'_i) and (u'_{j+1}, u'_j) with $i < j$, then u'_{i+1} is in X, u'_j is in $\overline{X}$, and u'_j is reachable from u'_{i+1} along an infinite capacity path that must go across the cut, contradicting the fact that the capacity of the cut is finite.

Lemma 4.3 *An uncapacitated flow problem on a graph of explicit width w can be reduced to a capacitated flow problem of cut width at most $2w$, without increasing the number of vertices and edges by more than a constant factor.*

Proof. We show a correspondence between flows in the uncapacitated flow problem on G and in the capacitated problem on G' constructed from it, in both directions. Given a flow on G, for each path P as above, if f_i is the flow out of vertex u_i with supply c_i (so that $0 \leq f_i \leq c_i$), we assign flow f_i to the new infinite capacity edges (u'_i, u_i), and assign flow $\sum_{j \leq i} f_j$ to the finite capacity edges (u'_{i+1}, u'_i). Demand vertices are handled similarly, so that if f_i is the flow into vertex u_i with demand d_i (so that $0 \leq f_i \leq d_i$), we assign flow f_i to the infinite capacity edges (u_i, u''_i) and assign flow $\sum_{j \geq i} f_j$ to the finite capacity edges (u''_i, u''_{i-1}). It is easy to check that the capacity bounds and flow conservation laws hold, and that the value of the flow in G' is the same as in G. This completes one direction of the reduction.

In the other direction, given a flow on G', we first perform a simple transformation on the flow; this transformation will ensure that the flow on G satisfies the supply and demand constraints. Let F_i be the flow along the edge (u'_{i+1}, u'_i), so that the flow along (u'_i, u_i) is $F_i - F_{i-1}$ by flow conservation (with $F_0 = 0$ by convention). Let $F'_r = F_r$ and $F'_{i-1} = \min(F'_i, C_{i-1})$ by induction, for $1 \leq i \leq r$. Now assign flow F'_i to the edge (u'_{i+1}, u'_i) and flow $F'_i - F'_{i-1}$ to the edge (u'_i, u_i); increase the flow along (u_i, u_{i+1}) by $F'_i - F_i$. One

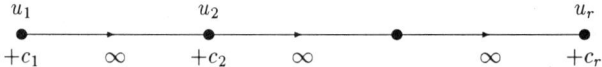

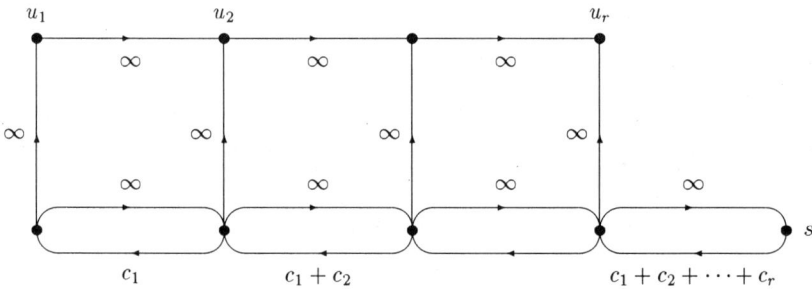

Figure 4.1: Providing flow at supply vertices.

can check by induction that $F_i \leq F'_i$, and directly that flow conservation and capacity bounds still hold. Furthermore, the flow along (u'_i, u_i) is $F'_i - F'_{i-1} = \max(0, F'_i - C_{i-1}) \leq C_i - C_{i-1} = c_i$. Thus if we remove all the u'_i vertices and let the flow out of supply vertex u_i be $F'_i - F'_{i-1}$, then this amount of flow will satisfy the supply constraint c_i as required. The analogous transformation for the demand vertices is as follows. Initially the flow along (u''_i, u''_{i-1}) is F_i and the flow along (u_i, u''_i) is $F_i - F_{i+1}$, where $F_{r+1} = 0$. We let $F''_1 = F_1$ and $F''_{i+1} = \min(F''_i, D_{i+1})$ by induction, assign flow F''_i to the edge (u''_i, u''_{i-1}) and flow $F''_i - F''_{i+1}$ to the edge (u_i, u''_i); we increase the flow along (u_{i-1}, u_i) by $F''_i - F_i$. The flow along (u_i, u''_i) is $F''_i - F''_{i+1} = \max(0, F''_i - D_{i+1}) \leq D_i - D_{i+1} = d_i$, so we can remove all the u''_i vertices and let the flow into demand vertex u_i be $F''_i - F''_{i+1}$, satisfying the demand constraint d_i. This completes the other direction of the reduction. □

Some flow algorithms use as a subroutine the computation of a blocking flow in a capacitated acyclic graph. A *blocking flow* is a flow such that every directed path from the source to the sink traverses a saturated edge. In the uncapacitated case, this means that if there is a directed path from a supply vertex to a demand vertex, then at least one of these two vertices must be saturated. The acyclicity assumption can be removed in the uncapacitated case, because the graph can be made acyclic by merging strong components and combining the corresponding supplies and demands, in $O(m)$ time [111].

The efficient implementation of flow algorithms often requires a data structure called dynamic trees, due to Sleator and Tarjan [105, 112]. This data structure is used to maintain a collection of vertex-disjoint rooted trees, with costs associated with the vertices, under the following operations. The maketree(v) operation creates a new tree containing the single vertex v, previously in no tree, with cost zero; findroot(v) returns the root of the tree containing vertex v; findcost(v) returns the pair (w,x) where x is the minimum cost of a vertex on the tree path from v to findroot(v) and w is the last vertex (closest to the root) on this path of cost x; addcost(v,x) adds x to the cost of every vertex on the tree path from v to findroot(v); link(v,w) combines the distinct trees containing vertices v and w by making v a child of w, where v must be a root; cut(v) divides the tree containing vertex v into two trees by deleting the edge joining v to its parent, where v must not be a root; and evert(v) makes v the root ot the tree containing v. In the algorithm of section 4.1.3, we shall also need to be able to perform the addcost operation separately for vertices at even and at odd depth in the tree. This modified version of the addcost operation can be incorporated without much difficulty into the dynamic tree data structure. This data structure makes it possible to execute an arbitrary sequence of any of these operations in time $O(\log t)$ per operation, where t is the size of the largest tree obtained during the execution.

4.1.2 Maximum Flow when the Width is Small

We now give a simple and efficient flow algorithm for the case of small width graphs. We describe the algorithm in the framework of capacitated flow problems (with edges of finite or infinite capacity).

The algorithm uses capacity scaling. Starting from a capacitated graph G, each finite capacity c is replaced by $\lfloor c/2 \rfloor$, and a maximum flow in this modified graph G' is obtained. By doubling this flow, we obtain a maximum flow in a graph G'' with capacities $2\lfloor c/2 \rfloor$. If we consider a corresponding minimum cut $X, \overline{X}$ in G'', then all the edges across the cut must be saturated, and must therefore have finite capacity. Therefore the number of such edges is at most the cut width w. We now restore the original capacities c of G. This increases all capacities by either 0 or 1, depending on whether c is even or odd. Therefore the capacity of the cut $X, \overline{X}$ increases by at most w. It follows that a maximum flow in G differs from the maximum flow in G'' by at most w units of flow. We may thus perform w augmentations, each running in $O(m)$ time, to obtain a maximum flow in G.

We have reduced a flow problem with optimum flow K to a flow problem with optimum flow at most $K/2$, by means of the reduction from G to G'. This reduction has time complexity $O(wm)$ and is referred to as a *scaling phase*. After $(\lg K) + 1$ such scaling phases, we are left with a problem whith maximum flow value 0; when this instance is reached, it is easily recognized in $O(m)$ time. Since each scaling phase takes $O(wm)$ time, this gives an $O(wm \log K)$ total running time. By Lemma 4.3, uncapacitated problems can be reduced to capacitated problems without affecting the size, width or maximum flow value of the given instance, provided that the notion of *explicit width* is used for the uncapacitated network.

Theorem 4.4 *A maximum flow of value K in a capacitated (uncapacitated) network with m edges and of cut width (resp. explicit width) w can be found in $O(wm \log K)$ time.*

4.1.3 Maximum Flow when the Optimum is Small

We are given a flow instance on a directed graph G with infinite capacities and integer supplies and demands. Our strategy for this problem will be the following. We first translate the problem into a flow problem in an auxiliary graph H, and give a simple algorithm for obtaining a maximum flow on H. We then show how the algorithm on H can be implemented efficiently without explicitly constructing the graph H. We assume that G is acyclic. This can be ensured via an $O(m)$ strong components computation [111], where strong components are replaced by single vertices and the corresponding supplies and demands are combined.

The auxiliary graph is a bipartite graph $H = (A \cup B, F)$. The set A consists of the supply vertices in G, the set B consists of the demand vertices in G (with supply and demand values inherited from G), and F contains a directed edge of infinite capacity (a, b) with $a \in A$ and $b \in B$ if and only if there is a directed path in G from a to b. The graph H is thus essentially the transitive closure of G. Clearly, a maximum flow in H corresponds to a maximum flow in the original graph G. For convenience, we assume that in the original graph G, the supply vertices have in-degree zero and the demand vertices have out-degree zero. This property can always be enforced for a supply vertex v by introducing a new vertex v' and an infinite capacity edge from v' to v, and moving the supply from v to v'; a similar transformation works for demands.

A maximum flow in H can be obtained by matching supplies in A to demands in B by a multiset M of edges in F: the number of occurrences of the edge (a, b) in M is the flow from a to b, and the number of edges in M involving a vertex $a \in A$ (resp. B) must be at most the supply (resp. demand) of the vertex. A maximum matching (one that maximizes the size of M) can be found with the Hopcroft and Karp matching algorithm, or alternatively with Dinits's flow algorithm [27, 60, 112]. This is done as follows. Suppose that some flow f in H is known. This flow induces a residual graph which consists of the same infinite capacity edges from A to B, but which also contains edges (b, a) of capacity c whenever f has a flow of value $c > 0$ along an edge (a, b) in H. Furthermore, supplies and demands have been updated in this residual graph in accordance with the flow f. (That is, the supply at a vertex $a \in A$ has been reduced by the amount of flow leaving a in f, and the demand at a vertex $b \in B$ has been reduced by the amount of flow entering b in f.)

From this residual graph, a layered graph is defined as follows. The vertices in A with positive supply in the residual graph are at level 0, and so are the vertices in B reachable in H from vertices of A at level 0. For $i > 0$, the vertices in A at level i are the vertices of A that are not at smaller levels and are reachable from vertices at level $i-1$ in B through a capacitated residual edge; the vertices in B at level i are the vertices of B that are not at smaller levels and are reachable in H from vertices of A at level i. The last level l is the

CHAPTER 4. OPTIMIZATION AND ENUMERATION   116

level that contains vertices in B with positive residual demand. The value l is called the *length* of the layered graph. The layered graph contains precisely those vertices that are at some level $0 \leq i \leq l$ and those edges of the residual graph that are either infinite capacity edges joining a vertex in A at level i to a vertex in B at level i for some i, or finite capacity residual edges joining a vertex in B at level $i-1$ to a vertex in A at level i. Therefore the layered graph is acyclic. The algorithm finds a blocking flow g in this layered graph, that is, a flow that cannot be increased by adding flow from a supply vertex to a demand vertex along unsaturated edges of the layered graph. This blocking flow can be added to the previous flow f to obtain a new flow $f' = f + g$. If we now construct the residual graph and the layered graph corresponding to f', it is known that the new layered graph must have length at least $l+1$. This in turn implies that the number of *phases*, i.e., the number of times that the flow f and the corresponding residual graph are updated before the final optimal flow is obtained, is at most $2\sqrt{K}$, where K is the value of the optimal flow (see below for a justification of this bound).

If we could implement each phase in $O(m)$ time, we would then have an $O(m\sqrt{K})$ algorithm, as desired. However, even if we put aside the complexity of obtaining H from G via a transitive closure computation, we face the difficulty that both the number of edges from A to B in H, and from B to A in the residual graphs, could be as large as m^2 (i.e., we could have G sparse but H dense), giving an $O(m^2)$ complexity for each phase. We shall handle this difficulty differently for the two types of edges: For the capacitated residual edges from B to A, we shall maintain a "forest" solution, thus keeping their number bounded by $O(m)$ (in fact by $n-1$). For the uncapacitated edges from A to B, we shall avoid computing the transitive closure explicitly, and work instead with edges in the original graph G, whose number is bounded by m. This will give the desired bound.

We first show how the number of capacitated edges from B to A in the residual graph can be kept within an $O(m)$ bound. The main idea to achieve this is to maintain a "forest" solution (sometimes known as a "spanning tree" solution). Suppose that the edges from B to A in the residual graph at the beginning of a phase form a forest (when viewed as undirected edges). We shall ensure that, after the blocking flow is obtained, the residual graph is updated so that this forest property is maintained. This means that the number of residual edges is in fact always at most $n-1$. A priori, each edge (a,b) from A to B along which the blocking flow g sends positive flow must be added as an edge (b,a) to the residual graph. It may happen that the addition of a new residual edge (b,a) completes a cycle among the edges from B to A in the residual graph (when viewed as undirected edges). If this happens, we perform the following transformation. Let δ be the minimum residual capacity of an edge (b',a') in this newly created cycle of capacitated residual edges. Note that the cycle has an even number of edges, since the residual graph is bipartite. The edges (b'',a'') on the cycle may thus be alternatively labelled even and odd, with the edge (b',a') of capacity δ labelled even by convention. We then decrease the flow f from A to B along (a'',b'') by δ if (b'',a'') is an even edge, and increase the flow by δ for an odd edge. This transformation preserves the validity of the flow f. Furthermore, the edge (b',a') is now no longer a residual edge, so the residual edges from B to A constitute once again a forest. Since this transformation creates no new residual edges, the length of the layered graph will still increase in each phase from l to at least $l + 1$, so the number of phases is still bounded by $2\sqrt{K}$. The trees in the forest can be maintained via dynamic trees. This data structure enables us to find the tree path from b to a (use *evert* to make a the root), to find the minimum capacity edge on this path (with *findcost* at b), to update the capacities of edges along the path (with *addcost* provided that even and odd depth costs are updated separately, as mentioned in section 4.1.1), and to link and cut trees in time $O(\log n)$ per operation. Thus, if R is the total number of capacitated residual edges created during all phases of the algorithm, then the complexity of maintaining a forest solution is $O(R \log n)$. We shall later see that $R = O(K \log n)$, giving an $O(K \log^2 n)$ time bound.

We now show how to compute a blocking flow in the layered graph efficiently. (See Fig. 4.2.) Recall that the layered graph can contain a large number of infinite capacity edges from A to B, but only $O(m)$ capacitated edges from B to A (by the forest property). We say that a vertex or an edge in G is at level i if i is the smallest number such that the vertex or edge can be reached in G from a vertex in A at level i. Note that if (a, b) is an edge in H with both endpoints at level i, then each path from a to b in G must be contained in level i. For given such a path, it is clear by definition that all vertices and edges along the path are at level i or smaller; if some such vertex or edge were at a level $j < i$, then this vertex or edge, and hence b itself, would be reachable in G from a vertex in A at level j, so b itself would be at level j or smaller. One can determine levels of vertices by giving length 0 to all infinite capacity edges in G, length 1 to all finite capacity edges in the residual graph, and performing a shortest path computation from the set of vertices in A at level 0, in $O(m)$ time.

We have partitioned the vertices and edges of G into disjoint levels; the edges (a, b) in H with a and b at level i are represented by paths in G within level i. We can discard the edges in G that join vertices in two different levels, since they do not represent any edge in H; after discarding such edges, we refer to the vertices and edges in a level i as the *level i graph*. If we now combine the level i graphs for all i with the capacitated edges joining two adjacent levels in the residual graph, we have an implicit representation of size $O(m)$ for the layered graph, with capacitated edges from B to A in the layered graph given explicitly, and infinite capacity edges in the layered graph from A to B given implicitly by paths in G within a level. We refer to this graph as the *representation of the layered graph*, illustrated in Fig. 4.2. This representation graph is acyclic, because each level is acyclic and capacitated edges only go from a level i to the next level $i+1$. An $O(m \log n)$ blocking flow algorithm [105, 106, 112] can be run directly in this representation graph. We sketch here such an algorithm, but with some modifications that lead to a better time bound for our problem. The reader is referred to [42, 105, 106, 112] for details.

The algorithm executes the following two steps in alternation: (1) repeatedly remove all vertices of outdegree zero other than the unsaturated demand vertices in B at level l, together with their incoming edges, and remove the edges of capacity zero as well; (2) traverse a path from an unsaturated supply vertex to an unsaturated demand vertex, and

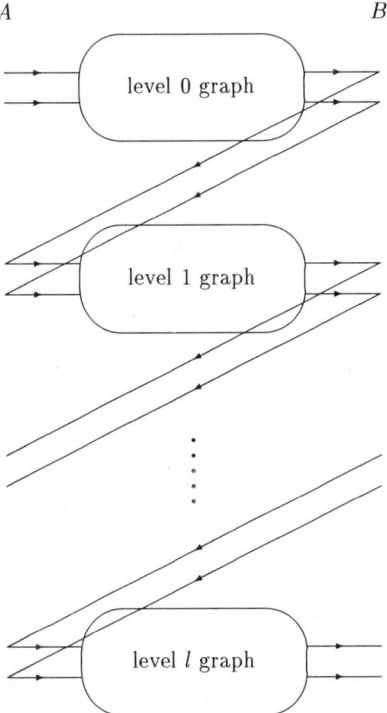

Figure 4.2: Representation of the layered graph.

send flow along this path until either the supply, the demand, or one of the capacitated edges along the path is saturated. Note that the path started at an unsaturated supply vertex can always be extended until an unsaturated demand vertex at level l is reached, since all the other vertices have nonzero outdegree and the graph is acyclic. An additional requirement is that whenever a path reaches a vertex that had previously been visited, it proceeds from this vertex along a previously visited edge, if there is any such edge left in the graph. As a result, at any point in time, there is at most one outgoing edge for each vertex that has been visited but has not been removed, so the visited edges remaining in the graph form a collection of in-trees. These trees can be maintained using dynamic trees, linking trees when a new edge is visited and cutting trees when an edge is removed. The edges on the path in step (2) that belong to dynamic trees are not visited, because whenever a path reaches a dynamic tree, it can proceed directly from the root of the tree. As a result, each edge is visited only twice, once the first time it is visited during step (2) at which point it

CHAPTER 4. OPTIMIZATION AND ENUMERATION

becomes part of the dynamic trees, and once when it is removed during step (1). Therefore the number of dynamic tree operations is $O(1)$ per edge, with an additional $O(1)$ per path created in step (2). The number of such paths is also $O(m)$, because each path saturates an edge or a vertex. This gives a total of $O(m)$ dynamic tree operations, for an $O(m \log n)$ time bound.

We introduce two modifications in this algorithm. First, we never link dynamic trees in different levels. Therefore the path in step (2) always visits the capacitated edges joining different levels and knows the vertices in A and B visited, hence the path in the (explicit) layered network used to increase the flow. Second, we enforce a fixed bound t on the dynamic tree size as in [42]. The cost of these modifications is the fact that certain edges are visited but not incorporated into the dynamic trees by the path in step (2). Each such edge costs a constant number of dynamic tree operations; we give an upper bound on the number of such edge. The edges not incorporated into the dynamic trees because of the first modification are the l capacitated edges joining different levels along the path. The edges not incorporated because of the second modification are edges whose endpoints belong to two dynamic trees, where one of the two trees has at least $t/2$ vertices (otherwise the bound t on the joint tree size would be met and the edge would be incorporated into the dynamic trees). The number of trees with at least $t/2$ vertices is at most $\lfloor 2n/t \rfloor$, and each such tree is charged at most twice, once by the edge where the path enters the tree, and once by the edge where the path leaves the tree at the root. Therefore at most $2\lfloor 2n/t \rfloor$ edges are visited because of the second modification. This gives an additional $O(l + \lfloor 2n/t \rfloor)$ dynamic tree operations per path. If the flow increase in a phase is h, then the number of paths in step (2) is at most h, giving a total of $O(m + (l + \lfloor 2n/t \rfloor)h)$ dynamic tree operations per phase. Each operation takes $O(\log t)$ time, so the time bound for the blocking flow computation at each phase is $O((m + (l + \lfloor 2n/t \rfloor)h) \log t)$.

To bound the total time complexity of the algorithm, we shall use the following fact, which is again known from standard matching algorithms (see, e.g., [27, 60, 112]): If the flow after a layered graph of length l has been obtained is K', then the optimum flow $K \geq K'$ differs from K' by at most K'/l. We can cover the possible values for l ($0 \leq l < n$) with $\lg n$ intervals of the form $L \leq l < 2L$ plus the additional value $l = 0$. If h is the amount of flow when the length is l, then $\sum_{L \leq l < 2L} h \leq K/L$ and $\sum_{L \leq l < 2L} lh \leq 2K$. Therefore $\sum_{0 \leq l < n} lh \leq 2K \lg n$. Note that the number of capacitated residual edges created in a phase is at most $(l+1)h$. Therefore the number R of such edges created in all phases is indeed $O(K \log n)$ as claimed, giving the $O(K \log^2 n)$ bound for maintaining a forest solution.

We shall choose $t = \max(2, \min(2n+1, K/l^2))$, so that the dynamic tree size bound is reduced to a constant as the length l of the layered graph approaches $\sqrt{K}$. By the bound in the preceding paragraph and the fact that $\log t \leq \log(2n+1)$, the $O(lh \log t)$ term of the time spent in each phase gives an $O(K \log^2 n)$ bound overall. The remaining term is $O((m + \lfloor 2n/t \rfloor h) \log t)$. For the phases with $l \geq L = \sqrt{K}$, we have $t = 2$ and

$\sum_{l \geq L} h \leq K/L = \sqrt{K}$. The number of such phases is as a result also at most $\sqrt{K}$. The term is then $O(m + nh)$ for each phase and $O(m\sqrt{K})$ for all phases $l \geq \sqrt{K}$.

For the phases with $1 \leq l < \sqrt{K}$, we again consider intervals of the form $L \leq l < 2L$, where L is of the form $L = \sqrt{K}/2^i$ with $i \geq 1$. Then

$$\sum_{L \leq l < 2L} (m + \lfloor \frac{2n}{t} \rfloor h) \log t \leq (Lm + 2n \cdot \frac{(2L)^2}{K} \frac{K}{L}) \log(\frac{K}{L^2})$$

$$\leq (m \frac{\sqrt{K}}{2^i} + 8n \frac{\sqrt{K}}{2^i})(2i).$$

Adding up this expression for all $i \geq 1$, we obtain an $O(m\sqrt{K})$ bound for all phases $1 \leq l < \sqrt{K}$.

As a last observation, note that the tree size bound $t = K/l^2$ depends on the unknown optimal flow value K. However, if the flow at the end of the first phase is K', then since $l \geq 1$ at that point, the flow in subsequent phases will increase by at most $K'/l \leq K'$, so $K/2 \leq K' \leq K$. Thus the flow K' at the end of the first phase provides an adequate estimate for K. The first phase itself is run with t growing from 1 to at most K'^2, letting $t = i^2$ for the ith path. The term for the first phase (which has $l = 0$ and $h = K'$) is thus $O((m + (\lfloor 2n/1^2 \rfloor + \lfloor 2n/2^2 \rfloor + \lfloor 2n/3^2 \rfloor + \cdots))(\log K')) = O((m + n) \log K') = O(m \log K)$.

Theorem 4.5 *A maximum flow of value K in an uncapacitated graph with m edges can be obtained in $O(m\sqrt{K} + K \log^2 m)$ time. This gives an $O(m\sqrt{K})$ time bound for $K \leq (m/\log^2 m)^2$.*

As presented, the algorithm will give the solution in terms of direct edges from supplies to demands, each of which corresponds to an infinite capacity path in the graph. If we want to know what the actual flow along the edges of the original graph is, we need to map each direct edge back to the path to which it corresponds. This path can be recovered (and the corresponding flow assigned to it) by using the dynamic tree structure that represented the path when the direct edge was first discovered. This structure may no longer exist at the end of the algorithm, and a re-execution of the algorithm to re-create the structure and assign flow to the corresponding paths will then be necessary. For most applications, knowledge of the actual infinite capacity path followed from a supply to a demand is not needed.

Note that if $K = O(m)$, then it is not necessary to maintain a forest solution to ensure an $O(m)$ bound on the number of residual edges, and furthermore the bound on the running time holds even if the $O(\log t)$ time used to implement dynamic trees is increased to $O(\sqrt{t}/\log^2 t)$.

In view of known bounds for network flow, an $O(m\sqrt{K})$ bound is only open for sparse graphs having $m/\log^2 m \leq \sqrt{K} \leq n \log n$.

4.1.4 Uncapacitated Blocking Flow

We now turn our attention to the problem of finding blocking flows in an uncapacitated graph. Our algorithm is based on the $O(m \log n)$ algorithm (for acyclic graphs) of [105, 106, 112] that uses dynamic trees. We also use the bounded tree size idea from the $O(m \log(n^2/m))$ algorithm of [42], but do not require finger search trees due to the fact that all capacities are infinite. Infinite capacities also make it possible to replace the parameter n by the potentially smaller explicit width w, thus giving an $O(m \log(w^2/m + 2))$ algorithm. The given graph G has infinite capacities, finite supplies and demands. Recall that we can assume without loss of generality that the graph is acyclic in the infinite capacity case. We are also given a set of w vertex-disjoint paths $P_1, P_2, \ldots, P_w$ in G that jointly cover all the vertices in G.

The algorithm proceeds by performing a series of depth-first searches on G, starting at different vertices of G. The depth-first searches will be used to send flow from supplies to demands; we shall first concentrate on the rules that guide the execution of these searches, and only later indicate how these searches are used to send flow. A depth-first search retreats from a vertex v only if it succeeds in saturating the demand at v (this saturation will be described later). Otherwise, the depth-first search stops (and the next depth-first search is started). As a result, if the depth-first search retreats from v, then the demand at v and at every vertex reachable from v is saturated, so no flow can be sent through v, and we can remove the vertex v together with its incoming edges from G. This also implies that a vertex v is removed from G only after all the vertices reachable from v have been removed from G. In particular, the vertices on a path P_i that have not been removed from G always constitute an initial segment of P_i. We denote the current initial segment by Q_i and the first and last vertices of Q_i by u_i and v_i respectively. When the depth-first search advances from a vertex v, it always proceeds along an edge visited by an earlier depth-first search, if such an edge exists. As a result, for every vertex v remaining in G, there is always at most one edge out of v that has been traversed so far. This implies that the visited edges form a collection of in-trees.

In choosing an edge out of a vertex v, we shall always give priority to the edge joining v to the vertex following v on the path Q_j that v belongs to, if $v \neq v_j$. As a result, there is always at most one visited edge out of Q_j, and this edge is in fact an edge (v, w) out of $v = v_j$. If w belongs to $Q_{j'}$ then we view this visited edge (v_j, w) as an edge joining Q_j to $Q_{j'}$. This implies that the visited edges joining two different Q_j form a collection of in-trees on the set of Q_j. At any point during the execution of a depth-first search, the edges already visited by the search that remain in the graph form a path R from the vertex where the search started to the current vertex; the remaining edges visited by the depth-first search have been removed from G because a retreat was performed through them. We always start a depth-first search at the vertex u_i of some Q_i; by the priority rule, the current path R will immediately and subsequently contain the initial segment Q_i. We can now specify how a vertex v is saturated before the depth-first search retreats. This is done simply by sending flow from the lowest unsaturated supply vertex u in Q_i to the vertex v, along R, so that either the supply at u or the demand at v is saturated. Therefore, when the depth-first search terminates because it cannot saturate the demand at v, all the supply vertices on P_i will have been either saturated or removed from G. The total number of depth-first searches performed is at most w, one for each P_i. By the time all these searches terminate, all supply vertices will have been either saturated or removed from G, and in either case no additional flow can be sent from them to an unsaturated demand, so we indeed have a blocking flow.

To implement this algorithm efficiently, we represent the in-trees joining the Q_j using dynamic trees. The nodes in the dynamic trees correspond to paths Q_j. Each path Q_j is represented separately by a doubly-linked list, and we also remember the end-vertex v_j of Q_j. When a depth-first search advances from a vertex v in a Q_j, it can always go directly to the vertex v_j, so the edges contained in Q_j are never visited when the search advances. In fact, the depth-first search can proceed directly to the vertex $v_{j'}$ of the root $Q_{j'}$ of the tree containing Q_j. Therefore, when the search advances, it only visits edges joining two different Q_j that were not visited by an earlier depth-first search, and adds each such edge to the dynamic tree structure so that later searches will not need to examine it when they advance. When the depth-first search retreats from the vertex v_j of a root Q_j, it removes v_j and its incident edges from G. An edge joining some $Q_{j'}$ to this root Q_j is of the form $(v_{j'}, w)$ with w in Q_j, and is removed from G only if $w = v_j$; therefore only some of the edges joining the root Q_j to its children $Q_{j'}$ are removed. The initial segment Q_j is also shortened when the last vertex v_j is removed, and v_j is updated accordingly. If the tree whose root is Q_j was entered by the depth-first search through a vertex v in some $Q_{j''}$, then the depth-first search retreats from the old v_j to the vertex $v_{j'}$ of the root $Q_{j'}$ of the tree containing $Q_{j''}$ after the updates (possibly $j' = j$).

Only a few dynamic tree operations are performed for each edge in G, namely those performed when the edge is seen for the first time by the advance of a depth-first search, and those performed when the edge is seen for the last time and removed from G. Therefore the number of dynamic tree operations is $O(m)$, and each operation takes $O(\log(w+1))$ time because the total number of nodes in the dynamic trees is at most w. This gives an $O(m \log(w+1))$ time bound.

To improve on this time bound, we introduce a limit t on the maximum dynamic tree size, and never link two Q_j if this linking would result in a dynamic tree with more than t nodes. Note that if a link operation is not performed, then one of the two trees involved must have at least $t/2$ nodes. We must now account for the traversal of an edge e by the advance of the depth-first search that does not result in a link. The edge e goes from the root Q_j of a tree to a node $Q_{j'}$ of a tree rooted at some $Q_{j''}$. If the depth-first search later retreats from the vertex $v_{j''}$ of $Q_{j''}$, then we charge the forward traversal of e to this retreat; otherwise we charge the traversal of e to one of the two trees involved, whichever is of size at least $t/2$. Note that in this second case the charged tree remains unchanged until the end of the depth-first search. There are at most $2w/t$ such trees, and each is charged by at most two edges, one where the depth-first search enters the tree and one where the depth-first search leaves the tree at its root. Therefore every such tree is

CHAPTER 4. OPTIMIZATION AND ENUMERATION 123

charged a bounded number of dynamic tree operations by the depth-first search, and we have an additional $O(w/t)$ dynamic tree operations per depth-first search for a total of $O(w^2/t)$ dynamic tree operations over all w searches. Each dynamic tree operation takes $O(\log(t+1))$ time, giving a total $O((m+w^2/t)\log(t+1))$ time bound. Letting $t = w^2/m + 1$ gives an $O(m\log(w^2/m + 2))$ time bound.

Theorem 4.6 *A blocking flow in an uncapacitated network with m edges and explicit width w can be found in $O(m\log(w^2/m + 2))$ time.*

Note that if $m = \Omega(w^2)$, then we can let $t = 1$ and still obtain a linear-time algorithm without dynamic trees. The efficiency of this $O(m + w^2)$ algorithm comes solely from the fact that we avoid traversing paths P_j by always jumping to the end-vertex of the path. This simplifies the implementation considerably.

Since the edges inside the paths P_j are skipped, the algorithm (with or without dynamic trees) remembers the flow only along edges that join two different paths P_j and $P_{j'}$. If at the end of the algorithm we need to know the flow for the edges inside a path P_j, we can just traverse P_j from its start-vertex and use flow conservation at each vertex to infer the flow along the edges of P_j. This can be done in $O(m)$ time.

4.2 Optimization on 2SAT instances

For the definition of 2SAT instances and some related constructions, see section 3.2.2. In order to reduce optimization problems for 2SAT instances to flow problems, we shall use a connection between 2SAT and vertex cover. A partial solution to a 2SAT instance is an assignment of values to a subset of the variables that can be extended to a complete solution by some assignment of values to the remaining variables. A partial assignment can be represented by a subset S of the vertices in the compatibility graph: If a literal u has been assigned the value 1 then the vertex u is included in S (thus $\bar{u}$ is included in S if u has been assigned the value 0); if u has not been assigned a value then both u and $\bar{u}$ are included in S. A vertex cover in an undirected graph is a set of vertices S with the property that at least one of the two endpoints of every edge is in S. Many results on the 2SAT problem depend implicitly on the following observation:

Lemma 4.7 *The partial solutions to a solvable 2SAT instance are the vertex covers of the compatibility graph of its transitive closure.*

Proof. Transitive closure in the implication graph corresponds to closure under the *resolution* rule in the compatibility graph. That is, if there is a clause $u \vee v$ and a clause $\bar{v} \vee w$, then there is also a clause $u \vee w$.

A partial solution to the 2SAT instance corresponds to a vertex cover of the compatibility graph because at least one literal of each clause involving two variables with assigned values in the partial solution is satisfied. To prove the converse, suppose that we have a vertex cover. Note that all vertex covers correspond to partial assignments because they include at least one of the two complementary literals $u, \overline{u}$ (given the presence of the edge $(u, \overline{u})$). We shall show that if we reduce the vertex cover to a minimal vertex cover (by repeatedly removing vertices from the cover while preserving the vertex cover property until no more vertices can be removed) then the corresponding extended partial assignment is indeed a complete assignment and hence a solution to the 2SAT instance (since all the clauses are satisfied by the vertex cover property). Therefore the partial assignment was indeed a partial solution. Suppose then, towards a contradiction, that we have a minimal vertex cover in which, for some literal v, both v and $\overline{v}$ are in this cover. Consider first the case where there is no self-loop edge (v, v) or $(\overline{v}, \overline{v})$. By minimality, v cannot be removed from the cover, so there must exist an edge (u, v) with u not in the cover. Similarly, $\overline{v}$ cannot be removed from the cover, so there is an edge $(\overline{v}, w)$ with w not in the cover. But then, by resolution, there is an uncovered edge (u, w), a contradiction. If there is one self-loop, say the edge $(\overline{v}, \overline{v})$, but not the edge (v, v), then we can still conclude that there is an edge (u, v) with u not in the cover, and then resolution yields edges $(u, \overline{v})$ and (u, u), so u must be in the cover, a contradiction. If both self-loops (u, u) and $(\overline{u}, \overline{u})$ are present, then the 2SAT instance has no solution. Therefore, if a solution exists, then every minimal vertex cover contains only one of $v, \overline{v}$ for each such pair, and hence defines a 2SAT solution. □

Note that the preceding proof also shows that the complete solutions to a solvable 2SAT instance are the *minimal* vertex covers of the compatibility graph of its transitive closure. The lemma implies that given a partial assignment to k variables of a transitively closed 2SAT instance, one can check whether this partial assignment is a partial solution by just checking the clauses involving only these k variables (at most $O(k^2)$ clauses). The reason is that the corresponding set S contains both u and $\overline{u}$ for every unassigned u, so all edges involving unassigned literals are automatically covered, and only edges involving the remaining $2k$ literals need to be checked. We let the implicit and the explicit *width* of a 2SAT instance be simply the implicit and the explicit width of its implication graph. Using the transitive closure algorithm from Lemma 4.2, we obtain the following:

Corollary 4.8 *Given a 2SAT instance with m clauses and explicit width w, one can determine, after $O(wm)$ preprocessing time, whether a query assignment to k of the variables is a partial solution, in $O(k^2)$ time.*

In section 4.2.1 we examine the complexity of an optimization problem for general 2SAT instances, which characterize the stable solutions for arbitrary nonexpansive networks. In section 4.2.2 we consider the case of bipartite 2SAT instances, which characterize the stable solutions to monotone nonexpansive networks.

4.2.1 The Non-Bipartite Case

An instance of the *minimum weight* 2SAT problem is a 2SAT instance with a nonnegative weight associated with each literal. The weight of a solution is the sum of the weights of the true literals (literals of value 1 in the solution). In the minimum weight 2SAT problem, the aim is to find a solution of minimum weight for a given weighted 2SAT instance. We can assume, for each variable x_i, that one of x_i and $\overline{x_i}$ has weight 0, by subtracting a constant from both weights if necessary; in fact we can assume that $\overline{x_i}$ has weight 0, by replacing x_i with its negation if necessary. We shall therefore assume that weights are associated with variables instead of literals. Given a weighted 2SAT instance, if we consider the compatibility graph of its transitive closure, with weights assigned to the vertices, and define the weight of a vertex cover to be the sum of the weights of vertices in the cover, then the minimum weight 2SAT problem can be viewed as a minimum weight minimal vertex cover problem.

Unfortunately, the minimum weight vertex cover problem is $\mathcal{NP}$-complete, even if all weights are 1. This also applies to the minimum weight 2SAT problem, because a graph on vertices x_i can be viewed as a 2SAT instance on the variables x_i, with clauses $x_i \vee x_j$ corresponding to edges (x_i, x_j) and where the variables x_i inherit their weight from the graph (the 2SAT instance is monotone and transitively closed). Therefore the minimum weight 2SAT problem is also $\mathcal{NP}$-complete.

On the other hand, there exist algorithms for approximating the minimum weight vertex cover within a factor of 2 of the minimum weight, and these algorithms give then solutions within a factor of 2 for the minimum weight 2SAT problem. See Bar-Yehuda and Even [12, 13], Clarkson [23], Gusfield and Pitt [53], Hochbaum [57, 58], Nemhauser and Trotter [84]. We shall describe one algorithm for the minimum weight 2SAT problem in terms of blocking flows, and this will enable us to use the efficient blocking flow algorithm from the last section.

The algorithm is as follows. Given a weighted 2SAT instance, we consider its implication graph. In this graph, we assign a supply to $\overline{x}$ equal to the weight of x and a demand to x equal to the weight of x, for each variable x; the edges are given infinite capacities. We now find a *symmetric blocking flow* in this uncapacitated network; symmetric means here that the flow out of supply vertex $\overline{x}$ equals the flow into demand vertex x. The set S of saturated demands x together with all vertices $\overline{x}$ gives then a partial solution to the 2SAT instance, which can be extended to a complete solution.

Theorem 4.9 *A solution to the minimum weight* 2SAT *problem with weight within a factor of 2 of the optimum can be obtained in* $O(m \log(w^2/m+2))$ *time for instances with m clauses and explicit width w.*

Proof. We first prove the correctness of the algorithm, and then discuss its implementation. If (x, y) is an edge of the compatibility graph of the transitive closure, then there is a path from $\overline{x}$ to y in the implication graph, with infinite capacity edges. Therefore a blocking flow must saturate the supply vertex $\overline{x}$ or the demand vertex y, and this means by the symmetry condition that either x or y must be saturated. Therefore the set S is a vertex cover of the compatibility graph of the transitive closure, and hence a partial solution by Lemma 4.7. Extending it to a complete solution means reducing S to a minimal vertex cover, and this can only reduce the weight of S.

The weight of S is at most the value of the blocking flow. The reason is that the weight of S is the sum of the saturated demands, i.e., the sum of the flow into saturated demand vertices, while the value of the flow is the sum of the flow into all demand vertices. On the other hand, let T be a solution to the 2SAT instance represented as a vertex cover of the compatibility graph of the transitive closure. Charge the amount of flow from a supply $\bar{x}$ to a demand y along some path to one or the other of these two vertices, depending on whether x or y is covering the edge (x,y) in T. The total flow is thus charged to supplies $\bar{x}$ or demands x such that x is in T, and neither is ever ever charged more than its corresponding supply or demand, which equals the weight of x. This shows that the total flow is never more than twice the weight of T. We have therefore proved that the weight of S is at most twice the weight of T, and so the solution obtained from S has weight at most twice the weight of any other solution, as desired.

To implement this algorithm, we use the blocking flow algorithm from the last section. To ensure the symmetry condition, whenever we send flow in the implication graph from supply $\bar{x}$ to demand y, we also send the same amount of flow from $\bar{y}$ to x (in the case $x \neq y$). This modification can be easily incorporated into the algorithm and does not affect the $O(m \log(w^2/m + 2))$ time bound. The saturated demands x define a partial solution. We can extend a partial solution to a complete solution using any $O(m)$ time algorithm for 2SAT (see section 3.2.2). □

An $O(nm)$ algorithm was given by Gusfield and Pitt [54].

If the given 2SAT instance is transitively closed, then we only need to send flow from a supply $\bar{x}$ to a demand y along a single edge (rather than a path). We can then enforce the symmetry condition by working directly with the undirected edge (x,y) of the compatibility graph rather than the two directed edges $(\bar{x},y)$ and $(\bar{y},x)$ of the implication graph. Each edge is considered once and assigned a flow that saturates either x or y, giving an $O(m)$ algorithm. This simple linear algorithm can be applied to the minimum weight vertex cover problem, because the corresponding 2SAT instance (see above) is always transitively closed. See also [12, 54].

4.2.2 The Bipartite Case

A 2SAT instance is *bipartite* if the corresponding compatibility graph is a bipartite graph $G = (U \cup V, E)$. Note that for each pair of complementary literals $u, \bar{u}$, one of them must be in U and the other one in V, since the edge $(u, \bar{u})$ is in G. The fact that G is bipartite also tells us that every edge (u, v) in the implication graph must join either two vertices in U or two vertices in V, and that implications $u \to v$ in U correspond to implications $\bar{v} \to \bar{u}$ in V. We shall therefore restrict the implication graph to the subgraph induced by the vertices in V, which always contains exactly one of each pair of complementary literals

$u, \overline{u}$. If $u \in U$, then setting $u = 0$ must be interpreted as setting $\overline{u} = 1$ in V (and setting $u = 1$ as setting $\overline{u} = 0$ in V).

The minimum weight solutions to a bipartite weighted 2SAT instance can be obtained as follows. For each variable $x \in U$, assign a supply to $\overline{x}$ (in V) equal to the weight of x. For each variable $y \in V$, assign a demand to y equal to the weight of y. Then find a maximum flow in the implication graph of the 2SAT instance with infinite capacities (in V). Now augment the 2SAT instance by adding the following constraints: If $\overline{x}$ is an unsaturated supply vertex, set $x = 0$. If y is an unsaturated demand vertex, set $y = 0$. If there is positive flow along an edge $u \to v$ in the implication graph, then add the constraint $v \to u$ to the 2SAT instance. The minimum weight solutions to the original 2SAT instance are then precisely the solutions to the modified 2SAT instance. Therefore a particular minimum weight solution can be obtained by solving the modified 2SAT instance. Note that the modified instance is simpler: Some variables have been replaced by constants, and the literals u and v for which a constraint $v \to u$ has been added must now satisfy $u = v$ and can therefore be replaced by a single literal. We shall prove that this algorithm is correct. With the time bounds of section 4.1, we have:

Theorem 4.10 *A minimum weight solution of weight K for a bipartite weighted 2SAT instance with m clauses and explicit width w can be found in $O(m\sqrt{K})$ time for $K \leq (m/\log^2 m)^2$ and in $O(wm \log K)$ for arbitrary K. In fact, a complete description of all minimum weight solutions can be found within this time bound.*

Proof. We only need to prove the correctness of the algorithm. Recall the definition from section 4.1.1 of uncapacitated flow problems in terms of capacitated problems with a source s, a sink t, and capacitated edges $(s, \overline{x})$ (resp. (y, t)) for supply vertices $\overline{x}$ (resp. demand vertices y). We can view the solutions to the 2SAT instance as cuts by setting $s = 1$ and $t = 0$, while using for the remaining vertices in V their values assigned as literals in the solution. Furthermore, the weight of a solution is the capacity of the corresponding cut. To see this, note that an edge (u, v) in the graph with $u = 1$ and $v = 0$ cannot be an infinite capacity edge, because it would then be an edge (u, v) in the implication graph and the constraint $u \to v$ in the 2SAT instance would be violated. For capacitated edges, we have $u = 1$ and $v = 0$ if and only if (u, v) is $(s, \overline{x})$ and $x = 1$, or (u, v) is (y, t) and $y = 1$. Therefore the true variables (variables set to 1) correspond to the edges across the cut, with the weight of the variable corresponding to the capacity of the edge, proving the claim.

Therefore the least possible weight for a solution is the weight of a minimum cut. By the max-flow min-cut theorem, given a maximum flow, the minimum cuts are the cuts such that if $(u, v) = (1, 0)$ then the edge (u, v) is saturated, and if $(u, v) = (0, 1)$ then the edge (u, v) has no flow [112]. For uncapacitated edges (u, v), this means that we cannot have $(u, v) = (1, 0)$, since these edges cannot be saturated; therefore all the implications $u \to v$ from the implication graph must be satisfied (implying that all minimum cuts are solutions to the 2SAT instance). Furthermore, if there is a positive flow along (u, v), then we cannot have $(u, v) = (0, 1)$, i.e., the additional constraint $v \to u$ must also be satisfied. For capacitated edges $(s, \overline{x})$ (which have $s = 1$), if this edge is not saturated (the supply $\overline{x}$ is not saturated), then we cannot have $\overline{x} = 0$, i.e., we must have $x = 0$. Similarly, for capacitated edges (y, t) (which have $t = 0$), if this edge is not saturated (the demand y is not saturated), then we cannot have $y = 1$, i.e., we must have $y = 0$. These conditions characterize the minimum cuts and therefore the minimum weight solutions to the 2SAT instance. □

The link between bipartite 2SAT and max flow can also be implicitly found in Irving, Leather and Gusfield [65]. By Lemma 4.7, it can also be viewed as the well-known link between vertex cover and matching for bipartite graphs (see Harary [55]).

We have shown that the minimum weight 2SAT problem can be solved in polynomial time for bipartite instances. A related problem on bipartite instances is the *balanced* 2SAT problem. Let $G = (U \cup V, E)$ be the bipartite compatibility graph. The aim in the balanced problem is to find a solution that minimizes the maximum of the sum of the weights of true variables in U and the sum of the weights of true variables in V. In a bipartite graph $G = (U \cup V, E)$, the *balanced vertex cover problem* is the problem of finding a vertex cover S that minimizes $\max(|S \cap U|, |S \cap V|)$. As in section 4.2.1, the balanced vertex cover problem can be reduced to the balanced 2SAT problem with all weights equal to 1. Garey and Johnson [39, 69] showed that the balanced vertex cover problem is $\mathcal{NP}$-complete. This implies that the balanced 2SAT problem is also $\mathcal{NP}$-complete. On the other hand, one can use here again the algorithm of section 4.2.1 to obtain an approximation result.

Theorem 4.11 *The balanced* 2SAT *problem is* $\mathcal{NP}$-*complete but can be approximated within a factor of 2 of the optimum in* $O(m \log(w^2/m + 2))$ *time for instances with* m *clauses and explicit width* w.

Proof. The only change with respect to the proof of Theorem 4.9 is that we strengthen the statement that the weight of S is at most the value of the flow to the statement that the weight of each of $S \cap U$ and $S \cap V$ is at most half the value of the flow. The reason for this is that the flow into demand vertices in U can come only from supply vertices $\overline{x}$ with x in V by the bipartite property, so this flow is the same as the flow out of supply vertices $\overline{x}$ with x in V, which in turn is the same as the flow into demand vertices in V by symmetry. The weight of $S \cap U$ is again at most the sum of the flow into demand vertices in $S \cap U$, which is exactly half of the total flow. Therefore the weight of each of $S \cap U$, $S \cap V$ is at most the weight of any solution T, and hence at most twice the maximum of the weights of $T \cap U$ and $T \cap V$. □

4.3 Enumeration on 2SAT instances

The last section looked at the problem of finding particular good solutions to a 2SAT instance. This section examines the problem of finding *all* solutions. Given a 2SAT instance, we can run a strong components algorithm to transform it into an acyclic instance (an instance with an acyclic implication graph) in $O(m)$ time. Two literals in the same strong component of the implication graph have the same value in all solutions, and can therefore be treated as a single literal. We assume that the 2SAT instance contains no implications of the form $u \to \overline{u}$. If it does, then we may set $u = 0$ and remove all occurrences of u from the 2SAT instance. This type of implication may remain present in a hidden form, as a chain of implications such as $u \to v \to \overline{u}$; detecting this would require executing a transitive closure algorithm. Fortunately, these hidden occurrences will not be of any consequence in the solution given below.

The *maximum degree* of a 2SAT instance in this *acyclic* form is the maximum d over all literals u of the number of clauses of the form $u \to v$. If the 2SAT instance has explicit width w, then we may assume that for each path P in the corresponding path cover, and each literal u, there is at most one literal v in P such that the clause $u \to v$ is present in the 2SAT instance. The reason is that if v' follows v on the path P, then the clause $u \to v'$ can be inferred from $u \to v \to v'$, and therefore does not need to be included in the 2SAT instance. Hence $d \leq w$.

It is well-known that a solution to a 2SAT instance with m clauses on the variables $x_1, \ldots, x_n$ can be found in $O(m)$ time (see section 3.2.2). Given such a solution, we may rename all variables for convenience so that the given solution is $x_i = 0$ for all $1 \leq i \leq n$ (the all-zero solution). Once this is done, all clauses must be of the type $x_i \vee \overline{x_j}$ or of the type $\overline{x_i} \vee \overline{x_j}$. The clauses of the first type can be viewed as implications $\overline{x_i} \to \overline{x_j}$. These implications form an acyclic graph, so we can perform a topological sort (in $O(m)$ time) and rename the variables to ensure that $i < j$ for all such clauses. The clauses of the second type are symmetric in x_i and x_j and can therefore always be written as implications $x_i \to \overline{x_j}$ with $i < j$.

We therefore assume that the 2SAT instance is given by implications of the form $\overline{x_i} \to \overline{x_j}$ or $x_i \to \overline{x_j}$ with $i < j$. We now observe the following property: Given a solution $x = x_1 x_2 \ldots x_n$ with at least one $x_i = 1$, if we change the last such x_i (i.e., $x_i = 1$ and $x_j = 0$ for all $j > i$) to $x_i = 0$, then we obtain another solution. This property holds because if the new assignment violates some clause, it must be a clause involving x_i which forbids $x_i = 0$. This can only be a clause $\overline{x_i} \to \overline{x_j}$ with $x_j = 1$ and $j > i$, contradicting the assumption that x_i whas the last variable equal to 1 in the given solution.

We refer to the solution obtained from a solution x by changing the last $x_i = 1$ to $x_i = 0$ as the *parent* of the given solution x. The solutions thus form a tree rooted at the known all-zero solution. We shall not build this solution tree explicitly; however, the execution of the recursive enumeration algorithm below shall correspond to a depth-first search started at the root of the tree.

Given a current solution x, the *index* is the largest l such that $x_l = 1$. By convention, the index is 0 if x is the all-zero solution. Note that the children of x in the solution tree are the solutions that can be obtained from x by setting $x_j = 1$ for a single j greater than the index of x. We say that an implication (either $\overline{x_i} \to \overline{x_j}$ or $x_i \to \overline{x_j}$ with $i < j$) is *active* if the antecedent (either $\overline{x_i}$ or x_i) has value 1 in the current solution x. We say that a variable x_j is *active* if $\overline{x_j}$ is *not* the consequent of any active implication.

Lemma 4.12 *The children of a solution x are precisely the solutions obtained from x by setting $x_j = 1$ for a single active x_j with j greater than the index of x.*

Proof. The fact that x_j must be active is clear: if not, then $\overline{x_j}$ is the consequent of an active implication whose antecedent equals 1, and setting $x_j = 1$ violates this implication. On the other hand, it x_j is active, then all implications with consequent $\overline{x_j}$ are inactive (antecedent equal to 0), and setting $x_j = 1$ does not violate these implications. No implication of the form $x_j \to \overline{x_k}$ can be violated either, because such an implication has $k > j$ and therefore $x_k = 0$ by the definition of index. □

Therefore, in order to determine the children of x, it is sufficient to maintain a list of the active variables x_j, ordered by the value of j. When we set $x_i = 1$ for some i, the implications of the form $x_i \to \overline{x_j}$ become active and those of the form $\overline{x_i} \to \overline{x_j}$ become inactive (the opposite happens if we set $x_i = 0$). We can maintain a count of the number of active implications that have $\overline{x_j}$ as a consequent, for each x_j. When this count becomes zero, the variable x_j becomes active and is added to the active list. When the count becomes nonzero, the variable x_j becomes inactive and is removed from the active list.

The algorithm maintains globally the current solution x, the list of active variables L, and a count for each x_j as explained above. Initially, x is the all-zero solution, and L and the counts are set appropriately. The algorithm starts with a call to *enumerate*$(0,0)$. (Ignore for now the output calls and the depth argument.) The two assignments to x_i within *enumerate* are meaningless in the top level call which has index=0 by convention, and they can be ignored in that special case.

procedure *enumerate*(index, depth);
begin
    set $i =$ index and $x_i = 1$;
    if depth is even then output the current x;
    update the active implications with antecedent x_i or $\overline{x_i}$;
    determine those x_j that have just become active or have just become inactive;
    remove from the active list L the x_j that have just become inactive, but
        remember their position in L (their predecessor in L) in an auxiliary local list N;
    let M be the ordered list of x_j that have become active;
    merge M into the ordered list L by traversing both lists in reverse order;
        as each variable x_j is inserted in the merged list, call *enumerate*$(j, \text{depth} + 1)$;
        stop when $j = i$ has been reached;
    if depth is odd then output current x;
    set $x_i = 0$, and restore the active list by removing the x_j from M and
        adding those from N;
    update the active implications with antecedent x_i or $\overline{x_i}$;
end;

The above procedure can be implemented so that it takes $O(d)$ time at the beginning and at the end of each call, plus an additional $O(1)$ time in between recursive calls (ignore for now the output calls). To see this, note that the number of active implications to be updated in the beginning is at most $2d$; the corresponding count update tells us then which variables should become active or inactive (again at most $2d$). The variables that become inactive can be removed from L in $O(1)$ time per variable if L is maintained as a doubly linked list. Therefore the operations before the merge take $O(d)$ time. As we merge the two lists, we do a recursive call for each element in the merged list, therefore spending $O(1)$ time between recursive calls. Since the merge is done starting from the end of the lists, the merged L is always correct from x_j on, and this is the only portion of L that is used inside the recursive call. The operations after the merge take again $O(d)$ time.

The reason for outputting the even depth solutions at entry time, and the odd depth solutions at exit time, is that after a solution x is output, the next solution to be output will be at most the third solution visited after x. This can be easily verified: Suppose that an even depth node e_1 has just been output. If e_1 has a child o_1, then either o_1 has no children and is immediately output, or o_1 has a child e_2 that is immediately output. If e_1 has no children, then we return to the parent o_2 of e_1, and again either o_2 has no children following e_1 and is immediately output, or its next child e_2 is immediately output. The other case is that of an odd depth node o_1 which has just been output. Then the algorithm returns to the parent e_1 of o_1. If e_1 has a next child o_2, then either o_2 has no children and is immediately output, or it has a child e_2 that is immediately output. If e_1 has no next child, then the algorithm returns to the parent o_3 of e_1, which again has either no next child and is immediately output, or has a child e_3 which is immediately output. Thus in all cases, the next node to be output is one of the next three nodes to be visited.

Since the algorithm spends $O(d)$ time at each node before moving to an adjacent node, the time between outputs is $O(d)$. Since adjacent solutions in the search tree differ in only one bit x_i, solutions that are consecutively output differ in at most three bits, and it is sufficient to output the values of the three bits that have changed (in constant time). The space used at each node is proportional to the number of clauses involving x_i for the current index i, and is therefore $O(m)$ overall. The information maintained globally also takes $O(m)$ space.

Theorem 4.13 *The solutions to a 2SAT instance with m clauses and maximum degree d can be enumerated after $O(m)$ pre-processing time in $O(d)$ on-line time per solution, using $O(m)$ space.*

4.4 Discussion

We have presented two algorithms for the uncapacitated maximum flow problem. The $O(m\sqrt{K})$ algorithm owes its efficiency to an implicit transitive closure representation via dynamic trees. The gradual reduction of the dynamic tree size ensures that the associated logarithmic cost averages out to a constant over the entire execution of the algorithm. The time bound currently applies up to values of K that are slightly larger than $(m/\log^2 m)^2$.

CHAPTER 4. OPTIMIZATION AND ENUMERATION

An open problem is to extend the bound to all $K \leq w^2$. This would yield, in combination with the second algorithm, a stronger $O(\min(\sqrt{K}, w)m \log(K/w^2 + 2))$ bound. The second algorithm runs in $O(wm \log K)$ time. This complexity is achieved by combining capacity scaling with the notion of width. An open question is whether the dependency on K can be significantly reduced (as in Ahuja, Orlin and Tarjan [2, 41] for example) or removed (as in the capacity scaling min cost flow algorithm of Orlin [89]) without increasing the main wm factor.

For the 2SAT problem, besides the optimization results that follow from the flow approach, we have studied the problems of recognizing partial solutions and of enumerating all solutions. The latter has an $O(m + dS)$ time complexity if the total number of solutions is S. The question of whether counting (a #$\mathcal{P}$-complete problem even in the bipartite case, see Provan and Ball [95]) is easier than enumerating remains open. The fact that consecutive solutions found by the algorithm differ in only a constant number of bits suggests that faster algorithms might be achievable.

The common element in the various algorithms is the use of the width of a graph, showing that combinatorial problems are sometimes easier in skinny graphs than in more general graphs. These graphs will be shown to arise naturally in the context of stable matching. The approach may well be applicable to other graph problems whose structure is the superposition of a simple collection of paths with a more complex structure.

Chapter 5

Stable Matching

In a stable matching problem, the aim is to match a set of people in pairs, subject to certain preference constraints. This problem was introduced by Gale and Shapley [37], who gave a solution for the marriage version of the problem. In this version, the people are men and women, and pairs must consist of a man and a woman. The more general roommates version, where arbitrary pairs of people are allowed, was solved by Irving [62]. Early work on this problem concentrated on the question of finding a single solution; since then, considerable attention has been given to the structure of the set of all solutions, and to algorithms that exploit this structure to look for solutions satisfying certain additional constraints. An introductory treatment of stable matching appears in Knuth [75], and in Polya, Tarjan, and Woods [94] A comprehensive treatment of structural and algorithmic aspects of stable matching may be found in Gusfield and Irving [51]. A study of stable matching from a game-theoretic point of view is presented in Roth and Sotomayor [102].

Subramanian [108] found a representation for stable matching as a network stability problem on networks of scatter-free gates, and used this representation to provide a new approach to stable matching. In this chapter, we present Subramanian's reduction to network stability in the framework of a more general problem, the stable arrangement problem. We then use the structural and algorithmic results from the previous three chapters to study the structure and complexity of stable matching.

5.1 The Stable Arrangement Problem

We begin by introducing a new problem, the stable arrangement problem. In the stable arrangement problem, the aim is to assign a set of people to tasks, so that certain preference constraints are satisfied. The stable matching problem is obtained from the stable arrangement problem by requiring that each task be listed by exactly two people.

People	Tasks
p	s t u v
q	u s t
r	v t

Table 5.1: The preference lists in a stable arrangements instance.

5.1.1 Stable Arrangements

An instance of the *stable arrangement* problem is defined on a set of *people* P and a set of *tasks* T. Each person $p \in P$ has a set of candidate tasks $T(p) \subseteq T$, together with a linear ordering $<_p$ on $T(p)$. The pair $(T(p), <_p)$ is the *preference list* of person p: If $s <_p t$, then we say that person p *prefers* task s to task t. We also introduce a special symbol $\bot$ and postulate $t <_p \bot$ for all $t \in T(p)$, thus extending the linear order $<_p$ to $T(p) \cup \{\bot\}$. We denote by t_{pi} the ith task in the preference list of person p, for $1 \leq i \leq |T(p)|$. Two characteristic parameters of an instance are the number of people $n = |P|$ and the *size* $m = n + \sum_{p \in P} |T(p)|$. Table 5.1 gives an instance on 3 people of size 12.

An *arrangement* is a mapping $\alpha : P \to T \cup \{\bot\}$ such that $\alpha(p) \in T(p) \cup \{\bot\}$ for all $p \in P$. If $\alpha(p) = t \in T(p)$ then we say that person p is *assigned* task t; if $\alpha(p) = \bot$ then person p is *unassigned*. An arrangement α is *stable* if it satisfies the following two conditions:

(a) *Partnership condition.* For every $p \in P$ with $\alpha(p) \in T$ there is some $q \in P$ with $q \neq p$ that satisfies $\alpha(q) = \alpha(p)$. That is, every person assigned to a task has at least one partner assigned to the same task.

(b) *Preference condition.* If $p \in P$ and $t \in T(p)$ with $t <_p \alpha(p)$, then every $q \in P$ with $q \neq p$ and $t \in T(q)$ satisfies $\alpha(q) <_q t$. That is, if a person prefers a certain task t to the proposed assignment, then all candidate partners have been assigned tasks that they prefer over t.

Intuitively, if the preference condition were violated, then we could improve the condition of person p while improving or leaving unchanged the condition of person q by making p and q partners in task t. (This might, however, cause a violation of the partnership condition for their partners.) Given an arrangement α, a *preference instability* is a pair (p, t) that violates the preference condition. A *partnership instability* is a pair $(p, \alpha(p))$ that violates the partnership condition with $\alpha(q) <_q \alpha(p)$ for all $q \neq p$ such that $\alpha(p) \in T(q)$. In words, all candidate partners for p are assigned tasks that they prefer over $\alpha(p)$. This does not cover all the possible violations of the partnership condition, because a violation can have $\alpha(p) <_q \alpha(q)$ for some $q \neq p$ with $\alpha(p) \in T(q)$ as well; that case, however, can be viewed as a violation of the preference condition with preference instability $(q, \alpha(p))$ with respect to p. Therefore α is stable if and only if it has no preference or partnership instabilities.

As an example, in the instance of Table 5.1, the arrangement $\alpha(p) = \alpha(q) = s$, $\alpha(r) = \perp$ is stable, while the arrangement $\beta(p) = \beta(q) = \beta(r) = t$ is unstable, with two preference instabilities (p,s) (with respect to q) and (q,s) (with respect to p).

The *stable arrangement* problem is the problem of finding a stable arrangement. The *stable matching* problem is a special case of the stable arrangement problem in which every task is listed by exactly two people, i.e., if $t \in T$ then $|\{p \in P : t \in T(p)\}| = 2$. In the *bipartite stable matching* problem, the set of people P is the disjoint union of a set of men M and a set of women W, and the two people that list a task must be a man and a woman, i.e., if $t \in T$ then $|\{p \in M : t \in T(p)\}| = |\{q \in W : t \in T(q)\}| = 1$. The *stable roommates* problem is a stable matching problem in which every pair of people lists at most one common task, i.e., if $p \neq q$ then $|T(p) \cap T(q)| \leq 1$. Note that in this case each task $t \in T(p)$ is uniquely determined by the partner $q \neq p$ with $t \in T(q)$, so the preference lists can be viewed as lists of candidate partners rather than lists of tasks. The *stable marriage* problem is a bipartite stable matching problem that is also a stable roommates problem. We say that a stable roommates problem has *complete preference lists* if $|T(p) \cap T(q)| = 1$ for all $p \neq q$; a stable marriage problem has complete preference lists if $|T(p) \cap T(q)| = 1$ for all $p \in M$ and $q \in W$.

This defines seven variations on the stable arrangement problem. The last four special cases, namely stable roommates and stable marriage with or without complete preference lists, are versions that have been studied in the literature. As we shall see, the model accomodates most naturally the first three versions, namely stable arrangements and bipartite or nonbipartite stable matching, where people are allowed to list more than one common task.

5.1.2 Reduction to Network Stability

An arrangement α can be described by an assignment x with boolean coordinates x_{pi} for all $p \in P$ and $0 \leq i \leq |T(p)|$ (the coordinates names pi stand for pairs (p,i), for brevity). A component x_{pi} of x equals 1 if and only if person p is not assigned any of the top i choices in $T(p)$. Thus $x_{p0} = 1$, and $x_{pi} = 1$ if and only if $t_{pi} <_p \alpha(p)$, for $1 \leq i \leq |T(p)|$, so that $x_{pi} \leq x_{p(i-1)}$. Conversely, if an assignment x satisfies $x_{p0} = 1$ for all $p \in P$, and satisfies the *consistency condition* $x_{pi} \leq x_{p(i-1)}$ for all $p \in P$ and $1 \leq i \leq |T(p)|$, then x describes an arrangement α, where $\alpha(p) = t_{pi}$ if i is the least value such that $x_{pi} = 0$, and $\alpha(p) = \perp$ if no such i exists.

Given $t \in T$, we let P_t be the set of all pairs (p,i) such that $t_{pi} = t$. Therefore the set P_t describes the people who list task t and the preference that they assign to t. We can use the sets P_t to state the stability conditions on an arrangement α as conditions on the corresponding assignment x. The condition stating that (p,t) must not be a partnership instability for α, where $t = t_{pi}$, is expressed in terms of x as follows: If $x_{pi} = 0$ and $x_{p(i-1)} = 1$, then $x_{q(j-1)} = 1$ for some $(q,j) \in P_t \setminus \{(p,i)\}$. This condition can be written as a single inequality $x_{pi} \geq x_{p(i-1)} \wedge (\bigwedge_{(q,j) \in P_t \setminus \{(p,i)\}} \overline{x_{q(j-1)}})$. The condition stating that (p,t) is not a

CHAPTER 5. STABLE MATCHING

preference instability for α, where $t = t_{pi}$, can also be expressed in terms of x: If $x_{pi} = 1$, then $x_{q(j-1)} = 0$ for all $(q,j) \in P_t \setminus \{(p,i)\}$. This, in conjunction with the consistency condition, can be written as a single inequality $x_{pi} \leq x_{p(i-1)} \wedge (\bigwedge_{(q,j) \in P_t \setminus \{(p,i)\}} \overline{x_{q(j-1)}})$. The combination of the partnership, preference and consistency conditions is therefore expressed by a single equation

$$x_{pi} = x_{p(i-1)} \wedge (\bigwedge_{(q,j) \in P_t \setminus \{(p,i)\}} \overline{x_{q(j-1)}})$$

for each choice of p and $t = t_{pi}$. If $P_t = \{(p_1, i_1), (p_2, i_2), \ldots, (p_k, i_k)\}$, then all the equations of this form involving a given task t can be written as a single equation using the X_k gate (see section 2.4), namely

$$x_{p_1, i_1} \ldots x_{p_k, i_k} = X_k(x_{p_1, (i_1-1)} \ldots x_{p_k, (i_k-1)}).$$

Summarizing, we have the following:

Lemma 5.1 *A assignment x describes an arrangement if and only if $x_{p0} = 1$ and $x_{pi} \leq x_{p(i-1)}$ for all $p \in P$ and $1 \leq i \leq |T(p)|$. A assignment x describes a stable arrangement if and only if $x_{p0} = 1$ for all $p \in P$ and $x_{p_1, i_1} \ldots x_{p_k, i_k} = X_k(x_{p_1, i_1-1} \ldots x_{p_k, i_k-1})$ for all $t \in T$, where the pairs (p_r, i_r) are the elements of P_t.*

We define now a network N with m coordinates pi, one for each $p \in P$ and each $0 \leq i \leq |T(p)|$. These coordinates are partitioned into $I(N) = \{pi : i = 0\}$, $O(N) = \{pi : i = |T(p)|\}$, and $L(N) = \{pi : 0 < i < |T(p)|\}$. The network N consists of $|T|$ gates g^t, one for each task $t \in T$. Each of these gates is an X_k gate, where $k = |P_t|$, and is defined by $I(g^t) = \{p(i-1) : (p,i) \in P_t\}$, $O(g^t) = \{pi : (p,i) \in P_t\}$, and $g^t(x) = y$ if $X_k(x_{p_1, i_1-1} \ldots x_{p_k, i_k-1}) = y_{p_1, i_1} \ldots y_{p_k, i_k}$, where the (p_r, i_r) are the elements of P_t.

If we now choose the input assignment for the network defined by $x_{p0} = 1$ for all $p \in P$, then the stable configurations for the network that are consistent with this input assignment are the configurations x that satisfy in addition the constraints $g^t(x_{I(g^t)}) = x_{O(g^t)}$ for each $t \in T$, and these are precisely the conditions stated in the lemma. This gives:

Theorem 5.2 *An instance of the stable arrangement problem can be viewed as a network with a given input assignment over a set of multivariate X gates. The stable arrangements correspond to the stable configurations of the network that are consistent with the input assignment.*

The network is illustrated in Fig. 5.1. Note that the coordinates $q0, q1, q2, \ldots$ for a person q form a path going through the X_k gates for the tasks listed by q. If an input coordinate $p0$ is assigned the value $x_{p0} = 0$ instead, then the definition of X_k gates implies that all coordinates pi will have value 0 in all stable configurations, for $0 \leq i \leq |T(p)|$. The stable configurations are then those of a stable arrangement problem in which p is simply

CHAPTER 5. STABLE MATCHING

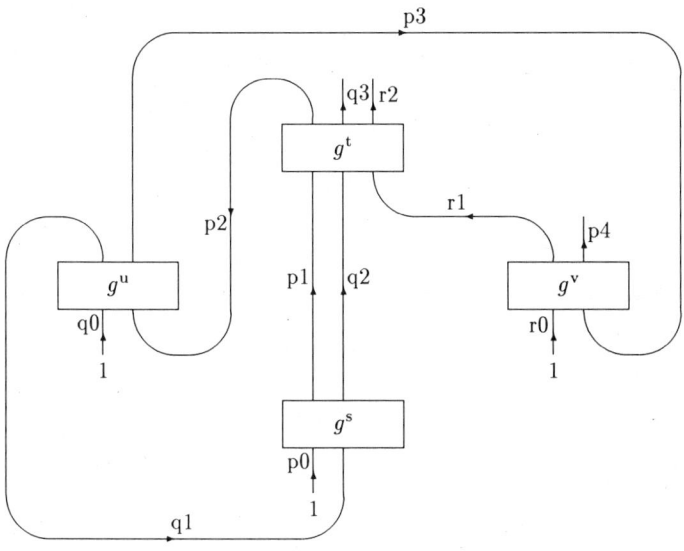

Figure 5.1: Network representation of a stable arrangement instance.

removed from the set of people; we might say that p declines to play the game. Note that the output value x_{pi} for $i = |T(p)|$ has value 1 if and only if p is playing the game and assigned no tasks, i.e., when $\alpha(p) = \bot$.

If a task t is listed by only one person so that $|P_t| = 1$, then the gate g^t is simply an $X_1 =$ ID gate; therefore t can be removed from the stable arrangements instance without affecting the stable configurations. In a stable matching problem, all tasks t have $k = |P_t| = 2$, so the network uses only regular (bivariate) X gates, with equations $y_{pi}y_{qj} = X(x_{p(i-1)}x_{q(j-1)})$ for $t \in T$ and $P_t = \{(p,i),(q,j)\}$. In a bipartite stable matching problem, these equations can be written using comparators as $y_{pi}\overline{y_{qj}} = C(x_{p(i-1)}\overline{x_{q(j-1)}})$, where $p \in M$ and $q \in W$. If each coordinate x_{qj} with $q \in W$ is replaced by its negation, then the resulting network uses comparators alone. This gives the following result of Subramanian [108]:

Corollary 5.3 *An instance of the stable matching problem can be viewed as a network over* $\{X\}$. *An instance of the bipartite stable matching problem can be viewed as a network over* $\{C\}$.

5.1.3 Instabilities and Local Search

The representation of the stable arrangement problem as a network stability problem makes it possible to give efficient algorithms for solving it, and we shall explore efficient algorithms for several related questions in section 5.2. In particular, the stable arrangment problem can be solved in linear time using the scatter-free algorithm of Mayr and Subramanian (see Theorem 3.17), or by a generalization of the stable matching algorithm of Irving [62]. In this section we present a less efficient stability algorithm which is still interesting because it illustrates the role of instabilities and the *local search* paradigm. A model for the study of local search problems was proposed by Johnson, Papadimitriou, and Yannakakis [70]. See also Papadimitriou, Schäffer, and Yannakakis [91]. In this model, we are given a set of candidate solutions to a problem; an objective function can be evaluated at each candidate solution in polynomial time. Furthermore, each candidate solution has a set of neighbor solutions, and there is a polynomial time algorithm that provides a neighbor with a better objective function value given a candidate solution, if such a neighbor exists. If no such neighbor exists, the candidate solution is a local optimum. The goal is to find such a local optimum.

The following neighborhood structure was proposed by Knuth [75] for the stable marriage problem with complete preference lists. The candidate solutions are the arrangements that satisfy the partnership condition. A neighbor of an arrangement α is obtained by selecting a pair of people p, q that violates the preference constraint, and replacing the pairs of matched people $(p, \alpha(p))$ and $(\alpha(q), q)$ with the pairs (p, q) and $(\alpha(q), \alpha(p))$. This represents an improvement from the point of view of p and q, but not necessarily from the point of view of $\alpha(p)$ and $\alpha(q)$. In the absence of an associated objective function, it is unclear whether a sequence of such interchanges will eventually lead to a stable solution. This question was settled negatively by Tamura [110], who exhibited stable marriage instances

CHAPTER 5. STABLE MATCHING

for which no sequence of interchanges from a given candidate solution will lead to a stable solution (although a stable solution always exists).

The reason for this failure can be attributed in part to the requirement that the partnership condition be maintained, which makes it necessary to match people that are left unmatched even if the two people dislike each other compared to their previous partners. We shall remove this condition, and obtain a simple algorithm for the general arrangement problem.

The set of candidate solutions is now the set of arrangements, which may violate both the partnership and the preference conditions. Each arrangement α has one neighbor for each instability (p, t) in it. If the instability is a partnership instability, then the neighbor is obtained by performing an *advance* operation for p which consists of replacing the task $\alpha(p)$ assigned to p with the item following $\alpha(p)$ in the preference list of p. If the instability is a preference instability, then the neighbor is obtained by performing a *retreat* operation for p which consists of replacing the item $\alpha(p)$ assigned to p with the task preceding $\alpha(p)$ in the preference list of p.

This neighborhood structure does not have an obvious objective function associated with it. However, we shall see that if we choose an instability at random and perform an advance or retreat operation accordingly, then the probability of reducing the Hamming distance to a stable configuration in the associated network is quite high. Therefore the Hamming distance can be used as a 'hidden objective function'. The *randomized local search* algorithm is then the following. Starting from an arbitrary arrangement α, choose an instability (p, t) uniformly at random from all the instabilities for α, then perform an advance or a retreat operation for p depending on whether (p, t) is a partnership or a preference instability, and repeat the process for the new α. The algorithm is intuitively appealing: If all candidate partners for p at task $\alpha(p)$ have assigned tasks that they prefer over $\alpha(p)$, then p should try the next item in his preference list; If there are candidate partners for p that would like p as a partner in a task t occurring before $\alpha(p)$ in the list of p, then p should retreat in the preference list $T(p) \cup \{\bot\}$, and especially so if there are many tasks t occurring before $\alpha(p)$ that provide a (p, t) preference instability.

Theorem 5.4 *The randomized local search algorithm finds a stable arrangement, if one exists, in expected $O(m^2)$ time, for instances of the stable arrangement problem of size m.*

Proof. Transform the network from Theorem 5.2 into a network with no inputs and no outputs, by assigning an absorption gate to each output and a constant gate that sends the value 1 to each input. Now the transition function of the network coincides with the internal function f. Let x be the configuration corresponding to a given arrangement α. If $f(x) = y$, then the value y_{pi} for $1 \leq i \leq |T(p)|$ is an output of an X_k gate. More precisely, if $t_{pi} = t$, then $y_{pi} = x_{p(i-1)} \wedge (\bigwedge_{(q,j) \in P_t \setminus \{(p,i)\}} \overline{x_{q(j-1)}})$. In the last section, we used this expression to show that $x_{pi} < y_{pi}$ if and only if (p, t) is a partnership instability, and $x_{pi} > y_{pi}$ if and only if (p, t) is a preference instability (we cannot have a violation of the consistency condition because all the configurations x considered correspond to arrangements). We are now in the same situation as for the randomized algorithm of section 3.1.6. That is, if a is the configuration corresponding to a stable arrangement, then $d(y, a) \le d(x, a)$ and so a random choice of an instability (p, t_{pi}), corresponding to a random choice of a coordinate where $x_{pi} \ne y_{pi}$, will have the property that $x_{pi} \ne a_{pi}$ with probability at least 1/2. Suppose that $x_{pi} \ne a_{pi}$. If $x_{pi} < y_{pi}$, then we have a partnership instability (p, t_{pi}), where $t_{pi} = \alpha(p)$, and the advance operation for p changes $x_{pi} = 0$ to $x_{pi} = 1$, thus decreasing $d(x, a)$. If $x_{pi} > y_{pi}$, then $x_{pi} = 1$ and $a_{pi} = 0$; if we consider the largest i' such that $x_{pi'} = 1$, then $a_{pi'} = 0$ as well by the consistency condition. Since (p, t_{pi}) is in this case a preference instability, we perform a retreat operation for p, setting $x_{pi'} = 0$ and thus decreasing $d(x, a)$ as well.

Therefore $d(x, a)$ changes by 1 at each iteration, and it decreases by 1 with probability at least 1/2. Under these assumptions, the analysis of the randomized algorithm of section 3.1.6 shows that a stable arrangement will be found in expected $O(m^2)$ time. □

This proof also shows that the number of instabilities is always $d(x, f(x))$. Unlike the algorithm of section 3.1.6, this algorithm does not always choose a coordinate where x and $f(x)$ differ (in the case $x_{pi} > y_{pi}$, when we perform a retreat operation, we change $x_{pi'}$, rather than x_{pi}, from 1 to 0, so if we had $y_{pi'} = 1$ then $d(x, y)$ would increase). Therefore the number of instabilities $d(x, f(x))$ may increase; only the probability of decreasing $d(x, a)$ for a stable configuration a is measured in the analysis.

5.1.4 Structural Properties

Several properties of the stable arrangement problem, some of them well-known, follow from its formulation as a nonexpansive network (in fact, a scatter-free network). We begin with a property proved by Irving and Leather [64].

Lemma 5.5 *Given a stable matching instance, suppose that persons p and q are assigned the same task t in some stable arrangement α. Then there is no stable arrangement β in which both p and q are assigned a task better than t.*

Proof. If x is the stable configuration corresponding to α in the associated network, and $t = t_{pi} = t_{qj}$, then the gate g^t has inputs $x_{p(i-1)} = x_{q(j-1)} = 1$ and outputs $x_{pi} = x_{qj} = 0$. A stable configuration y corresponding to β, on the other hand, would give inputs $y_{p(i-1)} = y_{q(j-1)} = 0$ and outputs $y_{pi} = y_{qj} = 0$. This contradicts the fact that the outputs of a gate uniquely determine the inputs of the gate over the set of stable configurations (Lemma 3.3). □

We give next a proof due to Subramanian [108] of the following theorem from Gale and Sotomayor [38].

Lemma 5.6 *In a stable arrangement problem, the set of assigned people is the same in all stable arrangements.*

CHAPTER 5. STABLE MATCHING

Proof. A person $p \in P$ is unassigned in an arrangement α if $\alpha(p) = \perp$, and this happens precisely when $x_{pi} = 1$ for $i = |T(p)|$, where x is the configuration of the network associated with the arrangement α. The coordinates pi with $i = |T(p)|$ are the outputs of the network, and since nonexpansive networks are convergent, the outputs must have the same value in all stable configurations. □

Recall the interpretation of a 1 input as indicating that the person is participating, and the interpretation of a 1 output as indicating that the person is unassigned. In a bipartite matching instance, when the network is made monotone, the variables x_{qj} for $q \in W$ are negated. Since monotone nonexpansive networks converge to monotone nonexpansive gates, this means that if a man changes his mind and decides to play the game by changing his input from 0 to 1, this can only affect who remains unassigned for at most one person. If the affected person is a man, then the man must become unassigned, changing his output from 0 to 1; if the affected person is a woman, then the woman must become assigned, changing her output from 0 to 1. Suppose now that we have a bipartite matching instance, where all participants are present except for four particular men m_1, m_2, m_3, m_4. Suppose also that some particular woman w_1 is currently unassigned. Is it possible that w_1 will become assigned if at least one of the three pairs of men $\{m_1, m_2\}$, $\{m_2, m_3\}$, or $\{m_3, m_4\}$ change their minds and decide to play the game, but in no other case? It turns out that this particular scenario cannot arise in any bipartite stable matching instance. If it did, then the output corresponding to w_1 would be given as a function of the inputs for the four men m_1, m_2, m_3, m_4 by $g(x_1 x_2 x_3 x_4) = (x_1 \wedge x_2) \vee (x_2 \wedge x_3) \vee (x_3 \wedge x_4)$. This means that the gate g would be the gate to which some comparator network converges. As mentioned in section 3.4.5, we have established that this is not possible, by running a large linear program. We lack however an intuitive understanding of why this is the case, since other similar-looking functions turn out to have a representation by means of comparator networks (and hence by stable marriage problems, see section 5.1.5).

Given three arrangements α, β and γ, we define their *median* $\text{med}(\alpha, \beta, \gamma)$ as the arrangement δ such that $\delta(p)$ is the item out of the three items $\alpha(p), \beta(p), \gamma(p)$ that holds an intermediate position in the preference list of person p. Thus $\delta(p)$ is obtained from these three items by discarding the one that occurs earliest and the one that occurs latest in the preference list of person p, and keeping the remaining task. It is a priori unclear whether for three stable arrangements α, β and γ, the median δ will satisfy either the partnership condition or the preference condition.

Lemma 5.7 *In a stable arrangement problem, the median of three stable arrangements is a stable arrangement.*

Proof. Consider the configurations x, y, z and w in the network corresponding to the arrangements α, β, γ and δ. From the definition of the median arrangement, we have $w_{pi} = 1$ for some $p \in P$ and $0 \leq i \leq |T(p)|$ if and only if at least two of the three values x_{pi}, y_{pi} and z_{pi} equal 1. That is, $\delta(p)$ is not among the i preferred tasks for p if and only if at least two of the three choices $\alpha(p), \beta(p)$ and $\gamma(p)$ are not among the i preferred tasks for p. By the definition of median in the hypercube, we have $w = \text{med}(x, y, z)$, and w must be a stable configuration as the median of three stable configurations. □

In the bipartite stable matching problem, the associated network is monotone since it only uses comparators. This means that there will always be a stable configuration, and in fact a zero-most stable configuration x^0 and a one-most stable configuration x^1. The fact that x^0 is zero-most indicates that every man m has as many x_{mi} equal to 0 as possible over all stable configurations, and thus m has his best possible assignment; for a woman w, since the variables x_{wi} were complemented to make the network monotone, the woman w has as many x_{wi} equal to 1 as possible over all stable configurations, and thus w has her worst possible assignment. Thus x^0 corresponds to a male-optimal/female-pessimal stable arrangment α^0. Similarly, x^1 corresponds to a female-optimal/male-pessimal stable arrangement α^1. These arrangements were shown to exist by Gale and Shapley when they introduced the subject [37]. If we denote by $\alpha \wedge \beta$ the arrangement that gives to every man m the best of the two choices $\alpha(m)$ and $\beta(m)$ in his preference list, and to every woman w the worst of the two choices $\alpha(w)$ and $\beta(w)$ in her preference list, then $\alpha \wedge \beta = \text{med}(\alpha, \beta, \alpha^0)$. Similarly, the arrangement $\alpha \vee \beta = \text{med}(\alpha, \beta, \alpha^1)$ gives to every woman w the best of the two choices $\alpha(w)$ and $\beta(w)$, and to every man m the worst of the two choices $\alpha(m)$ and $\beta(m)$. The lemma then gives as a special case the following well-known result of Conway [75]:

Corollary 5.8 *The set of stable arrangments in a bipartite stable matching problem forms a distributive lattice, with a man-optimal stable arrangement α^0, a woman-optimal stable arrangement α^1, and lattice operations $\alpha \wedge \beta$, $\alpha \vee \beta$ that give the better of the two choices from α and β to every man or to every woman respectively.*

The median structure gives in turn a 2SAT characterization for stable arrangements in terms of the boolean variables x_{pi}. In the bipartite stable matching problem, the network is monotone, so if we discard the variables x_{pi} in which all stable arrangements agree (i.e., those in which the man-optimal and the woman-optimal arrangements agree), the 2SAT instance for the remaining variables will be bipartite. In a monotone network, all the positive literals are on the same side of the bipartite compatibility graph of the 2SAT instance. In our case, since the variables for women have been complemented, the literals x_{mi} for $m \in M$ will be in one side together with the literals $\overline{x_{wi}}$ for $w \in W$, and the literals x_{wi} for $w \in W$ will be in the other side together with the literals $\overline{x_{mi}}$ for $m \in M$.

Lemma 5.9 *The stable arrangements in an arrangement problem are characterized by a 2SAT instance on the variables x_{pi} that indicate whether person p is assigned one of the first i choices in $T(p)$. In a bipartite stable matching problem, once the variables x_{pi} that have the same value in the male-optimal and female-optimal solutions have been discarded, the stable arrangements are characterized by a bipartite 2SAT instance on the remaining variables, with the variables x_{mi} and x_{wi} for men and women respectively on different sides of the bipartite compatibility graph.*

The structure of the stable roommates and stable marriage problem has been described in the literature in terms of a certain combinatorial object known as a partial order on a set of dual rotations. See Gusfield and Irving [62, 49, 51]. Without going into the algorithmic aspects of this combinatorial structure, we give a sketch of how it can be interpreted in the context of the stable networks approach. Given a network associated with a stable arrangement problem, there is an isomorphism g_σ on the space of internal assignments that coincides with the internal function of the network in the periodic set. The cycles of σ correspond to cycles in the underlying graph of the network. A cycle C of σ enters a gate g^t corresponding to task $t \in T$ along a coordinate $p(i-1)$ such that $(p,i) \in P_t$, and leaves g^t along a coordinate qj with $(q,j) \in P_t$ and $\sigma(qj) = p(i-1)$ or $\sigma(qj) = \overline{p(i-1)}$. The coordinates in a cycle C are equivalent variables over the set of stable configurations x. We can assume that $\sigma(qj) = p(i-1)$ if $p = q$ and $\sigma(qj) = \overline{p(i-1)}$ if $p \neq q$, that is, the negations along C occur precisely when switching from a person p to a different person q. Otherwise, if $\sigma(qj) = \overline{p(i-1)}$ with $p = q$ (and $i = j$), then no stable configuration x can have $x_{qj} = 1$ because then $x_{pi} = 1$ and $x_{p(i-1)} = 0$, contrary to the definition of the X_k gate; therefore $x_{qj} = 0$ for all the stable configurations x and we can assign constants to all the coordinates on C. Similarly, if $\sigma(qj) = p(i-1)$ with $p \neq q$, then no stable configuration x can have $x_{qj} = 1$ because then $x_{p(i-1)} = 1$, contrary to the definition of the X_k gate; therefore $x_{qj} = 0$ and we can assign constants to all the coordinates on C. As a result, the cycles C that are not replaced by constants have indeed the property that the negations along C occur when switching from some p to some $q \neq p$.

Let $(p_1, p_2, \ldots, p_r)$ be the cyclic sequence of people that occur as coordinates pi along cycle C. In a stable roommates or stable marriage problem, this cyclic sequence determines uniquely the cycle C: A segment p_{j-1}, p_j, p_{j+1} in this sequence corresponds to a segment of coordinates $p_j i$ on C such that the ith entry in the preference list of p_j is between p_{j-1} and p_{j+1} in that list (including p_{j-1} but excluding p_{j+1}). In a stable arrangement, the coordinates along C are equivalent, so all the coordinates involving a person p_j have the same value, and those involving p_{j-1} and p_{j+1} have the opposite value because of negations. The instance has a stable arrangement if and only if all these cycles are stabilizing cycles, and this happens when r is even, because switching from p_j to p_{j+1} introduces a negation. (In the marriage case, r is always even because the cyclic sequence of people alternates between men and women.) Say that a stable arrangement pairs p_j up with p_{j-1} if the coordinates on the chosen segment corresponding to p_j have value 0, and pairs p_j up with p_{j+1} if the coordinates for p_j have value 1. Then a pair (p_{j-1}, p_j) indicates that p_j has partner p_{j-1} or better, while a pair (p_j, p_{j+1}) indicates that p_j has partner p_{j+1} or worse. The two possible ways of pairing up a cyclic sequence are known in the literature as *dual rotations*. When the equivalent variables on each cycle C are combined into a single variable z_C, the 2SAT instance characterizing the stable arrangements becomes acyclic. The acyclic implication graph can then be viewed as a *partial order* on dual rotations, with the occurrences of z_C and $\overline{z_C}$ in this partial order corresponding to the dual rotations for $z_C = 1$ and $z_C = 0$ respectively.

A linear programming formulation of stable marriage [51] can also be obtained from nonexpansive networks, see section 3.4.6, where a parallel time upper bound close to $\sqrt{m}$ is obtained for stable matching as a special case via linear programming. It is of interest to see what the conditions defined by that linear program correspond to in the case of X gates and comparators. If an X gate in a stable matching instance has inputs x_1, x_2 and outputs y_1, y_2 then the linear inequalities on these variables are $y_1 = x_1 - \Delta$ and $y_2 = x_2 - \Delta$, with $\Delta \geq \max(0, x_1 + x_2 - 1)$. The reason is that we must compare distances between $X(x_1 x_2) = y_1 y_2$ and the four input assignments; now $X(10) = 10$ and $X(01) = 01$ give $1 - x_1 + x_2 \geq 1 - y_1 + y_2$ and $1 - x_2 + x_1 \geq 1 - y_2 + y_1$, or $x_1 - y_1 = x_2 - y_2$, the quantity we call Δ; also $X(00) = 00$ gives $x_1 + x_2 \geq y_1 + y_2$, or $\Delta \geq 0$, and $X(11) = 00$ gives $1 - x_1 + 1 - x_2 \geq y_1 + y_2$, or $\Delta \geq x_1 + x_2 - 1$. We may in fact extend the gate X to a gate whose inputs and ouputs are reals in $[0, 1]$, by letting Δ be a function of x_1, x_2 such that the resulting gate extends the X gate, and such that the partial derivatives of Δ are always between 0 and 1. Three possible choices for $\Delta(x_1, x_2)$ are $\max(0, x_1 + x_2 - 1) \leq x_1 x_2 \leq \min(x_1, x_2)$. Any choice of $\Delta(x_1, x_2)$ with the specified properties will define a real extension of X which is nonexpansive in the metric sense considered in Chapter 6. Furthermore, just like negating the second input and the second output of an X gate gives a comparator, replacing x_2 and y_2 by $1 - x_2$ and $1 - y_2$ gives an extension of the comparator. In the three examples given above, the first choice of Δ gives an extension that outputs $\min(x_1, x_2)$, $\max(x_1, x_2)$, the second choice gives the differentiable functions $x_1 x_2$, $x_1 + x_2 - x_1 x_2$, and the third choice gives $\max(0, x_1 + x_2 - 1)$, $\min(1, x_1 + x_2)$. For the stable arrangement problem, if each task is listed by k people, then X_k gates are used in the network representation. Each gate can be represented using about k^2 comparators and X gates, as shown in section 3.1.4, giving a parallel time upper bound close to $\sqrt{km}$.

5.1.5 Complexity of Several Arrangement Problems

We have seen that the stable arrangement problem can be viewed as an instance of network stability over multivariate X gates. The two special cases of stable matching and bipartite stable matching give networks over X gates and comparators respectively. This representation can then be used to study the structure of the stable arrangement and stable matching problems. In this section we go in the opposite direction and show, following the work of Subramanian [108], that networks over these various classes of gates can be viewed as instances of the stable arrangement and stable matching problems.

Given a network over X_k gates, we can decompose the edges of the underlying multigraph (corresponding to coordinates in the network) into paths as follows: If an edge is the rth input to an X_k gate, then the next edge in the path is the rth output of the gate. We would like each path to correspond to the preference list of a person in a stable arrangement problem. There are only a few difficulties in giving such an interpretation: (1) The path may cycle; (2) the path may not cycle but go through the same gate more than once; and (3) two paths may intersect more than once. The third difficulty is relevant only if we want to obtain a stable roommates or stable marriage problem, where two people can list each other only once.

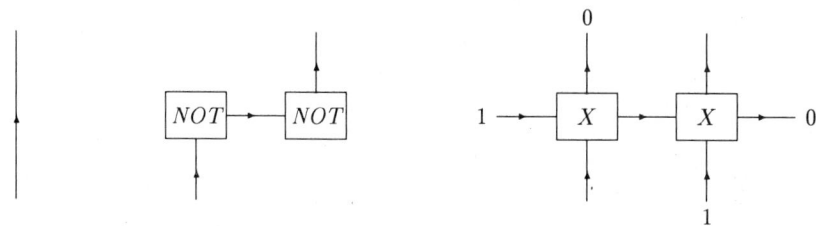

Figure 5.2: Breaking a path.

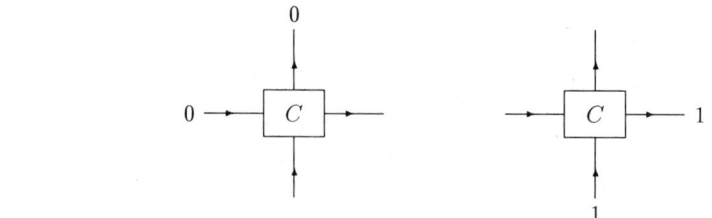

Figure 5.3: Breaking a path in the monotone case.

These difficulties are overcome by 'breaking the path', as illustrated in Fig. 5.2 (see [108]). First, two negations are introduced on each edge. The stable configurations of the resulting network correspond in an obvious manner to stable configurations in the original network. Second, each negation is viewed as an X gate (see Fig. 3.3 for the pictorial representation of X gates). The path of the edge entering the first gate now ends in an output edge leaving the gate, a short path starting at an input of the first gate ends at an output of the second gate, and a new path starting at an input of the second gate leaves at an output of the gate to proceed in the network. This substitution overcomes the three difficulties, i.e., paths can no longer cycle, meet again in the same gate, or meet more than once, because they are replaced by new paths at each edge of the original network. This gives a representation for networks of X_k gates as a stable arrangement problem, with each path corresponding to the preference list of a person and each gate corresponding to a task.

If the network contains only X gates, then the stable arrangement problem is a stable matching problem, because exactly two people meet at each task. In fact, it is a stable roommates problem, because two people meet at most once. If the network contains only comparators, then the simpler replacements from Fig. 5.3 makes it possible to break paths. Furthermore, we can make paths correspond to men and women so that the two paths meeting at a gate are a path for a man and a path for a woman: The replacement in Fig. 5.3 makes it possible to replace men with women and vice versa, as required. Therefore a network of comparators is a bipartite stable matching problem, and in fact a stable marriage problem as well.

Lemma 5.10 *Any network of X_k gates can be viewed as a stable arrangement problem. Any network of X gates can be viewed as a stable roommates problem. Any network of comparators can be viewed as a stable marriage problem.*

In fact, the roommates and marriage problems have preference lists of length at most 3, since paths consist of at most 4 edges (see also [108]). This characterization leads to the following complexity result. Recall that the bipartite stable matching problem is trivial as a decision problem, because the associated network is monotone and a stable arrangement always exists.

Theorem 5.11 *The seven versions of the stable arrangement problem have the same parallel complexity, as search problems, and as decision problems as well except for the bipartite versions, under logspace uniform $\mathcal{NC}^1$ reductions. The complexity of finding one stable arrangement is the same as that of finding a 2SAT characterization of all arrangements. Furthermore, these problems have the same complexity as the comparator circuit evaluation question, under $\mathcal{NC}^1$ reductions.*

This result follows directly from the complexity results of section 3.5.3, with the additional observation that the comparator circuit evaluation question and the X gate circuit evaluation question are equivalent to each other. See [108] for a proof of this fact and alternative proofs of some of the equivalences in the theorem.

A stable arrangement problem on a set of people P can be restricted to a problem on a subset $P' \subseteq P$ by removing the people in $P - P'$ and their preference lists from the instance.

Theorem 5.12 *Consider instances of the stable roommates problem, and suppose that we are given a subset P_0 of the set of people P. Then the problem of deciding whether there is a set P' with $P_0 \subseteq P' \subseteq P$ such that the problem restricted to P' has a stable arrangement is $\mathcal{NP}$-complete. In instances of the stable marriage problem, the problem of deciding whether there is a set P' with $P_0 \subseteq P' \subseteq P$ such that the problem restricted to P' matches some particular man/woman pair as partners is $\mathcal{NP}$-complete.*

Proof. One can remove a person p from a stable arrangement problem simply by assigning value 0 (instead of 1) to the input x_{p0}. This works because now every stable configuration must have $x_{pi} = 0$ for all $0 \leq i \leq |T(p)|$, and these values simplify all the X_k gates corresponding to tasks listed by p by removing the input and output for person p.

CHAPTER 5. STABLE MATCHING   147

A network of X gates can be viewed as a roommates problem. The reduction, given above, introduces some additional gates, inputs and outputs. Let P_0 be the set of people p corresponding to inputs $p0$ that were already present in the original network; thus a restriction to some P' with $P_0 \subseteq P'$ corresponds to a choice of an input assignment for the inputs in the original network, leaving the inputs added by the reduction with their input value 1 (see Fig. 5.3). The question of whether a network of X gates has a stable configuration for some choice of an input assignment is $\mathcal{NP}$-complete by Corollary 2.12, proving the first part of the theorem.

For the stable marriage problem, we consider the problem of deciding whether some input assignment for a comparator circuit gives outputs $x_a x_b = 10$ at some pair of chosen outputs. This problem is $\mathcal{NP}$-complete by Lemma 2.13. We can view the circuit as a marriage problem, and let the set P_0 take care of the constant inputs introduced in the reduction of Fig. 5.3 (a constant input 1 corresponds to man, a constant input 0 corresponds to a woman). The output x_a is the last entry in the preference list of a man m and the output x_b is the last entry in the preference list of a woman w. Thus $x_a x_b = 10$ means that neither m nor w are assigned before the end of their preference lists (since coordinates from W are negated in the comparator network representation). We can always introduce new men and women with the construction of Fig. 5.3, so we can assume that m and w have not listed each other in this marriage instance. If we add an additional comparator between these two outputs, and interpret it as m and w listing each other at the end of their lists, then m and w will be matched if and only if they are not matched earlier, and this happens precisely when $x_a x_b = 10$. □

Two variations of stable matching that have been studied are the stable roommates and stable marriage problems *with ties*. In these problems, a person may be indifferent to some of the possible choices, listing several people at the same level in the preference list. The preference condition now states that if a person p prefers some q to the assigned $\alpha(p)$, then q must not prefer p to $\alpha(q)$ (but could be indifferent with respect to these two choices). Note that the stable marriage with ties has solutions that can be obtained by breaking ties arbitrarily and solving the resulting instance without ties (guaranteed to have a solution by monotonicity). The $\mathcal{NP}$-completeness of stable roommates with ties was shown by Ronn [101].

Theorem 5.13 *The stable roommates problem with ties is $\mathcal{NP}$-complete. In the stable marriage problem with ties, the question of whether some particular man/woman pair is matched in some stable arrangement matching is $\mathcal{NP}$-complete.*

Proof.   Given a roommates instance and a subset $P_0 \subseteq P$, we add for each person $p \in P \backslash P_0$ two new people p', p''. Person p adds p'' at the beginning of its list. Person p' lists p'' alone. There is only indifference in the list of p'', who lists both p and p', with no preference.

A stable arrangement must choose a partner for p''. Both choices of partners p and p' could a priori lead to a stable matching, since both p and p' prefer p'' over everybody else.

Whether p will look for a partner in the original roommates instance or not (i.e., whether p is in the subset P' with $P_0 \subseteq P' \subseteq P$ of the previous theorem) depends on whether p'' is matched to p' or to p, respectively. Therefore the existence of a choice for each p'' leading to a stable arrangement depends on the existence of a set P' leading to a stable arrangement in the original instance, and this problem is $\mathcal{NP}$-complete.

The marriage result is proved similarly, using the $\mathcal{NP}$-completeness of deciding whether some particular man/woman pair is matched for some P' with $P_0 \subseteq P' \subseteq P$ from the previous theorem. □

Both Theorem 5.12 and 5.13 extend easily to the case of complete preference lists, by adding the remaining unlisted partners at the end of the preference lists in an appropriate fashion (i.e., without introducing cycles in the associated network).

Other versions of stable marriage with ties have also been studied. One of them strengthens the preference condition by requiring that if a person p prefers some q to the assigned $\alpha(p)$, then q must prefer $\alpha(q)$ to p (indifference between $\alpha(q)$ and p is no longer allowed). Here it is no longer guaranteed that a stable marriage must exist. For example, in the reduction given in the last proof, both choices for p'' would be unstable, and no stable matching would exist. Gusfield and Leather [51] have shown that even here, if a stable arrangement exists, then a male-optimal and a female-optimal arrangement exist as well, and can be found in polynomial time. The construction exhibits some similarity with respect to the monotone network stability problem with fanout, and we conjecture by analogy with Theorem 2.5 that the problem of deciding whether there is a stable arrangement other than the male-optimal and female-optimal ones is $\mathcal{NP}$-complete.

There are some other formulations of stable arrangement and stable matching problems that can be shown to be $\mathcal{NP}$-complete, both directly or by the network approach. One such version is a form of three-dimensional matching, where the partnership condition is strengthened by requiring that three people, rather than just two, be assigned to a task, and the set P is partitioned into three subsets P_1, P_2, P_3 rather than just two (see Knuth [75]). The $\mathcal{NP}$-completeness result, due to Ronn [100] and to Subramanian [108], can be explained in this case by the fact that the associated networks exhibit fanout [108]. A circular version of this problem, where a person in P_1 lists only preferences for people P_2 (and is indifferent to those in P_3), a person in P_2 lists only preferences for P_3, and a person in P_3 lists only preferences for P_1, is still open.

5.2 Optimization and Enumeration in Stable Matching

We examine next the complexity of optimization and enumeration questions in the case of stable matching. In order to obtain efficient algorithms, it will be necessary to give reductions between problems that are not only easy to perform, but also produce reasonably small instances of the target problem. The two parameters of size and width that we considered in the algorithms of the last chapter play an important role in this regard.

We focus on the stable matching problem rather than on the general stable arrangement problem in order to ensure that the values of these two parameters are suitably small.

5.2.1 Size and Width

Suppose that an instance of the stable matching problem has n people and size m. The corresponding network uses only X gates and has size m. Since the gatewidth is 2, we know that a 2SAT instance with $O(m)$ clauses on the variables x_{pi} for $p \in P$ and $0 \leq i \leq |T(p)|$ characterizing the stable arrangements can be found in $O(m)$ time, using the scatter-free algorithm of section 3.3.3. (Alternatively, one can find a single solution in $O(m)$ time by the stable matching algorithm of Irving [62] or the scatter-free algorithm of Subramanian [108], and then use the algorithm of section 3.3.2 for the general nonexpansive case.) The implications $x_{pi} \to x_{p(i-1)}$ hold by the consistency condition. Therefore the explicit width of the 2SAT instance is at most $2n$, with one path for the literals x_{pi} with p fixed, and one path for the literals $\overline{x_{pi}}$ with p fixed. If the stable matching problem is bipartite, then the network can be made monotone by using comparators instead of X gates, and so after identifying the variables x_{pi} that have the same value in the man-optimal and the woman-optimal solution (these are the trivial variables), the 2SAT instance on the remaining variables becomes bipartite.

Theorem 5.14 *The set of stable arrangements of a stable matching problem with n people and size m can be described by a 2SAT instance with $O(m)$ clauses and explicit width $O(n)$. If the stable matching problem is bipartite, then the 2SAT instance is bipartite. The 2SAT instance for a given stable matching problem can be found in $O(m)$ time.*

We show that the converse of this theorem is also true. This kind of correspondence, in the marriage case, was first obtained by Blair [16]; the existence of a polynomially bounded matching instance, with $O(m)$ people, was found by Gusfield, Irving, Leather, and Saks [52, 64].

Theorem 5.15 *Every 2SAT instance with m clauses and explicit width w characterizes the set of solutions of a stable matching problem of size $O(m)$ and with $O(w)$ people If the 2SAT instance is bipartite, then the stable matching problem is bipartite. The stable matching problem corresponding to a given 2SAT instance can be found in $O(m)$ time.*

Proof. We can assume that the 2SAT instance is acyclic, since any 2SAT instance can be made acyclic by a strong components algorithm, with the strong components corresponding to literals that have the same value in all solutions and can thus be treated as a single literal. Since the 2SAT instance has width w, we can in particular cover all the unnegated variables with w paths. Let p be such a path, of the form $x_{i_1} \to x_{i_2} \to \cdots \to x_{i_k}$. We denote by $\overline{p}$ the path $\overline{x_{i_k}} \to \cdots \to \overline{x_{i_2}} \to \overline{x_{i_1}}$ in the 2SAT instance. We thus have $2w$ paths $p, \overline{p}$ covering all the literals. We can assume that the only clauses involving two variables x_{i_j} in the same path are the clauses $\overline{x_{i_{j-1}}} \vee x_{i_j}$ included in the paths.

The corresponding stable matching problem is defined as follows. The set of people is the set P of $2w$ paths $p, \bar{p}$. The set of tasks is the set T of m clauses, with an additional $2w$ tasks, one for each path $p, \bar{p}$. The preference list of person q (either p or $\bar{p}$) corresponding to a path q of the form $u_1 \to u_2 \to \cdots \to u_k$ is the following. It starts with the task for q, then lists all the tasks for clauses $u_1 \vee v$ in some order, followed by the clause $\overline{u_1} \vee u_2$, followed by the tasks for the remaining clauses $u_2 \vee v$, followed by the clause $\overline{u_2} \vee u_3$, and so on. Eventually, the tasks for clauses $u_k \vee v$ are listed, and the list ends with the task for $\bar{q}$. The path for $\bar{q}$, used in constructing the preference list of $\bar{q}$, is given by $\overline{u_k} \to \cdots \to \overline{u_2} \to \overline{u_1}$.

Note that a clause $u \vee v$ is listed as a task precisely by the two people corresponding to the paths q, r that contain the literals u, v. The special tasks $q, \bar{q}$ are listed by the two people $q, \bar{q}$. Thus the arrangement problem is indeed a stable matching problem. Every person q is assigned a task in every stable arrangement, because the last task in the list of q is the task $\bar{q}$, and this task is the first task in the list of $\bar{q}$ so that q is guaranteed this task or better by the preference condition. The tasks listed by both q and $\bar{q}$ are $q, \bar{q}$, and the tasks for clauses $\overline{u_{i-1}} \vee u_i$ corresponding to edges along the paths. We claim that in every stable arrangement, q is matched to $\bar{q}$ in one of these tasks. Otherwise, person q is assigned some other task $u_i \vee v$, and person $\bar{q}$ must be assigned some other task $\overline{u_j} \vee v'$. Furthermore, the clause $\overline{u_i} \vee u_j$ must follow from the clauses in the 2SAT instance. To prove this last observation, it suffices to show that $i \le j$. If $i > j$, then the clause $\overline{u_{i-1}} \vee u_i$ precedes the clause $u_i \vee v$ assigned to person q in the preference list of q, and it also precedes the clause $\overline{u_j} \vee v'$ assigned to person $\bar{q}$ in the preference list of $\bar{q}$, thus violating the preference condition for q and $\bar{q}$. Therefore every pair $q, \bar{q}$ of people not matched to each other indicates the presence of three clauses $u_i \vee v$, $\overline{u_j} \vee v'$ and $\overline{u_i} \vee u_j$ either in the 2SAT instance or in its transitive closure (omitting the third clause in the case $i = j$). Note that if u_i occurs in these clauses, then so does $\overline{u_i}$, and in fact each of them occurs in precisely one of these clauses. If we construct the implication graph corresponding to these clauses, each of these vertices u_i will have in-degree 1 and out-degree 1, so the graph has a cycle, contradicting the fact that the 2SAT instance is acyclic. Thus the 2SAT instance can be acyclic only if there are no such clauses, i.e., if q and $\bar{q}$ are matched to each other for each pair $q, \bar{q}$.

If the path p of unnegated literals is of the form $x_{i_1} \to x_{i_2} \to \cdots \to x_{i_k}$, let the value $x_{i_j} = 1$ indicate the fact that person p is assigned task $\overline{x_{i_{j-1}}} \vee x_{i_j}$ or better in the preference list $T(p)$, with $x_{i_1} = 1$ indicating that p is assigned task p. The values of the x_{i_j} give the task that p is assigned, because p can only be assigned (1) tasks $\overline{x_{i_{j-1}}} \to x_{i_j}$ with $2 \le j \le k$, in which case $x_{i_{j'}} = 1$ if and only if $j' \ge j$, (2) task p, in which case $x_{i_j} = 1$ for all j, or (3) task $\bar{p}$, in which case $x_{i_j} = 0$ for all j. Note that the x_{i_j} also indicate the task assigned to $\bar{p}$, which is the same as the task assigned to p. In this assignment of boolean values x_{i_j}, the consistency condition is simply $x_{i_{j-1}} \to x_j$, and this implication is a clause in the 2SAT instance. The partnership condition is automatically satisfied, since p and $\bar{p}$ are partners. The preference condition for tasks listed by both p and $\bar{p}$ is automatically satisfied, since p and $\bar{p}$ list such tasks in opposite order. The preference condition for tasks $u \vee v$ listed by two noncomplementary people p, q, with u in the path for p and v in the path for q, indicates that either p has an assigned task better than $u \vee v$ or q has an assigned

CHAPTER 5. STABLE MATCHING

task better than $u \vee v$. This can be enforced with the clause $u \vee v$. For example, if the task is of the form $x_{i_j} \vee v$, and p is assigned a task better than $x_{i_j} \vee v$, then p must be assigned task $\overline{x_{i_{j-1}}} \vee x_{i_j}$ or better (or task p if $j = 1$), and this is indicated by the value $x_{i_j} = 1$ that satisfies the clause $x_{i_j} \vee v$. Therefore, the assigned values for the x_{i_j} characterize a stable arrangement if and only if they give a solution to the 2SAT instance, so the stable arrangements are indeed the solutions to the 2SAT instance.

The $O(m)$ time complexity of the reduction is immediate. The number of people is at most $2w$, and the number of tasks is $2w+m$, with each task occurring in two preference lists, so the size of the stable matching problem is $O(m)$. If the 2SAT instance is bipartite, then we can assume that all the clauses are of the form $x_i \vee \overline{x_j}$, so the two people listing the corresponding task are of the form p and $\overline{q}$ and can be viewed as a man and a woman. Therefore the stable matching problem is bipartite. The 2SAT instance that we have used to describe stable arrangements differs only trivially from the standard 2SAT instance that comes out of the representation of stable matching problems as network stability problems. More precisely, some of the variables x_{pi} have been negated, and the variables x_{pi} corrresponding to tasks in the preference lists of $p, \overline{p}$ that lie in between two tasks shared by p and $\overline{p}$ are equivalent and have been replaced by single variables x_i to give an acyclic 2SAT instance. We can therefore say that the theorem holds for the standard 2SAT instance associated with a stable matching problem. □

The last two theorems show that the number of people and size of stable matching problems correspond to the width and number of clauses of 2SAT instances, in both directions. The preceding theorem does not hold if we want to construct roommates and marriage instances, where a pair of people can have only one task in common. For example, if the 2SAT instance is a single path on n variables with $m = n - 1$ clauses, then the matching problem must have at least $\sqrt{m}$ people, much more than the constant width of the 2SAT instance. More generally, we can construct a 2SAT instance consisting of w disjoint paths, each with n/w variables, for a total of n variables and $m = n - w$ clauses. The corresponding matching problem will have $\Omega(\sqrt{n/w})$ people per path, giving an $\Omega(\sqrt{wn})$ lower bound on the number of people that may be required. This seems to indicate that the number of people needed is somewhere between the width w and the number of variables n.

Theorem 5.16 *Every 2SAT instance with n variables and m clauses characterizes the set of solutions of a stable roommates problem with $O(n)$ people and of size $O(m)$. If the 2SAT instance is bipartite, then the roommates problem is a marriage problem. The roommates or marriage problem corresponding to a given 2SAT instance can be found in $O(m)$ time.*

Proof. We use the same construction as in the preceding theorem, but with $2n$ paths instead of $2w$ paths, so that each path consists of a single literal u. Now the only people that have more than one task in common are people $p, \overline{p}$ who list both p and $\overline{p}$ as a common task, so these people share two tasks. The presence of common tasks can be overcome by 'breaking' the path p and introducing two new people, as in the proof of Lemma 5.10, using the construction of Fig. 5.2. The result is then a roommates problem, with four people (p, $\bar{p}$, and the two additional people) per variable. A bipartite 2SAT instance gives a bipartite matching problem as before, and this problem is now a marriage problem. □

This result can be extended to the case of complete preference lists by extending the lists arbitrarily (with all additions at the end of the lists). This works because every q is guaranteed a partner no worse than $\bar{q}$. The resulting roommates problem has size $O(n^2)$ (rather than $O(m)$).

The result of Provan and Ball [95] on the $\#\mathcal{P}$-completeness of counting solutions to bipartite 2SAT instances gives then the following result of Irving and Leather [64].

Theorem 5.17 *The problem of counting solutions to a stable marriage problem is $\#\mathcal{P}$-complete.*

5.2.2 Optimization

A form of weighted stable matching problem, known as the 'optimal' stable matching problem, was studied by Irving, Leather, and Gusfield [65]. In an instance of this problem, each person p assigns a nonnegative weight to each task in the list $T(p)$. The weights w_{pi} for tasks t_{pi} in the list of p are monotonically nondecreasing as a function of i, so that preferred tasks have smaller weights. The weight of a stable arrangement α is $\sum_{p \in P, \alpha(p) = t_{pi}} w_{pi}$. The aim in the weighted stable matching problem is to find a stable arrangement of minimum weight. The fact that the case $\alpha(p) = \bot$ has no assigned weight is not significant, since the set of p such that $\alpha(p) = \bot$ is the same over all stable arrangements. As a special case, the *egalitarian* stable matching problem is the weighted stable matching problem with $w_{pi} = i$.

When the stable matching problem is viewed as a network stability problem, each stable arrangement α is described by a stable configuration x. For each $p \in P$ with $\alpha(p) \neq \bot$ in a stable arrangement, assign to x_{pi} the weight $w'_{pi} = w_{p(i+1)} - w_{pi}$ for $0 \leq i < |T(p)|$, where $w_{p0} = 0$ by convention. Assign an arbitrary weight to the x_{pi} with $i = |T(p)|$. Then $\sum_{0 \leq j \leq T(p)} w'_{pj} x_{pj} = \sum_{j < i, t_{pi} = \alpha(p)} w'_{pj} = w_{pi}$, so the sum of the weights assigned to the x_{pi} such that $x_{pi} = 1$ is the same as the weight of the arrangement α. (The arbitrary weights w_{pi} with $i = |T(p)|$ do not matter since $x_{pi} = 0$.) In the egalitarian case, all weights can be set to $w'_{pi} = 1$.

Theorem 5.18 *The egalitarian stable roommates problem is $\mathcal{NP}$-complete.*

Proof. Every acyclic 2SAT instance can be viewed as a characterization of the stable arrangements for a stable roommates problem by Theorem 5.16. Ignore for now the fact that we had to introduce two additional people per pair $p, \bar{p}$ to make these two people share only one task. Each pair of people $p, \bar{p}$ has a single variable, call it x_p, in the correspondence between the 2SAT instance and the stable matching problem as defined in the proof of Theorem 5.15, with $x_p = 1$ if $p, \bar{p}$ are assigned task $p = t_{p1} = t_{\bar{p}r}$, and $x_p = 0$ if $p, \bar{p}$ are assigned task $\bar{p} = t_{\bar{p}1} = t_{ps}$. The contribution of $p, \bar{p}$ to the weight of a stable arrangement in the egalitarian case is thus $r+1$ if $x_p = 1$ and $s+1$ if $x_p = 0$. We can force $r+1 = 2n$ and $s+1 = n$ by introducing $2n$ additional pairs of people $q, \bar{q}$, forcing every stable arrangement to match q with $\bar{q}$ by having these two people list each other first in their preference lists, adding then some of these extra people to the preference lists of the original $p, \bar{p}$, in between the extreme choices $p, \bar{p}$, so as to force $p = t_{\bar{p}(2n-1)}$ and $\bar{p} = t_{p(n-1)}$.

Now each $x_p = 1$ contributes weight $2n$ and each $x_p = 0$ contributes weight n, so a minimum weight stable arrangement is a 2SAT solution with the fewest number of $x_p = 1$. We know from section 4.2.1 that the problem of finding such a solution to a 2SAT instance is $\mathcal{NP}$-complete. In fact, the proof involved a reduction from vertex cover that gives monotone 2SAT instances, and monotone 2SAT instances are always acyclic as required. It can be verified that the two additional people introduced to 'break a path' for each pair $p, \bar{p}$ simply increase the weight of every stable arrangement by 3, and therefore do not affect the characterization of the minimum weight solutions. Hence the problem of finding a minimum weight solution for a stable roommates problem in the egalitarian case is $\mathcal{NP}$-complete. □

It is possible to carry out the reduction from vertex cover carefully so as to preserve constant factor approximation guarantees in the reduction, in both directions. The best known constant approximation factor for vertex cover is 2. By Theorem 4.9, using the representation result of Theorem 5.14, we have:

Theorem 5.19 *A solution to the weighted stable matching problem of weight within a factor of 2 of the optimum can be obtained in $O(m \log(n^2/m + 2))$ time for instances with n people and of size m.*

In the case of complete preference lists, we have $m \approx n^2$, and hence an $O(m)$ algorithm. As mentioned in section 4.1.4, this case does not require dynamic trees, making the implementation quite simple. An algorithm running in $O(n^4)$ time was given by Gusfield and Irving [51].

For bipartite stable matching problems, the associated 2SAT instance is bipartite, and Theorem 4.10 with the representation of Theorem 5.14 gives the following:

Theorem 5.20 *A minimum weight stable arrangement, of weight K, for a bipartite weighted stable matching problem with n people and size m, can be found in $O(m\sqrt{K})$ time for $K \leq (m/\log^2 m)^2$ and in $O(nm \log K)$ for arbitrary K. In fact, a complete description of all minimum weight stable arrangements can be found within this time bound.*

In the egalitarian case, since the total length of the preference lists is m, we have $K \leq m$. Therefore the egalitarian bipartite matching problem (and the egalitarian marriage problem) can be solved in $O(m^{1.5})$ time. (Note that in this case, the algorithm of section 4.1.3 does not need to maintain a forest solution, since the number of capacitated edges in the residual graph cannot exceed m.) In the case of complete preference lists, with $m \approx n^2$, this gives an $O(n^3)$ time bound.

An $O(m^2 \log m)$ bound for the weighted marriage problem was known, due to Irving, Leather and Gusfield [65], with an $O(m^2)$ bound for the egalitarian case. An $O(m^{1.5}\sqrt{\log m})$ time bound for the egalitarian marriage problem was obtained by Ng [85]. Gusfield and Pitt [54] have given an algorithm for the nonbipartite problem, with the same approximation guarantee, running in $O(m^2)$ time.

Theorem 5.20 either matches or comes close to the bounds that we would expect from classical flow and matching results. For example, one can show using Theorem 5.16 that an improvement below $O(nm)$ for the weighted stable marriage problem would require at the very least a similar improvement for the uncapacitated max flow problem on graphs with n vertices and m edges; without such an improvement, this leaves open the possibility of reducing the $\log K$ factor to a factor depending on m alone. An improvement below $O(m^{1.5})$ for the egalitarian marriage problem would at least require a similar improvement for the *bipartite matching with bounds* problem. In this problem, a bipartite graph with n vertices and m edges is given, the vertices have associated *bounds* adding up to at most m, and the aim is to find a largest multiset F of edges from the graph with the constraint that the number of occurrences of a vertex in edges of F does not exceed its bound. (Thus the standard bipartite matching problem is the case where all bounds equal 1.) In both cases, it is the solution to the *dual* of the flow or matching problem that is actually needed, so we can say only that the weighted marriage problem is at least as hard as finding the *value* of an optimal flow or the *size* of an optimal multiset.

There are several other problems that can be solved efficiently after finding in $O(m)$ time the 2SAT instance for a given stable matching problem. The *minimum regret* stable matching problem asks for a stable arrangement α having the largest value i such that $\alpha(p) = t_{pi}$ for some $p \in P$ be as small as possible. Such a minimum regret arrangement can be found in $O(m)$ time by a careful execution of the stable matching algorithm of Irving [63] or the scatter-free network stability algorithm of Subramanian [108]. Alternatively one can use the structure of the set of stable solutions, as in Gusfield [48]. If we list the variables x_{pi} of the 2SAT instance in some order, so that the indices i are in nonincreasing order, then a minimum regret solution can be found by incrementally setting $x_{pi} = 0$ for the variables in order, until a variable x_{pi} that must have value 1 is found. Each assignment $x_{pi} = 0$ forces values for some subset of the remaining variables, using the implications in the 2SAT instance. The value $x_{pi} = 1$ will then be forced only if these implications starting with $x_{pi} = 0$ lead to a contradiction, because every contradictory partial assignment leads to a contradiction in the transitive closure (see the proof of Lemma 4.7).

The *lexicographic* bipartite stable matching problem asks for a stable arrangement α that gives the lexicographically smallest value of $(k_n, k_{n-1}, \ldots, k_2, k_1)$, where k_i is the number of people p such that $\alpha(p) = t_{pi}$. The problem of minimizing the number of people p that do not have $\alpha(p) <_p t_{pi}$ can be stated as a weighted problem by setting $w_{pj} = 1$ for $j \geq i$ and

CHAPTER 5. STABLE MATCHING

$w_{pj} = 0$ otherwise. This gives $w'_{p(i-1)} = 1$ and $w'_{pj} = 0$ for $j \neq i-1$. The minimum weight solution has $K \leq n_i$, where n_i is the number of people p for which there is a variable $x_{p(i-1)}$, with $\sum_i n_i = m$. This solution can be found in $O(m\sqrt{n_i})$ time. To solve the lexicographic problem, we solve this weighted problem for $i = n, n-1, \ldots, 2, 1$ in turn. Each problem solved adds constraints describing the set of all solutions to the 2SAT instance, and the successive 2SAT instances become simpler as i increases, so that the number of clauses does not increase if strong components are merged (see the proof of Theorem 4.10). The total running time is $O(m \sum_i \sqrt{n_i})$, and the worst case occurs when $n_i = m/n$, giving an $O(n^{1/2} m^{3/2})$ time bound. An $O(nm^2 \log^2 n)$ bound was previously known for this problem, by Irving, Leather, and Gusfield [65].

The *balanced* stable matching problem asks for a stable arrangement α that minimizes $\max(\sum_{p \in M, \alpha(p) = t_{pi}} w_{pi}, \sum_{q \in W, \alpha(q) = t_{qi}} w_{qi})$. Using Theorem 4.11, in conjunction with the reductions from section 5.2.1, one can show that the balanced stable matching problem is $\mathcal{NP}$-complete even in the marriage case, but can be approximated within a factor of 2 in $O(m \log(n^2/m + 2))$ time.

5.2.3 Enumeration and Partial Arrangements

We now turn to the enumeration question. From Theorems 4.13 and 5.14, using the fact that the degree is bounded by the explicit width, we obtain:

Theorem 5.21 *The stable arrangements of a stable matching problem with n people and of size m can be enumerated after $O(m)$ pre-processing time in $O(n)$ on-line time per solution, using $O(m)$ space.*

This result was previously known in the bipartite case, by a result due to Gusfield [48]. In the general case, the best known bound required $O(nm \log n)$ pre-processing time and $O(m)$ time per solution, using $O(m)$ space [49].

The *stable pairs* of a stable roommates or stable marriage problem are the pairs of people that can be paired up by some stable arrangement. Stable pairs can be viewed as partial assignments of one person to one task that can be extended to complete, stable arrangements.

Theorem 5.22 *Given a set of r people in a stable matching instance of size m, one can can determine, after $O(rm)$ pre-processing time, whether a query partial assignment of tasks to k of the r people extends to a stable arrangement, in $O(k^2)$ time.*

Proof. An arrangement assigns to person p the task t_{pi} if the corresponding assignment x in the associated network has $x_{pi} = 0$ and $x_{p(i-1)} = 1$. We must therefore check a partial assignment of values to $2k$ variables, and see whether this partial assignment can be extended to a complete solution. As in the proof of Corollary 4.8, if the corresponding 2SAT instance is transitively closed, it is sufficient to check the $O(k^2)$ clauses involving the $2k$ variables under consideration. Since the explicit width is bounded by the number of people n, the transitive closure can be obtained in $O(nm)$ time. If the set of people under consideration belongs to a restricted set of $r \le n$ people, then the corresponding variables x_{pi} belong to just r paths in the 2SAT instance, and the edges of the transitive closure that involve at least one vertex from a given path can be found in $O(m)$ time by Lemma 4.2, giving an $O(rm)$ time bound over all r paths. □

An algorithm for this problem with $O(m^2)$ pre-processing time was previously known, see Gusfield and Irving [51].

As a special case, the stable pairs of a roommates problem can be found in $O(1)$ time per pair because $k = 1$, and since only $O(m)$ candidate pairs need to be checked, the running time is dominated by the $O(nm)$ preprocessing time for all n people. By contrast, in the marriage case, Gusfield [48] has shown that the stable pairs can be found in $O(m)$ time. The difference between the two cases can be explained as follows. In order to assign person p to the task t_{pi}, there must exist a stable arrangement with $x_{pi} = 0$ and $x_{p(i-1)} = 1$. This condition can be written simply as $x_{pi} \ne x_{p(i-1)}$, because the consistency condition $x_{pi} \le x_{p(i-1)}$ holds for all stable arrangements. Suppose that we have found the 2SAT instance for the stable arrangements, its strong components, and a particular solution, in $O(m)$ time. Suppose also that we know the trivial variables of the 2SAT instance. If x_{pi} and $x_{p(i-1)}$ are nontrivial equivalent variables, then we can only have $x_{pi} \ne x_{p(i-1)}$ if this inequality holds for all stable arrangements, and in particular for the known stable arrangement. Similarly, if both variables are trivial, then it is sufficient to test the known stable arrangement. If one of them is trivial and one nontrivial, or both are nontrivial and inequivalent, then the inequality must hold for some stable arrangement. Therefore the stable pairs problem reduces to (and is in fact equivalent to, [47]) recognizing the trivial variables of the 2SAT instance.

In the bipartite case, the trivial variables are those that have the same value in the male-optimal and in the female-optimal solutions. These solutions can be found in $O(m)$ time. In the nonbipartite case, by contrast, one can show that a transitive closure computation is needed to find the trivial variables. Let the *partial transitive closure* problem be the problem of deciding, given a directed graph with n vertices and m edges, whether some n particular edges $e_i = (x_i, y_i)$ are in the transitive closure of the graph. Assume without loss of generality that the x_i are all distinct. We can view the vertices as Boolean variables and the edges as implications, and add implications $y_i \to \overline{x_i}$ for each edge e_i. Then the implication $x_i \to y_i$ is in the transitive closure precisely when the implication $x_i \to \overline{x_i}$ is in the transitive closure, that is, when x_i is a trivial variable. Thus recognizing trivial variables for a 2SAT instance reduces to the partial transitive closure problem. By Theorem 5.16, every 2SAT instance can be viewed as a stable roommates problem. Therefore a stable pairs algorithm with a time bound smaller than $O(nm)$ would require a partial transitive closure algorithm with a time bound smaller than $O(nm)$.

5.3 Discussion

The representation of stable matching problems as network stability problems on nonexpansive networks clarifies the study of the structure of stable matching. The usefulness of this representation comes from the associated Hamming metric on the space of configurations. This metric makes it possible to give simple proofs of old results, such as the invariance of the set of assigned people over all the stable arrangements, and new results, such as the median structure on the set of stable arrangements. A new randomized algorithm provides a simple way of defining a distribution on the space of stable configurations from a given distribution on the space of all configurations, by using the given distribution, e.g., the first preference for each person, as a starting point of the algorithm (see [108] for a different approach to obtaining distributions).

The average-case analysis of stable matching has received considerable attention in recent years [76, 92, 93]. Pittel showed that the expected number of solutions for a random stable marriage instance on n men and n women with complete preference lists is $e^{-1} n \log n$, but only $e^{1/2}$ for a random stable roommates instance [92, 93]. A similar discrepancy seems to arise when comparing monotone to non-monotone nonexpansive mappings in the hypercube. Thus, for example, if we consider distributions on the nonexpansive mappings on the m-cube which are invariant under isomorphisms, then each point is a fixed point with probability 2^{-m}, and the expected number of fixed points is again a constant 1. We would like to know whether networks can be used as a first approach to the analysis of less-understood questions on stable matching.

Chapter 6

Metric Networks and Product Graphs

The cartesian product of graphs has been extensively studied. Graphs have a prime factorization under the cartesian product operation; similarly, graphs have a canonical representation as *isometric* subgraphs of cartesian products with as many factors as possible. We study here the representation of graphs as *retracts* of cartesian products, as well as a new intermediate but computationally more tractable representation, namely the 2-*isometric* subgraphs of products. From the structural properties of 2-isometric subgraphs and retracts of cartesian products, the canonical representation result is extended to these two notions. The results are presented in the context of finite metric spaces, where the product of several metric spaces has points with one coordinate from each factor, and distances between points are obtained by adding the distances from each factor. When the finite metric spaces are those associated with connected graphs, the product operation on metric spaces corresponds to the cartesian product on graphs.

The starting point for the study of network stability on boolean nonexpansive networks, developed in Chapter 3, is the description of the space of configurations as the set of vertices in a hypercube under the associated Hamming metric. The distance function in the hypercube can be decomposed as the sum of distances in the different coordinates, where each coordinate belongs to the metric space consisting of two points 0 and 1 at distance 1 from each other. This formulation suggests extending the study of nonexpansive networks to products of more general metric spaces, with the product operation defined above. This chapter presents such a study on products of finite metric spaces.

Many of the proofs in the boolean case involve simple arguments about distances, and such arguments can be extended without difficulty to the more general metric case. Thus the basic structure of convergent networks leads to a polynomial time algorithm for stability in metric nonexpansive networks when the ratio between the largest and the smallest distance in each metric space is polynomially bounded. Similarly, the basic study of fixed cubes in the boolean case by means of the distance center can be directly extended to products of metric spaces.

The set of stable configurations for boolean nonexpansive networks is characterized in the hypercube in terms of median sets under the median operation. One cannot in general impose a median structure on arbitrary metric spaces; it turns out, however, that the *imprint* operation, introduced by Chung, Graham and Saks [21] for the case of products of cliques, can be suitably extended to all finite metric spaces, with the median operation on median graphs as a special case. Closure under medians in the boolean case gave the median sets; closure under imprints for products of metric spaces gives the 2-*isometric* subspaces. The connected median sets coincide in the boolean case with the retracts, a notion used to characterize the periodic configurations of nonexpansive networks. In the case of metric spaces, on the other hand, the retracts give a class of subspaces that contains, but does not coincide with, the 2-isometric subspaces. Indeed, the 2-isometric subspaces can be viewed as a first approximation to the notion of retract; one of the main advantages of such an approximation is that, in many ways, retracts are computationally intractable, while 2-isometric subspaces are not.

One of the significant properties in the boolean case was the characterization of the behavior of nonexpansive networks on the periodic configurations as, essentially, a permutation of coordinates. However, if several coordinates are combined and viewed as a single coordinate belonging to a larger metric space, then this characterization no longer holds. In order to expose the underlying permutation, the larger metric spaces have to be subdivided again into boolean coordinates. In general, given a product of metric spaces, one may hope to decompose each of the factors into several factors, and obtain in this way a *representation* that makes it easy to characterize the behavior of periodic points. For this purpose, we use the theory of isometric representations, developed by Graham and Winkler [46], and show that it can be extended to the main subspaces in our study, the 2-isometric subspaces and the retracts; this gives rise to the notions of a 2-isometric representation and a retract representation.

Sections 6.1 describes finite metric spaces and their products; section 6.2 incorporates the notion of a representation. Sections 6.3 examines the isometric representations. Section 6.4 studies the prime factorization in relation to the isometric representations. Section 6.5 introduces a new notion, that of a 2-isometric subspace, and uses it to examine the 2-isometric representations. Section 6.6 looks at the structure of retracts and retract representations. An application of this structural study to a dynamic location problem is described in section 6.7. Finally, section 6.8 shows how the study of metric spaces developed up to that point can be used to examine the behavior of metric nonexpansive networks; section 6.9 concludes with a discussion of the metric approach and some open problems.

6.1 Metric Spaces

For our purposes, a *metric space* $M = (V, d)$ is a nonempty *finite* set of points $V = V(M)$ equipped with a real-valued distance function $d = d_M$ satisfying the following conditions: (1) $d(x,y) = d(y,x) \geq 0$ for all points x, y; (2) $d(x,x) = 0$ for all points x; and (3)

$d(x, y) \leq d(x, z) + d(z, y)$ for all points x, y, z. The third condition is the *triangle inequality*. A *subspace* of a metric space M is a metric space L with $V(L) \subseteq V(M)$ and $d_L(x, y) = d_M(x, y)$ for all points x, y in L.

Given three points x, y, z in a metric space M, we say that z is *between* x and y if $d(x, y) = d(x, z) + d(z, y)$; we say that z is *strictly between* x and y if $z \neq x, y$ and z is between x and y. Two points $x \neq y$ are *adjacent* in M if there is no point z strictly between them. This notion of adjacency induces a graph structure on M: We let $G(M) = (V, E, w)$ be the undirected weighted graph on the set of vertices $V = V(M)$, with an edge $e = \{x, y\}$ in E for every pair of adjacent points x, y in M, and with a weight $w(e) = d_M(x, y)$ assigned to the edge. Given a weighted graph G, we say that a sequence $x = x^0, x^1, \ldots, x^r = y$ of points in G is a *path* from x to y if each $e^j = \{x^{j-1}, x^j\}$ is an edge of G. The e^j are the edges on the path, and the *length* of a path is the sum of the weights of the edges in it. A *shortest path* from x to y is a path from x to y of minimum length. Shortest paths can be used to recover M from $G = G(M)$, by means of the following lemma.

Lemma 6.1 *If M is a metric space, then the distance $d(x, y)$ between any two points x, y in M equals the length of a shortest path from x to y in $G(M)$.*

Proof. Given a path $x = x^0, x^1, \ldots, x^r = y$ in G with edges $e^j = \{x^{j-1}, x^j\}$, we have

$$d(x, y) \leq \sum_j d(x^{j-1}, x^j) = \sum_j w(e^j)$$

by the triangle inequality, so $d(x, y)$ is at most the length of a shortest path. On the other hand, we can construct a path from x to y of length $d(x, y)$ explicitly, by induction on the finitely many values $d(x, y)$ in M. If $x = y$, then the trivial path from x to y with no edges in it has length $0 = d(x, y)$. If x is adjacent to y, then the path consisting of the single edge $e = \{x, y\}$ has length $w(e) = d(x, y)$. If $x \neq y$ and x is not adjacent to y, then there is a point z strictly between x and y, so $d(x, y) = d(x, z) + d(z, y)$ and the distances in the right hand side are strictly smaller than the distance on the left hand side. By inductive hypothesis, there is a path from x to z of length $d(x, z)$, and a path from z to y of length $d(z, y)$; combining both paths produces a path from x to y of length $d(x, y)$. □

Given a connected weighted undirected graph G with nonnegative weights on the edges (or even a weighted multigraph, with self-loops and parallel edges of possibly different weights allowed), we let $M = M(G)$ be the metric space on the vertices of G with the distance $d_M(x, y)$ defined to be the length of a shortest path from x to y in G. By the lemma, if $G = G(M)$, then $M = M(G)$. The converse holds only if G satisfies the additional property that every edge in it constitutes a unique shortest path between its endpoints. Otherwise, we can repeatedly discard *unnecessary* edges in G, namely edges that do not constitute a unique shortest path between their endpoints, until no such edge remains; this removal does not affect $M = M(G)$, and the converse of the above statement must now hold for the reduced graph.

CHAPTER 6. METRIC NETWORKS AND PRODUCT GRAPHS

Given a set S and a collection of metric spaces M_i indexed by $i \in S$, we say that x is an *assignment* on S if x is a function that maps each i in $S = S(x)$ to a point $x(i)$ in M_i. An element $i \in S(x)$ is a *coordinate* of x, and the image $x(i)$ is its *value*. We denote by x_T the restriction of x to a subset $T \subseteq S(x)$. If $T = \{i\}$, then x_T is denoted by x_i; for convenience, we also identify x_i with its value $x(i)$. If x and y are assignments with $S(x) \cap S(y) = \emptyset$, then xy denotes the union of the two.

The *cartesian product* of a collection of metric spaces M_i indexed by $i \in S$ is the metric space $M = \square_{i \in S} M_i$ such that $V(M)$ is the set of assignments on S, and d_M is defined by

$$d_M(x,y) = \sum_{i \in S} d_{M_i}(x_i, y_i)$$

for all points x, y in M. If $S = \{1, 2, \ldots, k\}$, then $\square_{i \in S} M_i$ is written as $M_1 \square M_2 \square \cdots \square M_k$. The cartesian product can also be described in terms of the associated weighted graphs, as follows.

Lemma 6.2 *If $M = \square_{i \in S} M_i$, then $G(M)$ is the weighted graph on the set of assignments to S whose edges are those $e = \{x, y\}$ such that $x_i \neq y_i$ for precisely one value of $i \in S$ and $e_i = \{x_i, y_i\}$ is in $G(M_i)$, with $w(e) = w(e_i)$.*

Proof. Suppose that x and y differ in coordinate i. If x and y differ in some other coordinate, then $x_{S\setminus\{i\}} \neq y_{S\setminus\{i\}}$, and the fact that $d(x,y) = d(x_i x_{S\setminus\{i\}}, y_i x_{S\setminus\{i\}}) + d(y_i x_{S\setminus\{i\}}, y_i y_{S\setminus\{i\}})$ from the definition of the product implies that x and y are not adjacent. Therefore x and y can be adjacent only if $x_i \neq y_i$ for precisely one value of i. If $x_{S\setminus\{i\}} = y_{S\setminus\{i\}}$, then $d(x,y) = d(x_i, y_i)$. A point z is then strictly between x and y if and only if $d(x_i, y_i) = d(x_i, z_i) + d(z_i, y_i)$ and $z_{S\setminus\{i\}} = x_{S\setminus\{i\}}$, with $z_i \neq x_i, y_i$. Therefore x and y are adjacent in M if and only if x_i and y_i are adjacent in M_i. □

Given weighted graphs G_i for $i \in S$, we write $G = \square_{i \in S} G_i$ if G is the weighted graph on the set of assignments to S whose edges are those $e = \{x, y\}$ such that $x_i \neq y_i$ for precisely one value of $i \in S$ and $e_i = \{x_i, y_i\}$ is in G_i, with $w(e) = w(e_i)$. The lemma then says that if $G_i = G(M_i)$, then $G = G(M)$ for $G = \square_{i \in S} G_i$ and $M = \square_{i \in S} M_i$. One of the simplest examples of products is the case where each M_i consists of two points at distance 1; then $G = \square_{i \in S} G(M_i)$ is a hypercube, and the associated metric is the Hamming metric.

If $M = \square_{i \in S} M_i$, then the space M_i is the *ith factor* of M. Given an edge $e = \{x, y\}$ in $G(M)$, if i is the coordinate where x and y differ, then we say that e *belongs* to factor i. If $M = \square_{i \in S} M_i$, and $T \subseteq S$, then we denote $\square_{i \in T} M_i$ by M_T. More generally, if L is a subspace of M, then L_T is the subspace of M_T consisting of the points of the form x_T for some x in L. If $T = \{i\}$, then we denote L_T by L_i. A subspace L of the product M is an *irredundant* subspace if $L_i = M_i$ for all $i \in S$. A *subproduct* of the product $\square_{i \in S} M_i$ is a product $\square_{i \in S} L_i$, where each L_i is a subspace of the corresponding M_i. Note that if L is a subspace of the product $M = \square_{i \in S} M_i$, then L is an irredundant subspace of $\square_{i \in S} L_i$. Furthermore, L is an irredundant subspace of the product M if and only if the only subproduct of M with L as a subspace is M itself.

6.2 Representations

A mapping f from a metric space L to a metric space M is *nonexpansive* if $d(f(x), f(y)) \leq d(x,y)$ for all points x,y in L. We denote by $f(L)$ the subspace of M induced by the images under f of points in L. A mapping f from L onto M is an *isomorphism* if $d(f(x), f(y)) = d(x,y)$ for all points x,y in L. Consider two products $L = \square_{i \in S} L_i$ and $M = \square_{j \in T} M_j$ with $|S| = |T|$. Given a one-to-one correspondence ν from S to T, and isomorphisms g_i from $M_{\nu(i)}$ to L_i, we can define a one-to-one correspondence $\sigma(i, g_i(b)) = (\nu(i), b)$ from pairs (i,a) with $i \in S$ and $a \in L_i$ to pairs (j,b) with $j \in T$ and $b \in M_j$. We denote by g_σ the isomorphism from M to L defined by $g_\sigma(x) = y$ if and only if $y_i = g_i(x_{\nu(i)})$.

A *representation* for a metric space L in a cartesian product $M = \square_{i \in S} M_i$ is a nonexpansive mapping g from L to M such that (1) if x and y are adjacent in L, then their images $g(x)$ and $g(y)$ in M differ in at most one coordinate i; and (2) the image $g(L)$ is an irredundant subspace of the product M, each of whose factors M_i contains at least two points. In condition (1), if i is the coordinate where $g(x)$ and $g(y)$ differ, then we say that g maps the edge $\{x,y\}$ to factor i. Once condition (1) is satisfied, condition (2) can be easily enforced by first reducing the factors M_i to $g(L)_i$ as above so that $g(L)$ is an irredundant subspace of M, and then discarding the factors M_i with $|V(M_i)| = 1$. A partition of a set S is a collection of disjoint nonempty sets whose union is S. Let g be a representation for L in $M = \square_{j \in S} M_j$, and let g' be a representation L in $M' = \square_{i \in T} M'_i$. We say that the representation g *contains* the representation g' if there exists a partition of S consisting of sets S_i with $i \in T$, and representations h_i for M'_i in $M_{S_i} = \square_{j \in S_i} M_j$, such that the nonexpansive mapping h from M' to M defined by $h(x)_{S_i} = h_i(x_i)$ satisfies $h(g'(x)) = g(x)$. If two representations g and g' contain each other, then $|S| = |T|$ so that each S_i contains a single element $\nu(i)$, and each h_i maps M'_i to $M_{\nu(i)}$. In fact the mapping h_i is onto, since $g(L)$ is an irredundant subspace of M. Furthermore, given two points a, a' in M'_i, we can find points z, z' in $g'(L)$ with $z_i = a$, $z'_i = a'$ since $g'(L)$ is an irredundant subspace. Then $d(h(z), h(z')) = d(z, z')$ because of nonexpansiveness in both directions, so $d(h_i(z_i), h_i(z'_i)) = d(z_i, z'_i)$, proving that each h_i is an isomorphism. Therefore $h = g_\sigma$, where σ is defined by ν and the isomorphisms $g_i = h_i^{-1}$.

Given a metric space L and a relation ρ on the edges of $G(L)$, we let ρ^* denote the smallest equivalence relation containing ρ. Every equivalence relation ρ^* can be used to define a representation g^ρ for L in a product $M = \square_{i \in S} M_i$, as follows. Let the equivalence classes of ρ^* be the sets E_i with $i \in S$. Let G_i be the multigraph obtained from $G(L)$ by collapsing and discarding every edge of $G(L)$ that is *not* in E_i, thereby identifying any two vertices of $G(L)$ connected by a path none of whose edges lie in E_i. This defines a natural map g_i from vertices of $G(L)$ to vertices of G_i. We then let $M_i = M(G_i)$, and define the mapping $g = g^\rho$ from L to $M = \square_{i \in S} M_i$ by $g(x)_i = g_i(x)$. After discarding factors M_i containing a single point, we obtain the following:

Lemma 6.3 *The mapping $g = g^\rho$ is a representation for L in the cartesian product M.*

CHAPTER 6. METRIC NETWORKS AND PRODUCT GRAPHS

Proof. To prove that g is nonexpansive, it is sufficient to show that $d(g(x), g(y)) \le d(x, y)$ for points x, y that are adjacent in L; this property follows then inductively for points x, y that are not adjacent by introducing a point $z \ne x, y$ such that $d(x, y) = d(x, z) + d(z, y)$ and using the inductive assumption on x, z and on z, y. If x and y are adjacent then there is an edge $e = \{x, y\}$ in $G(L)$ and hence in some E_i. For $j \ne i$, this edge is collapsed in the definition of G_j, so $g_j(x) = g_j(y)$. In G_i, the length of a shortest path from $g_i(x)$ to $g_i(y)$ is at most $w(e) = d(x, y)$ since the edge e is not discarded, so $d_{M_i}(g_i(x), g_i(y)) \le d(x, y)$ and therefore $d(g(x), g(y)) \le d(x, y)$. Note that we have also shown that the images of adjacent points x and y differ in at most one coordinate i, so condition (1) in the definition of a representation is satisfied. Condition (2) is also satisfied, since all points in G_i come from points in $G(L)$ so that $g(L)$ is irredundant, and the spaces M_i with a single point have been discarded. □

Containment between relations gives reverse containment between representations.

Lemma 6.4 *If L is a metric space and ρ is a relation on the edges of $G(L)$, then every relation γ containing ρ defines a representation g^γ contained in g^ρ.*

Proof. The representation g^ρ for L in a product $M = \square_{j \in S} M_j$ is defined by the equivalence classes E_j of ρ^* with $j \in S$, and metric spaces $M_j = M(G_j)$; the representation g^γ for L in $M' = \square_{i \in T} M'_i$ is defined by the equivalence classes $E'_i = \bigcup_{j \in S_i} E_j$ of γ^* with $i \in T$, where the sets S_i form a partition of S, and metric spaces $M'_i = M(G'_i)$. The containment must be given by representations h_i for M'_i in $M_{S_i} = \square_{j \in S_i} M_j$. Note that the edges in G'_i are partitioned into equivalence classes E_j with $j \in S_i$, and that the graphs G_j with $j \in S_i$ can be obtained from G'_i by contracting edges not in E_j. If $G'_i = G(M'_i)$, then these contractions provide a representation h_i by the previous lemma. Otherwise, G_i differs from $G(M'_i)$ only in the presence of unnecessary edges, and each unnecessary edge will either be discarded or remain unnecessary after contracting edges not in E_j, so that the same metric space $M(G_j)$ is obtained even if we start from $G(M'_i)$ instead of G_i. □

Given a representation f for L in $M = \square_{i \in S} M_i$, consider the representation g^γ for L in $M' = \square_{i \in S} M'_i$ defined by the equivalence relation $\gamma = \gamma^* = \gamma(f)$ whose equivalence classes E_i consist of the edges $e = \{x, y\}$ such that $f(x)$ and $f(y)$ differ only in the factor M_i. (If $f(x) = f(y)$, put e in any equivalence class.)

Lemma 6.5 *Every representation f for a metric space L in a cartesian product M is contained in the representation g^γ for L in M', where $\gamma = \gamma(f)$ relates edges mapped to the same factor by f. The containment is given by nonexpansive mappings from factors M'_i onto corresponding factors M_i.*

Proof. Let $g = g^\gamma$. If $g_i(x) = g_i(y)$, then x and y are joined by a path of edges not in E_i, so $f(x)$ and $f(y)$ do not differ in factor i, i.e., $f_i(x) = f_i(y)$, where $f_i(x) = f(x)_i$. Therefore $f_i(x)$ depends only on $g_i(x)$; let h_i be the mapping from M'_i to M_i that takes $g_i(x)$ to $f_i(x)$. The mapping h_i is defined on all of M'_i since $g(L)$ is an irredundant subspace

of M', and it is onto since $f(L)$ is an irredundant subspace of M. If $d(g_i(x), g_i(y)) = d$, then x and y are joined by a path such that the edges in E_i along the path have weights adding up to d. If $e = \{x', y'\}$ is an edge on this path, then $d(f_i(x'), f_i(y')) = 0$ for $e \notin E_i$, and $d(f_i(x'), f_i(y')) \leq d(x', y')$ for $e \in E_i$ of weight $d(x', y')$ since f is nonexpansive, so $d(f_i(x), f_i(y)) \leq d$. This proves that h_i is nonexpansive, providing the required containment.
□

Let g be a representation for L in a product $M = \square_{i \in S} M_i$, and let T be a spanning tree of $G(L)$. Then every coordinate i must be such that the images under g of the endpoints of some edge in T differ in coordinate i, because otherwise all vertices of T (and hence all points in L) would be mapped to points that agree in coordinate i, contrary to condition (2) in the definition of a representation. If $G(L)$ has n vertices and m edges, then T has only $n - 1$ edges, so the number k of factors M_i in the product M is at most $n - 1$. (In fact $k = n - 1$ for some representation g for L if and only if $G(L) = T$.) When considering induced representations g^ρ, we shall only need to consider the equivalence classes E_i that appear on the edges of a given spanning tree T, at most $n - 1$ of them, since only those classes can give an edge of T whose endpoints are mapped by g^ρ to points that differ in coordinate i.

Algorithmically, given an equivalence class E_i of ρ^*, we can remove the edges in E_i from $G(L)$, then obtain the connected components of the resulting graph. Each connected component gives a vertex of G_i, and two vertices of G_i are connected by an edge if and only if there is an edge in E_i between two vertices in the two corresponding connected components. We can thus construct G_i and g_i from E_i in $O(m)$ time using $O(m)$ space. All the G_i and corresponding g_i can therefore be constructed from the equivalence classes E_i in $O(km)$ time using $O(m)$ space. Since every edge in each G_i is the image under g_i of some edge in E_i, the total space required to store all the factors G_i, together with the images $g_i(x), g_i(y)$ for edges $e = \{x, y\} \in E_i$ and all i, is $O(m)$. This provides an implicit description for the representation g^ρ that makes it possible to recover from the coordinates of $g(x)$ for some fixed point x in L the coordinates of $g(y)$ for any other y in L, simply by traversing a path from x to y in $G(L)$ and computing the image under g_i for the edges $e \in E_i$ along the path. (A more explicit description of g giving the k coordinates of $g(x)$ for each of the n vertices of G can be kept using $O(kn)$ space.) If we view each G_i as a description of the corresponding $M_i = M(G_i)$, we have the following.

Lemma 6.6 *If the equivalence classes $E_1, \ldots, E_k$ of ρ^* for some relation ρ are known, in a graph $G(L)$ with n vertices and m edges, then the representation g^ρ can be obtained from these equivalence classes in $O(km)$ time using $O(m)$ space, where $k \leq n - 1$.*

The following sections will examine the properties of various specific representations induced by the choice of a relation ρ. Algorithmically, we shall show in each case how the equivalence classes of ρ^* can be obtained efficiently from the definition of ρ. The properties of each representation studied will be proved in the general case, but we shall restrict the complexity analysis to the unweighted case, where $w(e) = 1$ for each edge e in $G(L)$ and where all metric spaces are specified by giving their underlying unweighted graphs. This makes the presentation cleaner, and also facilitates the comparison with the literature, where this assumption is usually made.

6.3 Isometric Representation

A representation g for a metric space L in a cartesian product M is *isometric* if it satisfies $d(g(x), g(y)) = d(x, y)$ for all points x, y in L. As an example, the 6-cycle, with edges of weight 1, can be isometrically represented in a 3-cube, as a cycle $(000, 001, 011, 111, 110, 100)$. Adjacent points in L must map to points that differ in only one factor, but not necessarily to adjacent points. For instance, a 5-cycle with one edge of weight 2 and the remaining edges of weight 1 can be embedded in the product of a path $0, 1, 2$ of length 2 and a path $0, 1$ of length 1, giving points $01, 00, 10, 20, 21$, where the two points 01 and 21 are not adjacent.

The property of being an isometric representation is preserved under containment:

Lemma 6.7 *Suppose that a representation g for L in a product $M = \square_{j \in S} M_j$ contains a representation g' for L in $M' = \square_{i \in M} M_i$, with the containment given by representations h_i for M_i' in M_{S_i}. Then the representation g is isometric if and only if the representation g' and the representations h_i are isometric.*

Proof. The nonexpansive property of representations gives, for all points x, y in L,

$$d_L(x,y) \geq d_{M'}(g'(x), g'(y)) = \sum_{i \in S'} d_{M_i'}(g'(x)_i, g'(y)_i)$$

$$\geq \sum_{i \in S'} d_{M_{S_i}}(h_i(g'(x)_i), h_i(g'(y)_i)) = d_M(g(x), g(y)) \leq d_L(x,y).$$

In this equation, the first inequality holds with equality for all choices of x, y precisely when g' is an isometric representation, the second inequality holds with equality for all x, y precisely when all h_i are isometric representations, and the third inequality holds with equality for all x, y precisely when g is an isometric representation. Furthermore, the third inequality holds with equality for all x, y if and only if the first two inequalities hold with equality for all x, y. Therefore g is isometric if and only if g' and all h_i are isometric. □

Graham and Winkler [46, 116] defined a relation θ and used it to obtain an isometric representation. Two edges $e = \{x, y\}$ and $e' = \{u, v\}$ in $G(L)$ satisfy $e \, \theta \, e'$ if

$$d(x, u) + d(y, v) \neq d(x, v) + d(y, u).$$

Note that if $e \, \theta \, e'$, then every isometric representation g must map the edges e and e' to the same factor. For if g maps these two edges to factors i and i' respectively, with $i \neq i'$, then $g_j(x) = g_j(y)$ for $j \neq i$ and $g_j(u) = g_j(v)$ for $j \neq i'$, so $d(x, u) - d(y, u) = d(g_i(x), g_i(u)) - d(g_i(y), g_i(u)) = d(g_i(x), g_i(v)) - d(g_i(y), g_i(v)) = d(x, v) - d(y, v)$, and therefore e and e' are not related by θ. Graham and Winkler's surprising result is that θ^* gives indeed an isometric representation g^θ, the *canonical* isometric representation.

A direct implementation of the relation θ requires $O(m^2)$ time, since all pairs of edges must be compared. We shall obtain a faster algorithm by defining a relation θ_1 contained in θ that defines the same embedding, and prove the result of Graham and Winkler using θ_1 instead of θ.

Let T be a spanning tree of the graph $G(L)$. Given two edges e and e' in $G(L)$, we let $e\,\theta_1\,e'$ if $e\,\theta\,e'$ and at least one of e and e' is in the spanning tree T. Although θ_1 is in general a proper subset of θ, it turns out that $\theta^* = \theta_1^*$, so that θ can be replaced by θ_1. We shall not prove this equality directly; instead, we shall work with θ_1 and establish its basic properties as was done in [46, 116] for θ. The equality between the equivalence relations θ^* and θ_1^* can be inferred indirectly from the fact that both lead to the same representation.

For ordered pairs $p = (x, y)$ and $q = (u, v)$ of vertices in $G(L)$, let

$$\mu(p, q) = d(x, v) - d(x, u) - d(y, v) + d(y, u).$$

Two edges $e = \{x, y\}$ and $e' = \{u, v\}$ in $G(L)$ can be viewed as ordered pairs by ordering their endpoints arbitrarily; we then have $e\,\theta\,e'$ if and only if $\mu(e, e') \neq 0$. If E and F are multisets of ordered pairs, we let $\mu(E, F) = \sum_{p \in E, q \in F} \mu(p, q)$. Given two ordered pairs (x, z) and (z, y), and a multiset of ordered pairs E, direct calculation shows that $\mu(\{(x, z), (z, y)\}, E) = \mu((x, y), E)$. More generally, if $p = (x, y)$, then we say that P is a directed path from x to y if P is a multiset of ordered pairs (not necessarily edges in a graph) given by $P = \{(z^0, z^1), (z^1, z^2), \ldots, (z^{r-2}, z^{r-1}), (z^{r-1}, z^r)\}$, where $z^0 = x$ and $z^r = y$. By the preceding observation, we now have $\mu(P, E) = \mu(p, E)$ for all multisets of ordered pairs E, all $p = (x, y)$, and all directed paths P from x to y.

Graham and Winkler [46, 116] proved the following lemma for θ. We give a proof for θ_1.

Lemma 6.8 *Let P_0 be a shortest path from x to y in $G(L)$, and let P be any path from x to y in $G(L)$. Then for each equivalence class E of θ_1^*, the sum of the weights of edges in P from E is at least as large as the sum of the weights of edges in P_0 from E.*

Proof. The paths can be viewed as multisets of directed edges, i.e., as directed paths as described above. Let $p = (x, y)$. For any directed edge $e = (u, v)$ in P_0, we have $\mu(p, e) = d(x, v) - d(x, u) - d(y, v) + d(y, u) = w(e) + w(e) = 2w(e)$, so $\mu(p, P_0 \cap E) = 2\sum_{e \in P_0 \cap E} w(e)$. Let P_T be a directed path from x to y contained in the spanning tree T. Recall that if e and e' are in two different equivalence classes of θ_1^*, and at least one of them is in T, then $\mu(e, e') = 0$. This implies, for any path P in $G(L)$ from x to y, that

$$\mu(p, P \cap E) = \mu(P_T, P \cap E) = \mu(P_T \cap E, P \cap E) = \mu(P_T \cap E, P) = \mu(P_T \cap E, p),$$

so $\mu(p, P \cap E) = \mu(p, P_0 \cap E) = 2\sum_{e \in P_0 \cap E} w(e)$. However, for each $e = (u, v) \in P \cap E$, we have $|\mu(p, e)| \leq |d(x, v) - d(x, u)| + |d(y, v) - d(y, u)| \leq w(e) + w(e) = 2w(e)$, so $|\mu(p, P \cap E)| \leq 2\sum_{e \in P \cap E} w(e)$ and hence $\sum_{e \in P \cap E} w(e) \geq \sum_{e \in P_0 \cap E} w(e)$. $\square$

This lemma can now be used to show that θ_1 defines a canonical isometric representation.

Theorem 6.9 *The relation θ_1 induces a canonical isometric representation for L in a cartesian product $M = \square_{i \in S} M_i$. The isometric representations for L are the representations contained in g^{θ_1}, and are given by the mappings g^γ with $\theta_1 \subseteq \gamma^*$.*

Proof. Let $g = g^{\theta_1}$ be the representation induced by θ_1. The distance between two points $g(x)$ and $g(y)$ in the cartesian product M equals the sum over all i of $d(g_i(x), g_i(y))$, and each such term is by the definition of M_i the minimum sum of weights of edges from the equivalence class E_i on a path from x to y in $G(L)$. By the lemma, this minimum is the sum of the weights of edges on a shortest path P_0 in $G(L)$ from x to y, so the sum of the terms over all i is the total sum of weights of edges in P_0, which is precisely $d_L(x, y)$. Therefore the distance between $g(x)$ and $g(y)$ in the product M equals the distance between x and y in L, and the representation is isometric.

Let f be an isometric representation. By Lemma 6.5, every representation f in a product M is contained in a representation $g = g^\gamma$ in a product M', where the containment is given by nonexpansive mappings h_i from M'_i onto M_i. In terms of the nonexpansive mapping h from M' to M defined by $[h(z)]_i = h_i(z_i)$, this means that $f(x) = h(g(x))$ for all $x \in L$. Given two points a, b in M'_i, choose points x, y in L such that $g_i(x) = a$, $g_i(y) = b$. If the representation f is isometric, we have $d(x,y) = d(f(x), f(y)) = d(h(g(x)), h(g(y))) \leq d(g(x), g(y)) \leq d(x, y)$, so $d(h(g(x)), h(g(y))) = d(g(x), g(y))$. Since each h_i is nonexpansive, this implies that $d(h_i(a), h_i(b)) = d(a, b)$. Therefore the h_i are isomorphisms, and the representation f is equivalent to g^γ. Since edges related by θ must map under the isometric representation g^γ to the same factor, we must have $\theta_1 \subseteq \gamma^*$. Conversely, if $\theta_1 \subseteq \gamma^*$, then g^{θ_1} contains g^γ by Lemma 6.4, and it follows from Lemma 6.7 that g^γ is an isometric representation. □

In unweighted spaces, the relation θ_1 can be obtained in $O(nm)$ time using $O(m)$ space. For each of the $n - 1$ edges $e = \{x, y\}$ in the spanning tree T, we compute distances from x and from y to all other vertices. This gives all the edges related to e under θ_1 in $O(m)$ time with $O(m)$ space, for a total $O(nm)$ time over all edges in the spanning tree. Whenever we determine that two edges are related under θ_1, we merge their equivalence classes. Since there are only m edges, we only need to perform at most $m - 1$ union and $m(n-1)$ find operations in a union-find algorithm to maintain the equivalence classes, so an implementation of union-find that takes $O(\log m)$ time per union and $O(1)$ time per find operation (see, e.g., Aho, Hopcroft, and Ullman [1]) will run in $O(nm)$ time, using $O(m)$ space. (For weighted spaces, the only increase in complexity occurs in the computation of distances.) Once the equivalence classes are found, Lemma 6.6 applies, so we have:

Theorem 6.10 *The canonical isometric representation g^{θ_1} for an unweighted graph with n vertices and m edges can be found in $O(nm)$ time and $O(m)$ space.*

A direct implementation of the relation θ requires $O(m^2)$ time, since all pairs of edges must be compared. We shall obtain a faster algorithm by defining a relation θ_1 contained in

θ that defines the same embedding, and prove the result of Graham and Winkler using θ_1 instead of θ. An algorithm for the canonical isometric representation with time complexity $O(nm)$ and space complexity $O(n^2)$ has been obtained by Aurenhammer and Hagauer [5].

6.4 Prime Factorization

A representation g for a metric space L in a product M is a *factorization* if g is an isomorphism. A *prime factorization* is a factorization that contains all factorizations. The results in the literature are generally concerned with the unweighted case. The existence of a prime factorization under the cartesian product was proven by Sabidussi [103], and later by Vizing [113]. (For a study of factorization under different products, see Feigenbaum [34].) The original proofs were difficult and did not provide efficient algorithms for finding the prime factorization. Polynomial time algorithms were eventually obtained independently by Feigenbaum, Hershberger and Schäffer [35], who succeeded in converting Sabidussi's original approach into an $O(n^{4.5})$ algorithm, and by Winkler [115], who used the canonical isometric representation described above, and whose algorithm can be implemented to run in $O(n^4)$ time [34]. Faster algorithms were obtained by Hochstrasser [59], who gave an $O(nm + n^2 \log^2 n)$ algorithm that uses $O(n^2)$ space, and by this author [31], with an $O(nm)$ running time using $O(m)$ space. Both algorithms are based on the canonical isometric representation. The fastest known algorithm has been obtained by Aurenhammer, Hagauer and Imrich [6], and runs in $O(m \log n)$ time using $O(m)$ space. This algorithm uses the basic properties of cartesian products directly, and does not depend on the isometric representation.

We shall describe the algorithm based on the isometric representation that runs in $O(nm)$ time in the unweighted case. Our main interest here is to show how the basic relation θ of the isometric representation can easily be extended to obtain other representations. A more complicated version of this idea will be used later to obtain the 2-isometric representation.

The analog of Lemma 6.7 is obvious in the case of the cartesian factorization.

Lemma 6.11 *Suppose that a representation g for L in a product $M = \square_{j \in S} M_j$ contains a representation g' for L in $M' = \square_{i \subset S'} M_i$, with the containment given by representations h_i for M'_i in M_{S_i}. Then the representation g is a factorization if and only if the representation g' and the representations h_i are factorizations.*

By Theorem 6.9, every isometric representation of a graph can be obtained by extending the relation θ_1 defined above. In particular, the factorizations must be given by some extension of θ_1, since every factorization is (trivially) an isometric representation. We give such an extension.

Two edges $e = \{x, z\}$ and $e' = \{z, y\}$ in $G(L)$ satisfy $e \, \tau \, e'$ if z is the unique common neighbor of x and y. If g is a factorization for L in a product M, we identify L with its

CHAPTER 6. METRIC NETWORKS AND PRODUCT GRAPHS   169

isomorphic image $g(L) = M$. We claim that two edges $e = \{x,z\}$ and $e' = \{z,y\}$ related by τ must belong to the same factor. For if e and e' belong to factors i and i' respectively, with $i \neq i'$, then $x_j = z_j = y_j$ for $j \neq i, i'$, with $x_i \neq z_i = y_i$ and $x_{i'} = z_{i'} \neq y_{i'}$. Let t be the point defined by $t_j = z_j$ for $j \neq i, i'$, with $t_i = x_i$ and $t_{i'} = y_{i'}$. Then $t \neq z$, and the edges $\tilde{e} = \{x,t\}$ and $\tilde{e}' = \{t,y\}$ belong to factors i' and i respectively. Therefore x and y have two common neighbors z and t, implying that e and e' are not related by τ.

The relation τ by itself does not give a factorization. In fact, the mapping g^τ will not even be one-to-one in general. We say that a representation h for L in a product M is a *strict representation* if h is one-to-one and $d(h(x), h(y)) = d(x,y)$ for all edges $\{x,y\}$ in $G(L)$. Clearly, all isometric representations and all factorizations are strict representations.

Theorem 6.12 *The factorizations of L are given by the representations g^γ for L in a cartesian product M such that g^γ is a strict representation and $\tau \subseteq \gamma^*$. Thus $g^{\theta_1 \cup \tau}$ is the prime factorization, and the factorizations of L are given by the mappings g^γ with $\theta_1 \cup \tau \subseteq \gamma^*$.*

Proof. Every factorization is an isometric representation, so by Theorem 6.9 it is induced by an equivalence relation $\gamma = \gamma^*$. We known that edges related by θ or by τ must map under g^γ to edges belonging to the same factor, so $\theta \cup \tau \subseteq \gamma$. Conversely, we shall show that if g^γ is a strict representation and $\tau \subseteq \gamma$, then g^γ is a factorization. If $\theta_1 \cup \tau \subseteq \gamma$, then these conditions are met, and g^γ is a factorization. In particular $\gamma = \theta_1 \cup \tau$ gives the prime factorization.

Suppose then that the mapping $g = g^\gamma$ is a strict representation for L in M, and that $\tau \subseteq \gamma$. For convenience, we identify $V(L)$ with its image $V(g(L)) \subseteq V(M)$, using the fact that g is one-to-one; the mapping g is then the identity mapping. To show that g is a factorization, we must now show that $L = M = \square_{i \in S} M_i$, or equivalently $G(L) = G(M)$. Note that there are products $G = \square_{i \in S} G_i$ such that G is a connected subgraph of $G(L)$: at least one such product can be obtained by selecting a single point x in L, and letting G_i consist of the single point x_i, for each i. We shall show that given such a product G, with G a proper subgraph of $G(L)$, there is another such product H that has G as a proper subgraph, and is also a subgraph of $G(L)$. This gives an increasing sequence of products G that are subgraphs of $G(L)$, so we must eventually obtain a product G that equals $G(L)$. This implies that $G(L) = \square_{i \in S} G_i$. Then $G(M_i) = G_i$ by the definition of the representation g^γ from γ, so $G(M) = G(L)$.

Let then $G = \square_{i \in S} G_i$ be any product such that G is a connected subgraph of $G(L)$. Let e be an edge in $G(L)$ that is not in G. Since $G(L)$ is connected, there is at least one such $e = \{x,y\}$ with x in G. This edge e joins vertices that differ in only one factor i. Let H_i be the graph G_i augmented with a single edge $e_i = \{x_i, y_i\}$ with $w(e_i) = w(e)$; the vertex x_i is already in G_i, while the vertex y_i may not be in G_i. Let $H_j = G_j$ for $j \neq i$. Then $H = \square_{j \in S} H_j$ is also a connected graph, and has G as a proper subgraph. We must show that H is also a subgraph of $G(L)$. The edges of H that are not in G are the edges

$e' = \{x_i u, y_i u\}$ with u a vertex of $G_{S\setminus\{i\}} = \square_{j \in S\setminus\{i\}} G_j$, as well as the edges $e'' = \{y_i u, y_i v\}$ with $\{u, v\}$ an edge of $G_{S\setminus\{i\}}$ if y_i is not in G_i. We shall show that both families of edges are in $G(L)$ as well, proving that H is a subgraph of $G(L)$.

Consider first the edges of the form $e' = \{x_i u, y_i u\}$ with u a vertex of $G_{S\setminus\{i\}}$. There is at least one such edge in $G(L)$, namely the edge $e = \{x, y\}$. Since $G_{S\setminus\{i\}}$ is connected, it suffices to show that if some such e' is in G, then every $\tilde{e}' = \{x_i v, y_i v\}$ with $\{u, v\}$ an edge of $G_{S\setminus\{i\}}$ is also in $G(L)$. The edge $\tilde{e}'' = \{x_i u, x_i v\}$ in G joins vertices that differ in some factor $j \neq i$. The two edges e' and $\tilde{e}''$ have an endpoint $x_i u$ in common, but are not related by τ since they are mapped to different factors. This is possible by the definition of τ only if the other endpoints $y_i u$ and $x_i v$ have a second common neighbor in $G(L)$, other than the vertex $x_i u$. Since these two vertices differ in coordinates i and j, the only other possible common neighbor is $y_i v$, so both $\tilde{e}' = \{x_i v, y_i v\}$ and $e'' = \{y_i u, y_i v\}$ are edges in $G(L)$. This proves that all edges of the form $\{x_i u, y_i u\}$ are in $G(L)$. Along the way, we have also proved that the edges of the form $\{y_i u, y_i v\}$ are in $G(L)$ as well, so all the edges in H are in $G(L)$. This completes the proof that every connected product G contained in $G(L)$ can be extended to a larger connected product H contained in $G(L)$, so we must eventually have $G(L) = G = \square_{i \in S} G_i = \square_{i \in S} G(M_i) = G(M)$. Therefore $L = M$ and g is a factorization. □

In the unweighted case, the relation $\theta_1 \cup \tau$ can be obtained in $O(nm)$ time using $O(m)$ space. First, determine edges related by θ_1 as before. To find edges related by τ, do the following. For each vertex x from $G(L)$ in turn, do the following. Build an adjacency vector for x indicating for each vertex z whether $\{x, z\}$ is an edge of $G(L)$ or not. Visit every vertex y and look up its neighbors z in the adjacency vector for x to determine whether there is exactly one neighbor z of y that is adjacent to x; if this is the case, then $\{x, z\}$ and $\{z, y\}$ are related by τ. There are n choices for x and m choices for $\{y, z\}$, giving an $O(nm)$ running time. The equivalence classes can again be maintained in $O(nm)$ time and $O(m)$ space by a union-find algorithm, since at most $m - 1$ union operations will be performed in $O(\log m)$ time per operation, and at most n^2 find operations (one per pair x, y) will be needed, taking $O(1)$ time per operation. The vector for x uses $O(n)$ space.

Theorem 6.13 *The prime factorization $g^{\theta_1 \cup \tau}$ for an unweighted graph with n vertices and m edges can be found in $O(nm)$ time using $O(m)$ space.*

One important difference between isometric representations and factorizations is that an isometric representation can have up to $k = n - 1$ factors, this maximum being achieved when G is a tree, while a factorization has at most $k = \lg n$ because a product of k nontrivial spaces has at least 2^k vertices. Therefore, for both representations, the best known time complexity is $O(km)$, where k is the upper bound on the number of factors, matching the complexity of obtaining the representation form the equivalence classes in Lemma 6.6. On the other hand, the bound in Lemma 6.6 can be improved to $O(m)$ for the equivalence classes of the prime factorization; it may be that a linear-time algorithm can be obtained for this representation.

6.5 The 2-Isometric Representation

The next step in our study consists of the definition of a new notion, that of q-isometric subspaces. The 1-isometric are arbitrary subspaces; the interesting case occurs for $q = 2$. In a hypercube, the 2-isometric subspaces coincide with the median sets. We shall first define an important function, the imprint function, for arbitrary metric spaces. The imprint function coincides with the median function in the case of median graphs, and thus constitutes a generalization of medians to arbitrary spaces. The study relies on a partial order associated with every metric space that views an arbitrarily chosen point as the smallest element of the partial order.

6.5.1 Closed Sets and the Imprint Function

Given a finite metric space M and a point x in M, we denote by $\leq_x$ the partial order on points in M defined by letting $y \leq_x z$ if y is between x and z, i.e., if $d(x,z) = d(x,y) + d(y,z)$, where $d = d_M$. Given a set of points U in M, we write $y \leq_x U$ if $y \leq_x z$ for all $z \in U$. If this is the case, we say that y is a *lower bound* of U with respect to x in M. The point y is a *maximal lower bound* of U with respect to x if $y \leq_x U$ and no point $y' \neq y$ with $y \leq_x y'$ satisfies $y' \leq_x U$. Note that a set U may have more than one maximal lower bound with respect to a given point x. By finiteness, every lower bound y with respect to x is less than or equal to some maximal lower bound. We say that y is a maximal lower bound of U in M if y is a maximal lower bound of U in M with respect to y itself; this is equivalent to the statement that y is a maximal lower bound of U in M with respect to some x in M. A set of points V is *closed under maximal lower bounds* with respect to x in M if for every nonempty subset $U \subseteq V$, all the maximal lower bounds of U with respect to x in M are in V. For brevity, we shall say that such a set V is x-*closed* in M. The following lemma gives a seemingly weaker yet equivalent characterization.

Lemma 6.14 *A set V of points in M is x-closed in M if and only if for every subset $U \subseteq V$ with $|U| = 2$, all the maximal lower bounds of U with respect to x in M are in V.*

Proof. The 'only if' direction is obvious. To prove the 'if' direction, suppose that V is not x-closed in G. Then there is a nonempty subset $U \subseteq V$ and a maximal lower bound y of U such that y is not in V. Pick a maximal such U. Let y' be a point in U closest to y. Since y is not in U, we have $y' \neq y$. By the definition of a lower bound, we must have $y \leq_x y'$, and by the definition of a maximal lower bound, there must exist some z in U such that $y' \leq_x z$ does not hold, with $y \leq_x z$ by the definition of a lower bound. Then y is a lower bound of $U' = \{y', z\}$, and there exists a maximal lower bound z' of U' with $y \leq_x z'$. We then have $z' \neq y'$, because $z' \leq_x z$. Since $y \leq_x z' \leq_x y'$ and $z' \neq y'$, the point z' is closer than y' to y, so z' is not in U. Since $y \leq_x z'$, the point y is also a maximal lower bound of $U \cup \{z'\}$, and this is impossible by the maximality of U unless z' is not in V. This shows that there is a set $U' \subseteq V$ with $|U'| = 2$ and a maximal lower bound z' of U' with respect to x such that z' is not in V. □

The *x-closure* of a set U in M is the smallest x-closed set V in M that contains U as a subset. The *imprint* of a nonempty set U with respect to x in M, denoted by $I_M(x, U)$, is the least element in the x-closure of U in M. This means that $I_M(x, U)$ is the point y in the x-closure V of U such that $y \leq_x z$ for all z in V. Such a point exists because V has a lower bound (the point x) and hence a maximal lower bound y that belongs to V because V is x-closed. It is also unique because two such y would be related to each other under $\leq_x$, contrary to the asymmetric property of partial orders.

For median graphs, one can show that given three vertices x, y, z, the vertex $\mathrm{med}(x, y, z)$ is a maximal lower bound of $U = \{y, z\}$ with respect to x, and that the set $\{\mathrm{med}(x, y, z), y, z\}$ is the x-closure of U. Therefore $\mathrm{med}(x, y, z) = I(x, \{y, z\})$, where I stands for the imprint function in the median graph. See the next section for details.

A set V is *M-closed* if it is x-closed in M for all points x in M. The *M-closure* of a set U is the smallest M-closed set V that contains U as a subset. The following lemma gives two useful alternative characterizations of M-closed sets.

Lemma 6.15 *The following statements are equivalent:*

1. *The set V is nonempty and M-closed;*

2. *For all points x in M, the point $I_M(x, V)$ is in V;*

3. *For all points x in M, there is a point y in V such that $y \leq_x V$.*

Proof. We give a cyclic chain of implications among the three statements. If V is M closed, then V is x-closed in M for all x in M, so $I_M(x, V)$ is in V. If $I_M(x, V)$ is in V for all x in M, then the point $y = I_M(x, V)$ in V satisfies $y \leq_x z$ for all $z \in V$. Finally, if for all points x in M there is a point $y \in V$ such that $y \leq_x z$ for all $z \in V$, then consider a nonempty subset $U \subseteq V$ and a maximal lower bound x' of U with respect to a point x. We have $x' \leq_x z$ for all $z \in U$, and a point $y \in V$ such that $y \leq_{x'} z$ for all $z \in V$, so $x' \leq_x y \leq_x z$ for all $z \in U$ and y is a lower bound of U with respect to x. Since x' is a maximal lower bound, we must have $y = x'$, so $x' \in V$. It follows that V is x-closed, for all points x in M, so V is M-closed. Therefore the three statements are equivalent. □

The notion of an M-closed set is quite strong. It implies in particular that if V is M-closed, then every point x on a shortest path joining two points y, z in V is also in V, because x is in that case a maximal lower bound of $\{y, z\}$. An important property of M-closed sets is the following.

Lemma 6.16 *If $U^1, U^2, \ldots, U^k$ are M-closed, and $U^i \cap U^j \neq \emptyset$ for all i, j, then $U^1 \cap U^2 \cap \cdots \cap U^k \neq \emptyset$.*

Proof. It is sufficient to consider the case $k = 3$; the result for arbitrary k follows inductively, by replacing the collection of U^i with a smaller collection $U^1 \cap U^2, U^3, \ldots, U^k$. For $k = 3$, let x^{ij} be a point in $U^i \cap U^j$ for $1 \leq i < j \leq 3$, and let y be a maximal lower

CHAPTER 6. METRIC NETWORKS AND PRODUCT GRAPHS   173

bound of $\{x^{13}, x^{23}\}$ with respect to x^{12}. Then y is in U^1 because it is between two points x^{12}, x^{13} in U^1; y is in U^2 because it is between two points x^{12}, x^{23} in U^2; and y is in U^3 because it is a maximal lower bound of two points x^{13}, x^{23} in U_3. □

The notions defined so far are all well-behaved with respect to cartesian products. Given a set U of points in a product, we denote by U_i the set of points x_i with x in U. Given sets U_i in the factors M_i of a product M, we call the set U of points x in M such that x_i is in U_i the *product* of the U_i.

Lemma 6.17 *Given a cartesian product $M = \square_{i \in S} M_i$, three points x, y, z in M satisfy $y \leq_x z$ in M if and only if $y_i \leq_{x_i} z_i$ in each M_i. If U is a set of points in M, and x, y are points in M, then y is a lower bound (maximal lower bound) of U with respect to x in M if and only if each y_i is a lower bound (resp. maximal lower bound) of U_i with respect to x_i in M_i. If V is the x-closure of U in M, then each V_i is the x_i-closure of U_i in M_i. In particular, the product of x_i-closed sets V_i is an x-closed set V. The imprint $y = I_M(x, U)$ has coordinates $y_i = I_{M_i}(x_i, U_i)$.*

Proof. Since $d(x, z) = \sum_i d(x_i, z_i) \leq \sum_i d(x_i, y_i) + d(y_i, z_i) = d(x, y) + d(y, z)$, we have $d(x, z) = d(x, y) + d(y, z)$ if and only if $d(x_i, z_i) = d(x_i, y_i) + d(y_i, z_i)$ for all i, so $y \leq_x z$ if and only if $y_i \leq_{x_i} z_i$ for all i. The characterization for lower bounds and maximal lower bounds follows immediately from this property, using the definitions in terms of $\leq_x$ and $\leq_{x_i}$. If V is the x-closure of U, then V is x-closed. This implies that every V_i must be x_i-closed, as follows. If y_i is a maximal lower bound of a subset U_i of V_i, then the points in U_i are the points u_i for u in some subset W of V, and so y_i can be extended to a maximal lower bound y of W by picking maximal lower bounds y_j of W_j with $j \neq i$ arbitrarily. Thus $y \in V$ and $y_i \in V_i$, proving that V_i is x_i-closed. Therefore each V_i contains the x_i-closure of U_i. On the other hand, using the characterization of maximal lower bounds, we see that the product over i of the x_i-closure of U_i is x-closed, so by minimality the x-closure V of U must be contained in this product, and V_i is contained in the x_i-closure of U_i, establishing equality. The statement about products follows from the fact that V is x-closed and the V_i are x_i-closed precisely when they equal their own x-closure and x_i-closure, respectively. The characterization of imprints follows immediately from this characterization of the x-closure and the definition of the imprint. □

The M-closed sets of cartesian products are particularly simple.

Lemma 6.18 *The M-closed sets of a cartesian product $M = \square_{i \in S} M_i$ are the sets U that are products of M_i-closed sets U_i.*

Proof. If U is M-closed, then each U_i is M_i-closed by the previous lemma. The product of M_i-closed sets U_i is M-closed by the previous lemma. We show that every M-closed set U is the product of the sets U_i. To prove this, it is sufficient to show that if x is a point in M, and x_i is in U_i for each i, then x is in U. This follows inductively from the fact that if x_S and x_T are in U_S and U_T respectively, then $x_{S \cup T}$ is in $U_{S \cup T}$. Finally, this fact holds because we have points of the form $x_S y_{T \setminus S}$ and $z_{S \setminus T} x_T$ in $U_{S \cup T}$, and the point $x_{S \cup T}$ is a maximal lower bound of these two. □

6.5.2 2-Isometric Subspaces

A subspace L of a metric space M is a *q-isometric* subspace if, for every point x in L and every set of points U in L with $1 \leq |U| \leq q$, if x is a maximal lower bound of U in L, then x is a maximal lower bound of U in M. This condition always holds if $|U| = 1$, so all subspaces are 1-isometric. However, not all subspaces are 2-isometric subspaces. For example, the space whose associated graph is a 6-cycle with edges of weight 1 is a subspace of the 3-cube. If we let x, y^1, y^2 be three points on the 6-cycle all at distance 2 from each other, then x is a maximal lower bound of $\{y^1, y^2\}$ with respect to x in the 6-cycle, but not in the 3-cube, showing that the 6-cycle is not a 2-isometric subspace of the cube. In general, the notions of a q-isometric subspace are all distinct and become stronger as q increases. For instance, consider the space M induced by the graph on $x, y^1, y^2, \ldots, y^q, z^0, z^1, z^2, \ldots, z^q$, with edges of weight 1 joining x to each z^i, and joining each z^i to each y^j with $j \neq i$, and let L be the subspace defined by all the points except z^0. Then L is a $(q-1)$-isometric subspace of M but not a q-isometric subspace of M, as the point x and the set $U = \{y^1, y^2, \ldots, y^q\}$ show.

The following lemma links 2-isometric subspaces with the notions from the last section.

Lemma 6.19 *Let L be a 2-isometric subspace of M, and let x be a point in L. Suppose that V is x-closed in M. Then $V \cap V(L)$ is x-closed in L.*

Proof. By Lemma 6.14, it is sufficient to show that if a point y in L is a maximal lower bound of U with respect to x in L, with $U \subseteq V \cap V(L)$ and $|U| = 2$, then $y \in V$. By the definition of a 2-isometric subspace, the fact that y is a maximal lower bound of U in L implies that y is also a maximal lower bound of U in M, and therefore a maximal lower bound of U in M with respect to x. Since V is x-closed, it follows that y is in V. □

We are mainly interested in irredundant 2-isometric subspaces of cartesian products. For such subspaces, the imprint function in the subspace has a simple characterization.

Lemma 6.20 *Let L be an irredundant subspace of a product $M = \square_{i \in S} M_i$. Suppose that for every point x in L and every x-closed set V in M, the set $V \cap V(L)$ is x-closed in L. Then for every x-closed set V in L, each V_i is x_i-closed in M_i. Furthermore, the imprint function I_M coincides with the imprint function I_L inside L.*

Proof. Suppose that V is x-closed in L. Two points y_i^1 and y_i^2 in V_i can be extended to points y^1 and y^2 in V. A maximal lower bound w_i of $U_i = \{y_i^1, y_i^2\}$ with respect to x_i in M_i can be extended to a point w in L, because L is irredundant. Since $w_i \leq_{x_i} y_i^1$, the set $W_i = \{w_i, y_i^1\}$ is x_i-closed in M_i, so the product W of W_i with the x_j-closed sets $V(M_j)$ for $j \neq i$ is x-closed in M by Lemma 6.17, and $W \cap V(L)$ is x-closed in L by hypothesis. It follows that there is a point $z \in W \cap V(L)$ such that $z \leq_x y$ for all $y \in W \cap V(L)$, and in particular $z \leq_x y^1$. We also have $z \leq_x w$, since $w \in W \cap V(L)$; hence $z_i = w_i$. A similar argument with $W_i = \{w_i, y_i^2\}$ gives a point z' with $z'_i = w_i$ and $z' \leq_x y^2$. Furthermore $z \leq_x z'$ and $z' \leq_x z$ because z and z' are in both of the sets W constructed for y^1 and y^2. Therefore $z = z'$ is a lower bound of $U = \{y^1, y^2\}$ with respect to x, with $z_i = w_i$. There

CHAPTER 6. METRIC NETWORKS AND PRODUCT GRAPHS   175

is therefore a maximal lower bound z'' of U with respect to x in L, with $z \leq_x z''$. Since V is x-closed in L, the point z'' is in V. Furthermore, z_i'' must be a lower bound of U_i with respect to x_i in M_i, and $w_i = z_i \leq_{x_i} z_i''$ for a maximal lower bound w_i of U_i, so $w_i = z_i''$. This shows that w_i is in V_i, so by Lemma 6.14 the set V_i is x_i-closed in M_i.

Given a point x and a set of points U in L, the imprint $t = I_L(x, U)$ is an element of the x-closure V of U in L with $t \leq_x y$ for all $y \in V$. Similarly, the imprint $t' = I_M(x, U)$ is an element of the x-closure V' of U in M with $t' \leq_x y$ for all $y \in V'$. Since V' is x-closed in M, the set $V' \cap V(L)$ containing U is x-closed in L by hypothesis, and must therefore contain the x-closure V of U in L. Therefore $t \in V'$ and $t' \leq_x t$. In the other direction, since V is x-closed in L, each V_i is x_i-closed in M_i, so the product W of the V_i is x-closed in M by Lemma 6.17, and must therefore contain the x-closure V' of U in M. Therefore $t' \in W$. Since $t \leq_x y$ for all $y \in V$, we have $t_i \leq_{x_i} y_i$ for all $y_i \in V_i$, so $t \leq_x y$ for all $y \in W$, and in particular $t \leq_x t'$. It follows that $t = t'$. □

The fact that imprints in M of points in L must also be in L gives a strong property for L.

Lemma 6.21 *Let L be an irredundant subspace of a product $M = \square_{i \in S} M_i$. Suppose that $I_M(x, U)$ is in L whenever x and U are in L and $|U| \leq q$. Then L is a q-isometric subspace of M.*

Proof. Only $q \geq 2$ is interesting. Given a point x in L and a set U with $|U| \leq q$ in L, suppose that x is not a maximal lower bound of U in M. Then x_i is not a maximal lower bound of U_i in M_i for some i, by Lemma 6.17. Let $w_i \neq x_i$ be a lower bound of U_i with respect to x_i in M_i. The point w_i in M_i can be extended to a point w in L, since L is irredundant. Let V be the set of points $I_M(x, \{w, y\})$ in L for $y \in U$, and consider the point $z = I_M(x, V)$ in L. Then $z \leq_x V$, so $z \leq_x I_M(x, \{w, y\}) \leq_x y$ for each $y \in U$. Therefore z is a lower bound of U with respect to x in L. For each $y \in U$, if $t = I_M(x, \{w, y\})$, then $t_i = I_{M_i}(x_i, \{w_i, y_i\}) = w_i$ because $w_i \leq_{x_i} y_i$, therefore $V_i = \{w_i\}$. Then $z_i = I_M(x, V_i) = w_i \neq x_i$, so $z \neq x$ and x is not a maximal lower bound of U in L. Therefore L is a q-isometric subspace of M. □

A set V of points in M is *closed under imprints* in M if for all points x, y^1, y^2 in V, the point $I_M(x, \{y^1, y^2\})$ is also in V. The preceding lemmas make it possible to give alternative characterizations for the 2-isometric irredundant subspaces of products.

Theorem 6.22 *Let L be an irredundant subspace of a cartesian product M. The following statements are equivalent.*

1. *L is a 2-isometric subspace of M;*

2. *L is a q-isometric subspace of M for all $q \geq 1$;*

3. *L is a subspace of M with the property that for every point x in L and every set of points V in M, if V is x-closed in M, then $V \cap V(L)$ is x-closed in L;*

4. L is closed under imprints in M.

Proof. The first property implies the third by Lemma 6.19. The third property implies the fourth by Lemma 6.20; in fact, it implies the stronger property that $I_M(x, U)$ is in L for all points x in L and sets U in L. This stronger property implies the second property by Lemma 6.21, which trivially implies the first. □

The existence of several equivalent characterizations seems to indicate that the 2-isometric irredundant subspaces of products constitute a natural class of subspaces. There are two alternative candidate characterizations that turn out to be too strong or too weak. The first one requires that given three points x, y^1, y^2 in L, every maximal lower bound of $\{y^1, y^2\}$ with respect to x in M must also be in L. This property is sufficient since it implies closure under imprints (property 4 above). However, not all 2-isometric irredundant subspaces of products satisfy this property. The following counterexample was found with Peter Winkler. Consider the product of two $K_{2,3}$ graphs, consisting of two points α, β, each adjacent to three points a, b, c, and at distance 1 of them, then the 2-isometric subspace consisting of the points x satisfying $(x_1 = x_2 \in \{a, b, c\}) \vee (x_1 = \alpha) \vee (x_2 = \alpha)$ does not enjoy the maximal lower bound property, since the point $\beta\beta$ is a maximal lower bound of $\{bb, cc\}$ with respect to aa, yet it is not in the subspace. The second candidate characterization requires the intersection of every M-closed set with L to be an L-closed set. This property is necessary by Lemma 6.19. On the other hand, is not sufficient, as illustrated in the product of two 5-cycles $(1, 2, 3, 4, 5)$, with edges of weight 1, by the subspace on the points $11, 23, 32, 44, 55$. However, we shall see in section 6.5.3 that this property becomes sufficient under the assumption that L is the range of an isometric representation.

We have seen that the intersection of an M-closed set with a 2-isometric subspace is an L-closed set. It turns out that if L is a 2-isometric irredudant subspace of M, then every L-closed set U can be represented as the interesection of an M-closed set V with L, where V is the product of the U_i. This intersection is L-closed, since V is M-closed because each $V_i = U_i$ is M_i-closed (using Lemmas 6.17 and 6.20). Furthermore, every point in this intersection must be in U. The proof is essentially the same as for Lemma 6.18, using the fact that if $x_{S \cup T}$ is in $L_{S \cup T}$, and the points x_S and x_T are in U_S and U_T respectively, then $x_{S \cup T}$ is in $U_{S \cup T}$ as a maximal lower bound of two points $x_S y_{T \setminus S}$ and $z_{S \setminus T} x_T$ in $U_{S \cup T}$.

As an example of 2-isometric subspaces, consider the case where M is the hypercube, so that each M_i consists of two points at distance 1. In this case, $I_{M_i}(x_i, \{y_i, z_i\})$ is the majority of x_i, y_i and z_i, so $I_M(x, \{y, z\}) = \text{med}(x, y, z)$ and the 2-isometric subspaces are the median sets by the fourth characterization. If each M_i is a clique (a set of points all at distance 1 from each other), then $I_{M_i}(x_i, \{y_i, z_i\})$ equals y_i if $y_i = z_i$, and equals x_i otherwise, so $I_H(x, \{y, z\})$ coincides with the imprint function of Chung, Graham, and Saks [21]. Note that already in this case, the imprint function is no longer symmetric in all three variables x, y, z.

In the hypercube case, we saw that the median sets are characterized by 2SAT instances. A similar characterization holds in general for 2-isometric subspaces and imprints.

Theorem 6.23 *A subspace L of a product $M = \square_{i \in S} M_i$ is a 2-isometric subspace of M if and only if L contains precisely those points x in M such that for all sets T with $|T| \leq 2$ the point x_T is in L_T, and each such L_T is a 2-isometric subspace of M_T.*

Proof. Suppose that L is a 2-isometric subspace of M. We show that L_T is a 2-isometric subspace of M_T, for all $T \subseteq S$. Suppose that x_T, y_T^1, y_T^2 are points in L_T, and that x_T is not a maximal lower bound of $U_T = \{y_T^1, y_T^2\}$ in M_T. Then there is a point z_T in M_T such that $z_T \neq x_T$ and z_T is a lower bound of U_T with respect to x_T in M_T. Extend x_T, y_T^1, y_T^2 to points x, y^1, y^2 in L, with x as close as possible to y^1, and extend z_T to a point z in M with $z_{S \setminus T} = x_{S \setminus T}$. Then $z \neq x$ and z is a lower bound of $U = \{y^1, y^2\}$ with respect to x in M. Since L is a 2-isometric subspace, it must have a lower bound $z' \neq x$ of $U = \{y^1, y^2\}$ with respect to x in L. Since z' is closer than x to y^1, it must have $z'_T \neq x_T$. Furthermore, z'_T is a lower bound of U_T with respect to x_T in L_T, proving that x_T is not a maximal lower bound of U_T in L_T. Therefore L_T is a 2-isometric subspace of M_T.

Clearly, for all points x in L, the point x_T is in L_T. We show that if x_T is in L_T for all T with $|T| \leq 2$, then x is in L. We do this by showing that $x_{T'}$ is in $L_{T'}$ for all $T' \subseteq S$, by induction on $|T'|$. The base case $|T'| \leq 2$ holds by assumption. If $|T'| > 2$, choose two distinct $i, i' \in T'$, and use the inductive assumption to infer the existence of points y^0, y^1, y^2 in L such that $y^0_{i,i'} = x_{i,i'}$, $y^1_{T' \setminus \{i\}} = x_{T' \setminus \{i\}}$, and $y^2_{T' \setminus \{i'\}} = x_{T' \setminus \{i'\}}$. Since L is an irredundant 2-isometric subspace of the product $M' = \square_{j \in S} L_j$, the imprint function $I_{M'}$ coincides with I_L in L, and imprints in M' can be decomposed into coordinate-wise imprints in each $M'_j = L_j$ by Lemma 6.17. Therefore $z = I_L(y^0, \{y^1, y^2\}) = I_{M'}(y^0, \{y^1, y^2\})$ has $z_j = I_{L_j}(y^0_j, \{y^1_j, y^2_j\}) = x_j$, since two out of the three points y^0_j, y^1_j, y^2_j agree with x_j, for all $j \in T'$. This shows that $x_{T'} = z_{T'}$ is in $L_{T'}$, completing the induction and the 'only if' part of the lemma.

In the other direction, suppose that L contains precisely those points x in M such that for all sets T with $|T| \leq 2$ the point x_T is in L_T, and that each such L_T is a 2-isometric subspace of M_T. We want to show that L is a 2-isometric subspace of M. We first show that L is a 2-isometric subspace of $M' = \square_{j \in S} L_j$. Since L is an irredundant subspace of M', it is sufficient to show that L is closed under imprints in M'. If $x = I_{M'}(y^0, \{y^1, y^2\})$ with y^0, y^1, y^2 in L, then $x_T = I_{M'_T}(y^0_T, \{y^1_T, y^2_T\})$ is in L_T for all T with $|T| \leq 2$, since each such L_T is an irredundant 2-isometric subspace of M'_T and therefore closed under imprints in M'_T. This implies, by assumption, that x is in L, proving that L is closed under imprints in M' and therefore a 2-isometric subspace of M'.

We finally show that the fact that L is a 2-isometric subspace of $M' = \square_{j \in S} L_j$ and each L_j is a 2-isometric subspace of M_j implies that L is a 2-isometric subspace of M. If y is a maximal lower bound of some set $\{z^1, z^2\}$ in L, where all three points are in L, then it is also a maximal lower bound in M', since L is a 2-isometric subspace of M'. Therefore y_i is a maximal lower bound of $\{z_i^1, z_i^2\}$ in $M'_i = L_i$ for all i, by Lemma 6.17. Since L_i is a 2-isometric subspace of M_i, it follows that y_i is a maximal lower bound of $\{z_i^1, z_i^2\}$ in M_i as well, for all i, and therefore y is a maximal lower bound of $\{z^1, z^2\}$ in M. This shows that L is a 2-isometric subspace of M, completing the 'if' direction as well. □

A shorter proof of the fact that if x_T is in L_T for all T with $|T| \leq 2$, then x is in L, is the following. Let U^j be the set of points y in M with $y_j = x_j$. The set U^j is a product of M_i-closed sets U_i^j, hence M-closed. Therefore $V^j = U^j \cap V(L)$ is L-closed. Furthermore, $V^j \cap V^{j'}$ is nonempty since $x_{\{j,j'\}}$ is in it. Therefore $\{x\} \cap V(L) = \bigcap_{j \in S} V^j$ is nonempty, and x is in L. Note that this proof uses only the assumption that the intersection of M-closed sets with L is L-closed, which is weaker than the assumption that L is a 2-isometric subspace of M.

The connection in the previous lemma between closure under imprints and the property that membership in L can be reduced to testing two coordinates at a time, can be further generalized.

Given a set S and a subset T of S^k, let $T_{i_1, i_2, \ldots, i_k}$ be the subset of S^n consisting of those elements $x = (x_1, x_2, \ldots, x_n)$ satisfying $(x_{i_1}, x_{i_2}, \ldots, x_{i_k}) \in T$. Suppose that U is the intersection of such subsets, obtained from different choices for $1 \leq i_1, i_2, \ldots, i_k \leq n$. We say that U satisfies the l-Helly property if given an assignment of values to some r of the x_i, if for every choice of l out of the r assigned x_i, some element y of U assigns the same values to those l chosen x_i, then some element x of U gives those preassigned values to all r of them (see also Quillot [97]). Given S, we say that a subset T of S^k satisfies the l-Helly property under intersections if for every subset U of some S^n, obtained as an intersection of subsets obtained from T as above, the subset U satisfies the l-Helly property.

Consider now the following property for T. There exists a mapping $f : S^{l+1} \to S$ such that (1) if all but at most one of the $l+1$ arguments of f have the same value a, then the value of f is also a, and (2) given $l+1$ k-tuples $x^j = (x_1^j, x_2^j, \ldots, x_k^j)$, for $1 \leq j \leq l+1$, if each of these k-tuples is in T, then the k-tuple $y = (y_1, y_2, \ldots, y_k)$ is also in T, where $y_i = f(x_i^1, x_i^2, \ldots, x_i^{l+1})$.

It turns out that the property that given S, the subset T of S^k satisfies the l-Helly property under intersections is equivalent to the existence of and appropriate mapping f as above. In fact, we may use several different T for the intersections, and the l-Helly property under intersections is then equivalent to the existence of a common mapping f with the properties stated above for the sets T. A basic example for $S = \{0, 1\}$ consists of the subsets T of S^2, which are just the 2SAT clauses, in which case U is the solution set of a 2SAT instance and satisfies the 2-Helly property. Then the mapping f is just majority, $f(a, b, c) = \text{maj}(a, b, c)$, and the application of f to each coordinate is then the median function. More generally, imprints play the same role with regards to 2-isometric subspaces.

The link between the l-Helly property and imprint-like functions, and some of its computational implications, is studied in Feder and Vardi [33]. We sketch here a proof of the equivalence stated above. The fact that the existence of an f with properties (1) and (2) implies the l-Helly property is a straightforward extension of the argument for medians and more generally for imprints; the case $l = 2$ is the most natural one. We show the converse. Suppose that we treat each of the $|S|^{l+1}$ values for f as a variable z_i. Condition (2) defines a set of values for the z_i which is obtained from T under intersections. Condition (1) assigns values to specific z_i. But if we only require l of these prescribed value assignments, then they correspond to l choices of at most one of the $l+1$ argument positions for f, namely the one, if any, that appears as the exceptional one in condition (1). Therefore some argument i is not the exceptional argument in any of the l chosen assignments, so f may return the value of argument i and satisfy condition (2), as well as condition (1) for the l choices of prescribed values. But then, by the l-Helly property, there must exist values for f that satisfy all conditions.

6.5.3 2-Isometric Representations

A representation g for a metric space L in a cartesian product M is q-isometric if g is an isometric representation and $g(L)$ is a q-isometric subspace of M. Since all subspaces are 1-isometric, the 1-isometric representations are just the isometric representations. Since $g(L)$ is an irredundant subspace of M by the definition of a representation, it is a q-isometric subspace for any given $q \geq 2$ if and only if it is a 2-isometric subspace, so the q-isometric representations with $q \geq 2$ are the same as the 2-isometric representations. A subspace L of a space M is a *connected subspace* if the intersection of $G(L)$ and $G(M)$ is a connected graph. The following lemma characterizes the 2-isometric representations as the isomorphisms from L to $g(L)$ such that $g(L)$ is a connected 2-isometric irredundant subspace of a product M.

Lemma 6.24 *The 2-isometric representations for a metric space L in a product $M = \Box_{i \in S} M_i$ can be defined as the isomorphisms g from L to a 2-isometric irredundant subspace $L' = g(L)$ of M satisfying a chosen condition from the following:*

1. *Adjacent points in L are mapped by g to points that differ in only one coordinate;*

2. *The graph $G(L')$ is a subgraph of $G(M)$;*

3. *The intersection of $G(L')$ and $G(M)$ is connected.*

Proof. Condition (1) coincides with the definition of a representation. Suppose that condition (1) holds. If x, y are adjacent in L, then $x' = g(x)$, $y' = g(y)$ differ in only one coordinate i. If x' and y' are not adjacent in M, then there is a point z'_i strictly between x'_i and y'_i in M_i, and z'_i can be extended to a point z' in L' by irredundancy. The imprint $I_M(x', \{y', z'\}) = I_{L'}(x', \{y', z'\})$ is then a point $t' = g(t)$ in L' strictly between x' and y', because $t' \leq_{x'} y'$ and $t'_i = I_{M_i}(x'_i, \{y'_i, z'_i\}) = z'_i$ is different from x'_i and from y'_i. This implies that t is strictly between x and y, contradicting the fact that x and y are adjacent in L, so x' and y' must be adjacent in M as well. This proves condition (2). Suppose instead that condition (2) holds. Then the intersection of $G(L')$ and $G(M)$ is just $G(L')$, and all graphs induced by metric spaces are connected, so condition (3) holds. Suppose finally that condition (3) holds. Let x and y be two points adjacent in L, and let $x' = g(x)$,

$y' = g(y)$. Consider a path $x' = z^0, z^1, \ldots, z^r = y'$ from x' to y' in the intersection of $G(L')$ and $G(M)$, which exists by connectedness. Consecutive points z^j, z^{j+1} on the path are adjacent in $G(M)$ and therefore differ in only one coordinate. As a result, the points $t^j = I_M(x', \{y', z^j\}) = I_L(x', \{y', z^j\}) = g(u^j)$ in L have the property that consecutive points t^j, t^{j+1} differ in at most one coordinate, since imprints in M can be decomposed into imprints in each M_i separately. By the definition of imprint, each t^j is between x' and y', with $t^0 = x'$ and $t^r = y'$. But if t^j is strictly between x' and y', then u^j is strictly between x and y in L, contradicting adjacency. Therefore all t^j are equal to one of x', y', with $t^j = x'$ and $t^{j+1} = y'$ for some value of j. This implies that x' and y' differ in only one coordinate, so condition (1) holds. □

Given a subspace L' of a product M, let G' be the subgraph of $G(M)$ induced by the points in $G(L')$. Note that G' is a subgraph of $G(L')$. If condition (2) in the above lemma holds, then $G' = G(L')$. On the other hand, if we require in the lemma that g be an isomorphism from L to $M(G')$ instead of an isomorphism from L to L', then the intersection of $G(L')$ and $G(M)$ must be connected since G' is connected because it is the same as $G(L)$ under the isomorphism g, so condition (3) must hold for L', and the lemma applied to the identity mapping on L' gives condition (2) as well. Then $G(L') = G'$ and $L' = M(G')$ so that g is indeed an isomorphism from L to L', and g is a 2-isometric representation.

For subspaces L' of M that come from isometric representations, the condition that requires the intersection of M-closed sets with L' to be L'-closed is sufficient to prove that L' is a 2-isometric subspace.

Lemma 6.25 *Let g be an isometric representation for L in a product $M = \Box_{i \in S} M_i$, let $L' = g(L)$, and suppose that the intersection of every M-closed set with L' is L'-closed. Then the representation is 2-isometric.*

Proof. We claim that given a point x in L', a coordinate i, and two points b, c in M_i such that $b \leq_{x_i} c$, the unique points y and z in L' with $y_i = b$ and $z_i = c$ closest to x satisfy $y \leq_x z$. The points y, z are guaranteed to exist since the points in M with $y_i = b$, and those with $z_i = c$, are M-closed, so their intersection with L' is L'-closed, and also nonempty by irredundancy. This fact implies the lemma, because of the following. Given a point x in L' and a set $U = \{y^1, y^2\}$ of points in L', if x is not a maximal lower bound of U in M, then there is a lower bound t of U with respect to x with $t \neq x$. Then $t_i \neq x_i$ for some i, and $t_i \leq_{x_i} U_i$, so letting w, z^1, z^2 be the points in L with $w_i = t_i$, $z_i^1 = y_i^1$ and $z_i^2 = y_i^2$ closest to x, we have $w \leq_x z^1 \leq_x y^1$ and $w \leq_x z^2 \leq_x y^2$ by the claim, so $w \neq x$ is a lower bound of U with respect to x in L', proving that x is not a maximal lower bound of U in L' either.

To prove the claim, we can assume that b and c are adjacent in M_i. Otherwise, we can introduce a point b' strictly between b and c in M_i, assume inductively that the claim holds for b, b' and b', c, and prove from this that the claim holds for b, c. Assume then that b and c are adjacent in M_i. The proof is by induction on $d(x, y)$, where y and z are the two unique points in L' with $y_i = b$ and $z_i = c$ closest to x. Consider a shortest path joining

CHAPTER 6. METRIC NETWORKS AND PRODUCT GRAPHS   181

y and z in L'. There must be two consecutive points y', z' on this path such that $y'_i \neq z'_i$. By condition (1) in the definition of a representation, we must have $y'_{S\setminus\{i\}} = z'_{S\setminus\{i\}} = u$ for some u. Since there is no point strictly between b and c in M_i, we must have $y'_i = b$ and $z'_i = c$. Let V be the set of points w' in M such that w'_i is in the M_i-closure V_i of $\{b, c\}$. Since V is M-closed, the intersection $V' = V \cap V(L)$ is L'-closed, and V' contains y and z. If x is not in V', then the unique point x' in V' closest to x satisfies $x' \leq_x y$ and $y' \leq_x z$ with $x' \neq x$, so replacing x by x' gives an instance of the claim with $d(x', y) < d(x, y)$, and the claim follows by induction.

If x is in V', then x_i is in V_i. The set of all points of the from vu with v in M_i is M-closed, so its intersection W with L' is L'-closed. It follows that W_i is M_i-closed, because a point closest to some w'_i in W_i can be obtained from a point closest to w' in W, where w' is a point in L' consistent with w'_i. Since $\{b, c\} \subseteq W_i$, the M_i closure of $\{b, c\}$ satisfies $V_i \subseteq W_i$, so x_i is in W_i, and $x_i u$ is in W, hence in L'. Since $z \leq_x z'$, the point $z_{S\setminus\{i\}}$ is between $x_{S\setminus\{i\}}$ and u, so the fact that both x and $x_i u$ are in L' implies that the point z'' closest to z in L' satisfying $z''_i = x_i$ must have $z''_{S\setminus\{i\}} = z_{S\setminus\{i\}}$. Since y_i is between x_i and z_i, and both $x_i z_{S\setminus\{i\}}$ and z are in L', the point y'' closest to y in L' satisfying $y''_{S\setminus\{i\}} = z_{S\setminus\{i\}}$ must have $y''_i = y_i$. Then $y \leq_x y'' \leq_x z$. □

Unlike the case of the previous lemma, it is not sufficient here to assume that L is isomorphic to $M(G')$, where G' is induced by the points in L'; a counterexample is in that case provided by the non-isometric representation for a 6-cycle in the product of two 3-cycles, where the edges of the 6-cycle alternate between the two factors.

The analog of Lemma 6.7 holds for 2-isometric representations as well.

Lemma 6.26 *Suppose that a representation g for L in a product $M = \square_{j \in S} M_j$ contains a representation g' for L in $M' = \square_{i \in S'} M'_i$, with the containment given by representations h_i for M'_i in M_{S_i}. Then the representation g is 2-isometric if and only if the representation g' and the representations h_i are 2-isometric.*

Proof. We can restrict our attention to isometric representations by Lemma 6.7, since all 2-isometric representations are isometric. Under an isometric representation, every space is isomorphic to its image, so we can identify the two and let the representation mapping be the identity mapping. Assume then that L is an irredundant subspace of $M = \square_{j \in S} M_j$ and of $M' = \square_{i \in S'} M'_i$, and that each M'_i is an irredundant subspace of M_{S_i}. Since L is an irredundant subspace of M', the space M'_i is just the subspace L_{S_i} of M_{S_i}. We must show that L is a 2-isometric subspace of $M = \square_{i \in S'} M_{S_i}$ if and only if L is a 2-isometric subspace of $M' = \square_{i \in S'} L_{S_i}$ and each L_{S_i} is a 2-isometric subspace of M_{S_i}. If L is a 2-isometric subspace of M, then it is a 2-isometric subspace of M' since M' is a subspace of M containing L. The fact that L_{S_i} is a 2-isometric subspace of M_{S_i} was shown at the beginning of the proof of Theorem 6.23. The implication in the other direction was shown at the end of the proof of Theorem 6.23. □

A *canonical* 2-isometric representation for a metric space L in a product is a 2-isometric representation for L that contains all others. The idea for constructing such a representation is once again to extend the relation θ of the canonical isometric representation with an additional relation κ. Given two edges e and e' in $G(L)$, let $e \kappa e'$ if there exist points x, y, z, t, u in L with $e = \{x, y\}$ and $e' = \{z, t\}$, such that the point x is a maximal lower bound of $\{t, u\}$ in L, while the point y satisfies $z \leq_y t \leq_y u$.

Theorem 6.27 *The relation $\theta \cup \kappa$ induces a canonical 2-isometric representation for L in a cartesian product $M = \square_{i \in S} M_i$. The 2-isometric representations for L are the representations contained in $g^{\theta \cup \kappa}$, and are given by the mappings g^γ with $\theta \subseteq \gamma^*$.*

Proof. Since 2-isometric representations are isometric representations, they must be of the form $g = g^\gamma$ with $\theta \subseteq \gamma^*$. Any such representation is in turn isometric, so we only need to consider isometric representations in the proof. We can then identify L with its isomorphic image $g(L)$ in M. We show that if g is 2-isometric, then we must have $\kappa \subseteq \gamma^*$ as well. Suppose that e and e' are related by κ but belong to different equivalence classes of γ^*, and therefore to different factors i and j respectively in M. We use the names for edges e, e' and vertices x, y, z, t, u from the definition of κ. Since $z \leq_y t \leq_y u$, we have $z_j \leq_{y_j} t_j \leq_{y_j} u_j$. Furthermore $z_j \neq t_j$ since e' belongs to factor j, so $y_j \neq t_j$. This implies $x_j \neq t_j$ since x and y only differ in factor $i \neq j$ and must therefore have $x_j = y_j$. Consider the point w in M given by $w_{S \setminus \{j\}} = x_{S \setminus \{j\}}$ and $w_j = t_j$. Note that $x_j \neq w_j$. From these observations, we get $w_j \leq_{x_j} t_j$ and $w_j \leq_{x_j} u_j$; furthermore, $w_{S \setminus \{j\}} \leq_{x_{S \setminus \{j\}}} t_{S \setminus \{j\}}$ and $w_{S \setminus \{j\}} \leq_{x_{S \setminus \{j\}}} u_{S \setminus \{j\}}$ trivially. Therefore $w \leq_x t$ and $w \leq_x u$ with $w \neq x$, proving that x is not a maximal lower bound of $\{t, u\}$ in M. Since L is a 2-isometric subspace of M, it follows that x is not a maximal lower bound of $\{t, u\}$ in L either. But this is contrary to the condition $e \kappa e'$, so the assumption was false and the edges e and e' must indeed belong to the same equivalence class. Therefore $\theta \cup \kappa \subseteq \gamma^*$.

In the other direction, we show that if $\theta \cup \kappa \subseteq \gamma$, then $g = g^\gamma$ is a 2-isometric representation. Recall that g is an isometric representation. Let v, u^1 and u^2 be three points in L, and suppose that v is not a maximal lower bound of u^1 and u^2 in M. Then there is some point $w \neq v$ in M such that $w \leq_v u^1$ and $w \leq_v u^2$. Choose such a w closest to v, so that w is adjacent to v in M. Let i be the factor such that $w_i \neq v_i$; note that w_i is adjacent to v_i. Since L is an irredundant subspace of M, there is a point t in L with $t_i = w_i$. Choose such a t closest to v. Note that t_i is adjacent to v_i with $t_i \leq_{v_i} u_i^1$ and $t_i \leq_{v_i} u_i^2$, and that $t \neq v$. If $t \leq_v u^1$ and $t \leq_v u^2$, then v is not a maximal lower bound of u^1 and u^2 in L either, satisfying then the condition in the definition of a 2-isometric subspace.

Suppose that $t \leq_v u^1$ does not hold. The minimality condition in the choice of t implies that the last edge on a shortest path from v to t in $G(L)$ must be an edge $e' = \{z, t\}$ that belongs to factor i, because otherwise z would be a choice for t closer to v. Furthermore, $z \leq_v t$, so $z_i \leq_{v_i} t_i$ and therefore $z_i = v_i$, because v_i and t_i are adjacent and $z_i \neq t_i$. This determines z uniquely, so $e' = \{z, t\}$ is the last edge on every shortest path from v to t, and therefore every point $x' \neq t$ in L satisfying $x' \leq_v t$ must also satisfy $x' \leq_v z$. Also $t_i \leq_{z_i} u_i^1$

because $z_i = v_i$, so we must have $t \leq_z u^1$, since z and t agree in all coordinates other than i. Let x be a maximal lower bound of $\{t, u^1\}$ in L with $x \leq_v z$; choose such an x closest to z. Such an x exists, because we can let x be a maximal lower bound of $\{t, u^1\}$ with respect to v in L, and then $x \neq t$ since $t \leq_v u^1$ does not holds by assumption, implying that $x \leq_v z$. Note that $x \neq z$, since $t \leq_z u^1$ so that z is not a maximal lower bound of $\{t, u^1\}$ in L. Let $e = \{x, y\}$ be the first edge on a shortest path from x to z in $G(L)$, so that $x \leq_v y \leq_v z \leq_v t$, and in particular $z \leq_y t$. Furthermore $v_i = x_i = y_i = z_i$ since $v_i = z_i$, so e does not belong to factor i. Consider a maximal lower bound t' of $\{t, u^1\}$ with respect to y in L. Then $t' \leq_v t$, so if $t' \neq t$ then we must have $t' \leq_v z$, and t' satisfies the conditions defining x but is closer than x to z, contradicting the minimality of x. Therefore $t' = t$ and $t \leq_y u^1$. This gives $z \leq_y t \leq_y u^1$. The five points x, y, z, t, u^1 now satisfy the conditions for $e \kappa e'$. This is not possible because e and e' are not related by γ^*, since e' belongs to factor i and e does not. Therefore the assumption that $t \leq_v u^1$ does not hold was false. A similar argument applies to u^2, so the condition in the definition of a 2-isometric subspace is indeed satisfied. □

As in the case of θ, it is possible to avoid the obvious implementation of $\theta \cup \kappa$ and thus obtain an algorithm for the canonical 2-isometric representation in the unweighted case that runs in $O(nm)$ time and uses $O(m)$ space. One application of this representation algorithm is the recognition of median graphs, which are just the graphs mapped by the canonical 2-isometric representation into a hypercube. Another application will be mentioned in section 6.7. This representation algorithm, together with those for the canonical isometric representation and the cartesian prime factorization, can be parallelized so that these representation problems belong to the class $\mathcal{NC}$. See [31] for details on all these algorithms. In the following sections, we shall examine a class of subspaces whose structure allows a canonical representation theorem, but where by contrast, finding such a representation is computationally hard.

6.6 Retracts

A mapping f from a metric space M to itself is nonexpansive if $d(f(x), f(y)) \leq d(x, y)$ for all points x, y in M. A nonexpansive mapping f on a metric space M is a *retraction* if $f^{(2)} = f$, i.e., if $f(f(x)) = f(x)$ for all points x in M. This means that all the elements $y = f(x)$ of the range $f(M)$ satisfy $f(y) = y$ and are *fixed points* of f, so that the range and the fixed point set of f coincide. A *retract* of a metric space M is a metric space L such that $L = f(M)$ for some retraction f. The study of retractions and retracts for graphs was initiated by Banaschewski, Bruns [8], and Hell [56]. The connection with retractions and retracts on metric spaces was made explicitly by Quillot [98]. For a study of retracts on a different notion of product, see Jawhari, Misane, and Pouzet [67]. The problem of recognizing the retracts of graphs has been studied in some special cases (see, e.g., Quillot [96]).

As an example, if we consider the graph $K_{2,3}$ consisting of two points α, β adjacent to three points a, b, c, then the subspace of $K_{2,3} \square K_{2,3}$ defined in section 6.5.2, consisting of points satisfying $(x_1 = x_2 \in \{a, b, c\}) \vee (x_1 = \alpha) \vee (x_2 = \alpha)$, is a retract. The corresponding retraction has $f(x_1\beta) = \alpha x_1$ for $x_1 \in \{a, b, c\}$, $f(\beta x_2) = x_2\alpha$ for $x_2 \in \{a, b, c\}$, $f(x) = x$ for all x in the subspace, and $f(x) = \alpha\alpha$ for all remaining points.

A *retract representation* for a metric space L is an isometric representation g for L in a cartesian product M such that $g(L)$ is a retract of M.

Before examining the retracts of metric spaces, we shall briefly examine a notion stronger than the q-isometric subspaces but weaker than retracts, namely the subspaces without q-holes, and use it to study the distance center of subspaces of cartesian products. This notion was introduced by Nowakowski and Rival [88] in their study of retracts of graphs.

6.6.1 Subspaces without Holes and the Distance Center

Given a subspace L of a metric space M, a *q-hole* of L in M is a set U of points in L with $|U| \leq q$ and a point z in M such that there is no point z' in L with $d(z', y) \leq d(z, y)$ for all $y \in U$. The metric space L is a subspace of M without q-holes if there is no q-hole of L in M. A subspace *without $(q+1)$-holes* is always a connected q-isometric subspace. For if x in L is not a maximal lower bound in M of some set U of points in L with $|U| \leq q$, then there is a point z in M with $d(x, z) > 0$ and $d(x, y) = d(x, z) + d(z, y)$. Since $\{x\} \cup U$ is not a $(q+1)$-hole, there must be a point z' in L with $d(x, z') \leq d(x, z)$, and therefore $d(x, z') = d(x, z)$ since $d(x, y) \leq d(x, z') + d(z', y)$. This shows that x is not a maximal lower bound of U in L either. Connectivity follows from the fact that points adjacent in L are also adjacent in M, since otherwise we would have a 2-hole. A subspace *without holes* is a subspace without q-holes for all q. This condition is equivalent to the statement that there exists a mapping f from M to L such that $d(f(z), y) \leq d(z, y)$ for all points z in M and y in L, since we get such a function when $q = |V(L)|$. A retract is in particular a subspace without holes, because if f is the corresponding retraction, then $d(f(z), y) = d(f(z), f(y)) \leq d(z, y)$ for all points z in M and y in the retract $L = f(M)$.

The *distance center* of a metric space M is the set U of points x in M that minimize $\sum_{y \in V(M)} d(x, y)$.

Theorem 6.28 *Let L be a subspace without holes of a product space $M = \square_{i \in S} M_i$, and let U be the distance center of L. Then U is a subproduct of M.*

Proof. Let V be the set of points x in M that minimize $\sum_{y \in V(L)} d(x, y)$. Note that

$$\sum_{y \in V(L)} d(x, y) = \sum_{y \in V(L)} \sum_{i \in S} d(x_i, y_i) = \sum_{i \in S} \left(\sum_{y \in V(L)} d(x_i, y_i) \right),$$

so the set V is the product of sets V_i, where V_i consists of the points x_i in M_i that minimize $\sum_{y \in V(L)} d(x_i, y_i)$. If f is the mapping associated with L as a subspace without holes, then $\sum_{y \in V(L)} d(f(x), y) \leq \sum_{y \in V(L)} d(x, y)$ for all $x \in V(M)$, and the inequality is strict if x does not belong to L because in that case we have $d(f(x), y) = 0 < d(x, y)$ for $y = f(x)$.

CHAPTER 6. METRIC NETWORKS AND PRODUCT GRAPHS   185

Therefore all the points in M that minimize the sum defining V are in L, so the distance center U satisfies $U = V$ and is therefore a product. □

We seek a subproduct with stronger properties than the distance center.

Lemma 6.29 *Let L be a 2-isometric subspace of a product $M = \Box_{i \in S} M_i$ and let $U \subseteq V(L)$ be a subproduct of M. Then the L-closure of U is a subproduct of M.*

Proof. We can assume without loss of generality that L is an irredundant subspace of M, simply by replacing M_i with L_i. Let V be a maximal subproduct containing U and contained in V', the L-closure of U. We claim that $V = V'$. If $V \neq V'$, then V is not L-closed, so by Lemma 6.14 there is a point $x \in V(L) - V$ and points $y^1, y^2 \in V$ such that x is a maximal lower bound of $\{y^1, y^2\}$ in L. Since x is not in V, and V is a product, there is a coordinate i such that $x_i \notin V_i$. Note that x_i is a maximal lower bound of y_i^1, y_i^2 in M_i by Lemma 6.17. Let $z^1, z^2 \in V$ be any two points with $z_i^1 = y_i^1$, $z_i^2 = y_i^2$, and $z_{S \setminus \{i\}}^1 = z_{S \setminus \{i\}}^2$ for $j \neq i$. Then the point $t = I_L(x, \{z^1, z^2\}) = I_M(x, \{z^1, z^2\})$ has $t_i = x_i$ and $t_{S \setminus \{i\}} = z_{S \setminus \{i\}}^1 = z_{S \setminus \{i\}}^2$, and furthermore t is in the L-closure V'. This shows that V is not a maximal subproduct containing U and contained in V', because we can replace V_i with $V_i \cup \{x_i\}$ and thus obtain a larger subproduct. Therefore $V' = V$, and the L-closure V' is a subproduct. □

The M-closed sets have strong properties inside M.

Lemma 6.30 *Let U be an M-closed set. Then the subspace of M induced by U is a retract of M.*

Proof. Since U is M-closed, every point $x \in V(M)$ has a unique point $y = f(x) \in U$ such that $y \leq_x z$ for all $z \in U$. Clearly $f(x) \in U$ for all $x \in V(M)$, and $f(x) = x$ for all $x \in U$. Furthermore, given two points $x, y \in V(M)$, say with $d(x, f(x)) \leq d(y, f(y))$, we have $d(y, x) + d(x, f(x)) \geq d(y, f(x)) = d(y, f(y)) + d(f(y), f(x))$, so $d(f(x), f(y)) \leq d(x, y)$. Therefore f is nonexpansive, and hence and a retraction with the subspace induced by U as retract. □

Combining these three results, we have:

Theorem 6.31 *Given a subspace without holes L of a product M, the subspace of L induced by the L-closure of the distance center of L is a subproduct of M and a retract of L.*

6.6.2 Partial Mappings and Projections

The main difference between retracts and the other kinds of subspaces that we have studied so far is that the retract condition is a global condition, and this makes it much harder to determine whether a subspace is a retract. As a simple example of this difficulty, we have the following:

Lemma 6.32 *Given an unweighted graph H_1 and a vertex $b \in V(H_1)$, consider the product $H = H_1 \square K_2$ of H_1 with the unweighted graph K_2 consisting of a single edge on $V(K_2) = \{0, 1\}$. Let G be the subgraph of H induced by $V(G) = \{x = x_1 x_2 : (x_1 = a) \vee (x_2 = 0)\}$. Then the question of whether $M(G)$ is a retract of $M(H)$ for some given H_1 and b is $\mathcal{NP}$-complete.*

Proof. A retraction on $M(H)$ that gives $M(G)$ as a retract is a nonexpansive mapping mapping f from $M(H)$ to $M(G)$ such that $f(x) = x$ if $x_1 = a$ or $x_2 = 0$, with $f(x) \in V(G)$ if $x_1 \neq a$ and $x_2 = 1$. Note that a nonexpansive mapping in the unweighted case is just a mapping that maps adjacent vertices to vertices that are either adjacent or equal. To specify f, we only need to determine $f(x)$ in this second case $x = x_1 1$ with $x_1 \neq a$. Such an x is adjacent to $x_1 0 \in V(G)$, so $f(x)$ is adjacent or equal to $x_1 0$, and therefore $f(x) \neq a1$. It follows that $f(x) = y_1 0$ for $x = x_1 1$ and $x_1 \neq a$, and to specify f we only need to specify the mapping $y_1 = f'(x_1)$ that gives the first coordinate of $f(x)$ for $x = x_1 1$. Expressed in terms of f', the conditions on f state that f' must be a nonexpansive mapping on H_1, with $f'(x_1)$ adjacent or equal to x_1 and $f'(x_1) = a$ if x_1 is adjacent to a in H_1. (The last condition expresses the fact that if $x_1 \neq a$ is adjacent to a, then $x_1 1$ is adjacent to $a1$ and therefore $y_1 0 = f(x_1 1)$ must be adjacent or equal to $a1$ in $V(G)$, and the only such vertex has $y_1 = a$.)

Given an instance of SAT with variables v_i and clauses c_j, we construct a graph H_1. The vertices in $V(H_1)$ are the clauses c_j, the variables v_i, their negations $\overline{v_i}$, and an additional vertex a. The set of edges is the set $E(H_1) = E_1 \cup E_2 \cup E_3 \cup E_4$. The set E_1 has edges joining every pair of clauses c_j, c'_j. The set E_2 has edges joining every clause c_j to the literals v_i or $\overline{v_i}$ that occur in c_j. The set E_3 has edges joining every pair of literals with the exception of complementary pairs $v_i, \overline{v_i}$. The set E_4 has edges joining every literal v_i or $\overline{v_i}$ to the vertex a.

For this graph H_1, the mapping f' must have the following properties. First, we must have $f'(v_i) = a$ and $f'(\overline{v_i}) = a$ for all v_i, since both v_i and $\overline{v_i}$ are adjacent to a. Second, we must have $f'(c_j) = u$, where u is a literal that occurs in c_j, because every literal that occurs in c_j maps to a, forcing c_j to map to a neighbor of a adjacent to c_j. Third, a pair of literals $f'(c_j)$ and $f'(c_{j'})$ from different clauses $j \neq j'$ cannot be complementary literals v_i and $\overline{v_i}$, because the adjacent vertices c_j and $c_{j'}$ map to adjacent literals, and the nonadjacent literals are precisely the complementary literals. If these three conditions are satisfied, then f' does indeed map the graph H_1 properly in the sense that the conditions on f' are satisfied. Such an f' is simply a choice of a literal from each clause, with the property that at most one of each pair of complementary literals is chosen. Given f', we can assign the value 1 to the chosen literals, and assign other variables arbitrarily, thus obtaining a satisfying truth assignment. Conversely, given a satisfying truth assignment, we can choose a literal with value 1 from each clause, thus defining an appropriate f'. Therefore such an f' exists, so that $M(G)$ is a retract of $M(H)$, precisely when the instance of SAT has a satisfying truth assignment. □

It is a priori unclear, given a representation g for a metric space L in a product M, whether one can decide in $\mathcal{NP}$ if g is a retract representation, since the metric space M that must be retracted to $g(L)$ may have exponentially many vertices (consider, for instance, a path as a retract of a hypercube). The study below will give a positive answer to this question, and show that despite their computational hardness, retracts of product spaces have a fairly simple structure.

In order to study retracts of product spaces, we shall need to consider partial mappings as well, i.e., mappings that are defined for only a subset of the possible points. A *partial nonexpansive mapping* on a metric space M is a partial mapping f on M such that $d(f(x), f(y)) \leq d(x, y)$ for all points x, y in M such that f is defined on both x and y. Let f be a partial nonexpansive mapping on a product $M = \square_{i \in S} M_i$, so that f computes $y = f(x)$ with x, y in M. If we partition S into two sets T and $\overline{T} = S \backslash \{T\}$, and view M as the product $M = M_T \square M_{\overline{T}}$, then f computes a mapping $f(x_T x_{\overline{T}}) = y_T y_{\overline{T}} = g_{x_T}(x_{\overline{T}}) h_{x_T}(x_{\overline{T}})$, with x_T, y_T in M_T and $x_{\overline{T}}, y_{\overline{T}}$ in $M_{\overline{T}}$. A *periodic point* of $h = h_u$ with u in M_T is a point v in $M_{\overline{T}}$ such that $h^{(p)}(v) = v$ for some $p \geq 1$. The least such p is the *period* of v. Note that h may not have a periodic point, since f and h are only partial mappings. The *projection* of f under T is the partial mapping f_T on M_T defined by letting $f_T(u) = g_u(v)$ if v is a periodic point of h_u, and leaving $f_S(u)$ undefined if h_u has no periodic point. As the next lemma shows, the projection f_S is well-defined and preserves nonexpansiveness.

Lemma 6.33 *If f is a partial nonexpansive mapping on $M = \square_{i \in S} M_i$, and $T \subseteq S$, then f_T is a partial nonexpansive mapping on M_T. If f is total, then so is f_S. If f is a retraction and $L = f(M)$ is the corresponding retract of M, then f_T is a retraction and the corresponding retract $f_T(M_T)$ of M_T is L_T.*

Proof. We write $M = M_T \square M_{\overline{T}}$ and $f(uv) = g_u(v) h_u(v)$. We first show that f_T is well-defined. Given u in M_T, suppose that $h = h_u$ has two periodic points v, v' in $M_{\overline{T}}$, and let p be a common multiple of the periods of v and v'. Since f is nonexpansive, the mapping h_u is also nonexpansive and so

$$d(v, v') \geq d(h(v), h(v')) \geq d(h^{(p)}(v), h^{(p)}(v')) = d(v, v'),$$

implying that $d(h(v), h(v')) = d(v, v')$. Then

$$\begin{aligned} d(g_u(v), g_u(v')) + d(h_u(v), h_u(v')) &= d(f(uv), f(uv')) \\ &\leq d(uv, uv') = d(v, v') = d(h_u(v), h_u(v')), \end{aligned}$$

proving that $d(g_u(v), g_u(v')) = 0$ and $g_u(v) = g_u(v')$. Therefore the value $g_u(v)$ is independent of the choice of a periodic point v, and $f_T(u) = g_u(v)$ is well-defined.

To show that f_T is nonexpansive, suppose that f_T is defined on two points u, u'. Then the mappings h_u and $h_{u'}$ each have at least one periodic point, say v and v' respectively.

Choose such a pair v, v' with $d(v, v')$ minimum. Then

$$\begin{aligned} d(g_u(v), g_{u'}(v')) + d(h_u(v), h_{u'}(v')) &= d(f(uv), f(u'v')) \\ &\leq d(uv, u'v') = d(u, u') + d(v, v') \\ &\leq d(u, u') + d(h_u(v), h_{u'}(v')), \end{aligned}$$

where the last inequality follows from the minimality of $d(v, v')$. Therefore $d(f_T(u), f_T(u')) = d(g_u(v), g_{u'}(v')) \leq d(u, u')$, proving that f_T is nonexpansive.

If f is defined everywhere then h_u has a periodic point for all u in M_T so that f_T is defined everywhere as well.

Suppose that f is a retraction and $L = f(M)$ is the corresponding retract. Given u in L_T, there is a point uv in L with v in $L_{\overline{T}}$, so that $f(uv) = uv$. Then $h_u(v) = v$, so v is a fixed point and hence a periodic point of h_u, and therefore $f_T(u) = g_u(v) = u$ and u is a fixed point of f_T. Given any u, v, the point $f(uv) = g_u(v)h_u(v)$ is in L, so $g_u(v)$ is in L_T and $f_T(u) = g_u(v)$ is in L_T. Therefore f_T is a retraction and L_T is the corresponding retract. □

Using this lemma, we can once again prove an analog of Lemma 6.7.

Lemma 6.34 *Suppose that a representation g for L in a product $M = \Box_{j \in S} M_j$ contains a representation g' for L in $M' = \Box_{i \in S'} M'_i$, with the containment given by representations h_i for M'_i in M_{S_i}. Then the representation g is a retract representation if and only if the representation g' and the representations h_i are retract representations.*

Proof. As in the 2-isometric case (Lemma 6.26), We must show that L is a retract of $M = \Box_{i \in S'} M_{S_i}$ if and only if L is a retract of $M' = \Box_{i \in S'} L_{S_i}$ and each L_{S_i} is a retract of M_{S_i}. If L is a retract of M, then L is a retract of M' since M' is a subspace of M containing L, and each L_{S_i} is a retract of M_{S_i} by the preceding lemma. In the other direction, given a retraction f' on M' with $f'(M') = L$, and retractions f_i on M_{S_i} with $f_i(M_{S_i}) = L_{S_i}$, the mapping f on M defined by $f(x) = y$ if $y = f'(z)$ with $z_{S_i} = f_i(x_{S_i})$ is a retraction on M with $f(M) = L$. □

The analogue of Theorem 6.23 also holds.

Theorem 6.35 *A subspace L of a product $M = \Box_{i \in S} M_i$ is a retract of M if and only if L contains precisely those points x in M such that for all sets T with $|T| \leq 2$ the point x_T is in L_T, and each such L_T is a retract of M_T.*

Proof. If L is a retract of M, then L is a 2-isometric subspace of M, so by Theorem 6.23 the space L contains precisely those points x in M such that for all sets T with $|T| \leq 2$ the point x_T is in L_T, and by Lemma 6.33 each such L_T is a retract of M_T.

In the other direction, suppose that L contains precisely those points x in M such that for all sets T with $|T| \leq 2$ the point x_T is in L_T, and that each such L_T is a retract of M_T.

We show how to construct from the retractions f_T on M_T with $f_T(M_T) = L_T$ for $|T| \leq 2$ a retraction f on M with $f(M) = L$. The retractions f_T can be viewed as retractions g_T on $M = M_T \square M_{\overline{T}}$ with $g_T(M) = L_T \square M_{\overline{T}}$, by letting $g_T(x_T x_{\overline{T}}) = f_T(x_T) x_{\overline{T}}$. Given a point x in M, if x_T is not in L_T, then there is some point y in L such that $y_T = f_T(x_T) \neq x_T$, and so $d(g_T(x), y) = d(y_T x_{\overline{T}}, y) < d(x, y)$. Furthermore, $d(g_T(x), y) \leq d(x, y)$ for all x in M and y in L. Therefore, when x is replaced by $g_T(x)$, the distance to each point y in L does not increase, and this distance actually decreases for some point y in L if x_T is not in L_T. Let g be the composition of the mappings g_T in some arbitrary order, and let $f = g^{(r)}$ for some very large r. Note that $g(x) = x$ and $f(x) = x$ if x is in L. Let $d_s(x) = \sum_{y \in V(L)} d(g^{(s)}(x), y)$. If $z = g^{(s)}(x)$ is not in L, then z_T is not in L_T for some T with $|T| \leq 2$ by assumption, so when we apply g to z we get $d(g(z), y) < d(z, y)$ for some y in L, and therefore $d_{s+1}(x) < d_s(x)$. Since $d_s(x)$ can only take finitely many possible values because all spaces are finite, there is some value s such that $d_{s+1}(x) = d_s(x)$, and then $z = g^{(s)}(x)$ is in L. If we pick r larger than the value s associated with each x in M, then $f(x) = g^{(r)}(x)$ is in L for all x, proving that g is a retraction. $\square$

Given a representation g for L in a product $M = \square_{i \in S} M_i$, possibly exponentially large, with only the factors M_i and the mapping g given explicitly, we can check in $\mathcal{NP}$ whether g is a retract representation, as follows. First, we verify that g is a 2-isometric representation. To verify that $g(L)$ is a retract of M, we use the preceding lemma. The first condition is automatically satisfied since $g(L)$ is a 2-isometric subspace. The second condition only requires guessing retractions on spaces M_T with $|T| \leq 2$, and these spaces are small (quadratic in the size of the M_i). In combination with Lemma 6.32, we obtain:

Corollary 6.36 *The problem of deciding whether a given representation g is a retract representation is $\mathcal{NP}$-complete.*

6.6.3 Canonical Retract Representation

Our aim is to show that there is a retract representation that contains all the retract representations. In essence, what we must show is that if two equivalence relations γ_1 and γ_2 induce retract representations, then so does $\gamma_1 \cap \gamma_2$. The following special case will turn out to contain all the necessary ingredients for the general case. For convenience, we use the set notation $i_0* = \{ij : i = i_0\}$ and $*j_0 = \{ij : j = j_0\}$ for indices, as often used to denote rows and columns in matrices. We shall also use parentheses as in $g_{(T)}$ when we simply intend to describe an indexed family of functions, to avoid confusion with the notation f_T denoting the projection of f under T.

Lemma 6.37 *Let L be a connected 2-isometric irredundant subspace of a product*

$$M = M_{11} \square M_{12} \square M_{21} \square M_{22} = M_{1*} \square M_{2*} = M_{*1} \square M_{*2}.$$

Suppose that L is a retract of $L_{1*} \square L_{2*}$ and of $L_{*1} \square L_{*2}$. Then L is a retract of M.

Proof. By assumption, we have a retraction f^1 on $L_{1*} \square L_{2*}$ with L as retract, and a retraction f^2 on $L_{*1} \square L_{*2}$ with L as retract. If we had retractions $g_{(i*)}$ on M_{i*} with L_{i*} as retract, for $i = 1, 2$, then we could extend f^1 to a retraction $\tilde{f}^1$ on M with L as retract by letting $\tilde{f}^1(x) = f^1(g_{(1*)}(x_{1*})g_{(2*)}(x_{2*}))$. Then the projections $\tilde{f}^1_{*j}$ would be retractions on M_{*j} with L_{*j} as retract, for $j = 1, 2$, by Lemma 6.33. In the perpendicular direction, starting from retractions $g_{(*j)}$, we could construct retractions $\tilde{f}^2_{i*}$. In general, if L turns out to be a retract of M, there may be more than one choice of retractions $g_{(i*)}$, leading to possibly different $\tilde{f}^1_{*j}$, and similarly in the perpendicular direction. Furthermore, if we go in one direction from $g_{(i*)}$ to $g_{(*j)} = \tilde{f}^1_{*j}$ then back to $\tilde{f}^2_{i*}$, we may obtain retractions different from the ones we started with. The key to obtaining specific retractions turns out to be the additional imposed requirement that the retractions so obtained match the original ones, i.e., that $\tilde{f}^2_{i*} = g_{(i*)}$.

Let $\mathcal{S}$ be the system of two equations $(g_{(*1)}, g_{(*2)}) = (\tilde{f}^1_{*1}, \tilde{f}^1_{*2})$ and $(g_{(1*)}, g_{(2*)}) = (\tilde{f}^2_{1*}, \tilde{f}^2_{2*})$ in two unknowns $(g_{(1*)}, g_{(2*)})$ and $(g_{(*1)}, g_{(*2)})$, where the mappings $\tilde{f}^1$ and $\tilde{f}^2$ are defined by $\tilde{f}^1(x) = f^1(g_{(1*)}(x_{1*})g_{(2*)}(x_{2*}))$. and $\tilde{f}^2(x) = f^2(g_{(*1)}(x_{*1})g_{(*2)}(x_{*2}))$.

Lemma 6.38 *The system of equations $\mathcal{S}$ has a unique solution over the space of retractions $g_{(T)}$ on M_T with retract $g_{(T)}(M_T) = L_T$, with T taking the four values $1*, 2*, *1, *2$.*

Proof. The equations in $\mathcal{S}$, and in particular the projections on $\tilde{f}^1$ and $\tilde{f}^2$, make sense, from our definitions, for partial nonexpansive mappings. The idea of the proof is to assign partial nonexpansive mappings $g^k_{(T)}$ on M_T to the unknowns $g_{(T)}$, and then use the two equations in $\mathcal{S}$ to obtain new values $g^{k+1}_{(T)}$ for the unknowns in the left side of the equations. We start with the four partial mappings $g^0_{(T)}$ defined by $g^0_{(T)}(u) = u$ for u in L_T, and undefined elsewhere. These partial mappings are clearly nonexpansive, and consistent with any retraction with retract L_T. By Lemma 6.33, if the $g^k_{(T)}$ are nonexpansive, then so are the $g^{k+1}_{(T)}$, because nonexpansiveness is preserved under projection and composition of functions. If the $g^k_{(T)}$ are consistent with some solution (given by retractions) to $\mathcal{S}$, then so are the $g^{k+1}_{(T)}$, because extensions of mappings give extensions under projection and composition. The range of each $g^k_{(T)}$ is contained in L_T: For instance, the range of f^1 is contained in L, so the range of $\tilde{f}^1$ is contained in L, and therefore the range of $g^k_{(*1)} = \tilde{f}^1_{*1}$ is contained in L_{*1}. The mapping $g^1_{(T)}$ is an extension of $g^0_{(T)}$: For instance, if u is in L_{*1}, then $x_{*1} = u$ for some x in L, and so $g^1_{(*1)}(u) = \tilde{f}^1_{*1}(u) = u = g^0_{(*1)}(u)$ since

$$\tilde{f}^1(x) = f^1(g^0_{(1*)}(x_{1*})g^0_{(2*)}(x_{2*})) = f^1(x_{1*}x_{2*}) = f^1(x) = x.$$

If the $g^k_{(T)}$ are extensions of the $g^{k-1}_{(T)}$, then the $g^{k+1}_{(T)}$ are extensions of the $g^k_{(T)}$, because extensions of mappings give extensions under projection and composition.

Summarizing, we have an increasing sequence of partial nonexpansive mappings $g^k_{(T)}$ on M_T in the sense that each $g^{k+1}_{(T)}$ is an extension of $g^k_{(T)}$, the mappings have range L_T

with each element of L_T as a fixed point, and are consistent with every solution (given by retractions) to $\mathcal{S}$. Since the spaces M_T are finite, this increasing sequence must reach a stage where $g_{(T)}^{k+1} = g_{(T)}^k$ for each T, thus giving partial nonexpansive mappings $g_{(T)}$ that satisfy the equations in $\mathcal{S}$. If we can show that each $g_{(T)}$ is a total mapping (defined on all of M_T), then each $g_{(T)}$ will be a retraction on M_T with retract L_T, thus giving a solution consistent with all solutions, hence a unique solution. To show that each $g_{(T)}$ is a total mapping, we shall associate a nonnegative *potential* with each point in each M_T, and show that each $g_{(T)}^k$ is at the very least defined at every point of M_T that has potential π_k or less, where π_k is the $(k+1)$th smallest value of $d(y,z)$ for points y,z in L. The potential of each point will be such a distance $d(y,z)$. Thus, once π_k reaches a value equal to the largest of all such distances, hence at least as large as each potential, the mappings $g_{(T)}^k$ will indeed be total mappings.

The potential of a point u in M_T is defined to be the minimum of the distance $d(y,z)$ over all y,z in L such that u is between y_T and z_T in M. Such points y,z do exist, because if $T = \{a,b\}$, then u is of the form $u = u_a u_b$, and we could pick any y,z with $y_a = u_a$ and $z_b = u_b$ since L is an irredundant subspace of the product M. The proof that the $g_{(T)}^k$ are defined on points of potential at most π_k is by induction on k. If $k = 0$, then a point u in M_T of potential $\pi_0 = 0$ has $d(y,z) = 0$ so that $y = z$ and $y_T = z_T = u$ with y in L, proving that u is in L_T so that $g_{(T)}^0$ is indeed defined at u. Assume now that the inductive assumption holds for all the $g_{(T)}^{k-1}$, and consider a point u in M_T of potential at most π_k, with the corresponding points y,z in L such that $d(y,z) \leq \pi_k$. If the potential of u is at most π_{k-1}, then $g_{(T)}^{k-1}(u)$ is defined by inductive assumption, so $g_{(T)}^k(u)$ is also defined since g^k extends g^{k-1}. So we can assume that the potential of u is exactly π_k. We do the proof for the case where T is $*1 = \{11, 21\}$; the other three cases are identical. Recall that $g_{(*1)}^k = \tilde{f}_{*1}^1$, where $\tilde{f}^1(x) = f^1(g_{(1*)}^{k-1}(x_{1*}) g_{(2*)}^{k-1}(x_{2*}))$. We shall write $\tilde{f}^1(uv) = g_u(v) h_u(v)$, with $u, g_u(v)$ in L_{*1} and $v, h_u(v)$ in L_{*2}. Given the choice of u and the two corresponding y,z with $d(y,z) = \pi_k$, we let $V = \{v \text{ in } L_{*2} : v \text{ is between } y_{*2} \text{ to } z_{*2} \text{ in } M_{*2}\}$, and show that the mapping $h = h_u$ is defined at every point $v \in V$ and has $h(v) \in V$ for all $v \in V$. As a result, the mapping h has a periodic point in V, and so the projection $g_{(*1)}^k(u) = \tilde{f}_{*1}^1(u) = g_u(v)$ with v a periodic point of h_u is defined, completing the induction.

We must first show that h is defined at every point $v \in V$, or equivalently that $\tilde{f}^1$ is defined at $x = uv$. We obtain this fact by showing that x_{1*} and x_{2*} both have potential at most π_{k-1}. This implies that the mappings $g_{(1*)}^{k-1}$ and $g_{(2*)}^{k-1}$ are defined at these two points respectively, by inductive assumption, so $\tilde{f}^1$ is indeed defined at x. We prove that the potential of x_{1*} is at most π_{k-1}; the case for x_{2*} is identical. Note that x is between y and z in M, since u is between y_{*1} and z_{*1} in M_{*1} and v is between y_{*2} and z_{*2} in M_{*2}. Therefore x_{1*} is between y_{1*} and z_{1*}, showing that the potential of x_{1*} is at most π_k. Suppose that y' is a neighbor of y in $G(L)$ with $d(y', z) < d(y, z)$. The edge $e = \{y, y'\}$ belongs to some factor. If e belongs factor 12 or factor 22, then $y'_{*1} = y_{*1}$ and so u would be between y'_{*1} and z_{*1} in M_{*1}, contradicting the minimality of $d(y,z)$ in the definition of the potential of

u. If the edge e belongs to factor 21, then $y'_{1*} = y_{1*}$, showing that the potential of x_{1*} is smaller than π_k and hence at most π_{k-1} as claimed. It remains to consider the case where every such e belongs to factor 11, and where by symmetry every $e' = \{z, z'\}$ where z' is a neighbor of z in $G(L)$ with $d(y, z') < d(y, z)$ also belongs to factor 11. This implies that $y_{11} \neq z_{11}$, because if $y_{11} = z_{11}$ then we could walk from y to z in $G(L)$ along a shortest path, without visiting any edges in 11. Then u_{11} differs from one of these two coordinates, say $u_{11} \neq y_{11}$. Let t be a point in L with $t_{11} = u_{11}$ (since L is an irredundant subspace). Then t_{11} is between y_{11} and z_{11}, with $t_{11} \neq y_{11}$, so y_{11} is not a maximal lower bound of $\{t_{11}, z_{11}\}$ in M_{11}, and therefore y is not a maximal lower bound of $\{t, z\}$ in M. Since L is a 2-isometric subspace of M, it follows that y is not a maximal lower bound of $\{t, z\}$ in L either, so there is a point $y'' \neq y$ in L such that $y'' \leq_y t$ and $y'' \leq_y z$. Choose such a y'' closest to y, so that y'' is adjacent to y in $G(L)$. The edge $e'' = \{y, y''\}$ must belong to factor 11 since $d(y'', z) < d(y, z)$ and we are assuming that such edges belong to factor 11. So $u_{11} = t_{11}$ is between y''_{11} and z_{11}, and therefore u is between y'_{*1} and z_{*1}, contradicting again the minimality of $d(y, z)$.

Therefore the only possibility is that the potential of x_{1*} is at most π_{k-1} as claimed, and similarly for x_{2*}, so $\tilde{f}^1$ is indeed defined at $x = uv$, and $h = h_u$ is defined at all $v \in V$. Since x is between y and z in M, the image $\tilde{f}^1(x) = g_u(v) h_u(v)$ is also between y and z in M, because y and z are fixed points of $\tilde{f}^1$, and $\tilde{f}^1$ is nonexpansive. Therefore $h_u(v)$ is between y_{*2} and z_{*2}, and so $h_u(v) \in V$ for all $v \in V$. It follows that $h = h_u$ has a periodic point in V, and so $g^k_{(*1)} = \tilde{f}^1_{*1}$ is defined at u, proving as claimed that $g^k_{(*1)}$ is defined at all points of potential at most π_k; a similar argument gives this result for the remaining $g^k_{(T)}$. If we choose k such that π_k is the largest potential, then $g_{(T)} = g^k_{(T)}$ is a total mapping, and therefore a retraction on M_T with retract L_T, consistent with all retractions that solve the system $\mathcal{S}$, and thus giving uniqueness as well. $\square$

The mappings $\tilde{f}^1$ and $\tilde{f}^2$ from the solution of the system $\mathcal{S}$ are then retractions of M with retract L, consistent with the partial mappings on M given by f^1 and f^2 respectively. The existence of such retractions proves Lemma 6.37. $\square$

This lemma extends to products of factors M_{ij}. We write $\overline{i_0 j_0} = \{ij : (i \neq i_0) \vee (j \neq j_0)\}$, $\overline{i_0 j_0} = \{ij : (i \neq i_0) \wedge (j \neq j_0)\}$, and $\overline{\{i_0 j_0, i_1 j_1\}} = \overline{i_0 j_0} \cap \overline{i_1 j_1}$.

Lemma 6.39 *Let L be a connected 2-isometric irredundant subspace of a product*

$$M = \square_{i,j} M_{ij} = \square_i M_{i*} = \square_j M_{*j}.$$

Suppose that L is a retract of $\square_i L_{i*}$ and of $\square_j L_{*j}$. Then L is a retract of M.

Proof. To prove that L is a retract of M, it is sufficient to prove that the two conditions in Theorem 6.35 are satisfied. The first condition is satisfied by Theorem 6.23, since L is a 2-isometric subspace of M. We must therefore show that the second condition holds, namely that L_T is a retract of M_T for all sets T with $|T| \leq 2$. The case $|T| = 1$ is clear,

CHAPTER 6. METRIC NETWORKS AND PRODUCT GRAPHS 193

since $L_{ij} = M_{ij}$ because L is an irredundant subspace of M. We therefore consider the case $|T| = 2$.

We first observe that for all i, j, the space L is a retract of $L_{ij} \square L_{\overline{ij}}$. To see this, consider the space $M' = M'_{11} \square M'_{12} \square M'_{21} \square M'_{22} = L_{ij} \square L_{i\overline{j}} \square L_{\overline{i}j} \square L_{\overline{ij}}$. We repeatedly use the fact that if L is a retract of some space, it is a retract of any subspace of this space containing L. Note that L is a retract of the subspace $L_{i*} \square L_{\overline{i}*}$ of $M'_{1*} \square M'_{2*}$, since L is a retract of $\square_{i'} L_{i'*}$. Similarly, L is a retract of the subspace $L_{*j} \square L_{*\overline{j}}$ of $M'_{*1} \square M'_{*2}$. By Lemma 6.37, L is then a retract of M', and therefore a retract of $L_{ij} \square L_{\overline{ij}}$.

Next, we observe that for all $ij \neq i'j'$ (that is, $i \neq i'$ or $j \neq j'$), the space L is a retract of $L_{ij} \square L_{i'j'} \square L_{\overline{\{ij,i'j'\}}}$. We write this product as $M'' = M''_{11} \square M''_{12} \square M''_{21} \square M''_{22} = L_\emptyset \square L_{ij} \square L_{i'j'} \square L_{\overline{\{ij,i'j'\}}}$. We observe that L is a retract of the subspace $L_{ij} \square L_{\overline{ij}}$ of $M''_{1*} \square M''_{2*}$, and also a retract of the subspace $L_{i'j'} \square L_{\overline{i'j'}}$ of $M''_{*1} \square M''_{*2}$. Therefore, again by Lemma 6.37, L is a retract of M'' and therefore a retract of $L_{ij} \square L_{i'j'} \square L_{\overline{\{ij,i'j'\}}}$. Since projections of retracts are retracts by Lemma 6.33, we can conclude that $L_{\{ij,i'j'\}}$ is a retract of $L_{ij} \square L_{i'j'} = M_{ij} \square M_{i'j'}$, completing the proof. □

This lemma gives then a canonical representation theorem for retracts.

Theorem 6.40 *There is a canonical retract representation g^ρ for a metric space L in a product M such that the retract representations are precisely the representations contained in g^ρ, given by mappings of the form g^γ with $\rho \subseteq \gamma^*$.*

Proof. All retract representations are of the form g^γ, where γ is an equivalence relation on the edges of $G(L)$, by Theorem 6.9 We show that if γ_1, γ_2 are equivalence relations such that g^{γ_1} and g^{γ_2} are retract representations, then g^γ is a retract representation for the equivalence relation $\gamma = \gamma_1 \cap \gamma_2$. Note that g^γ is a 2-isometric representation by Theorem 6.27, because $\theta \cup \kappa$ is contained in both γ_1 and γ_2 and therefore in γ as well. If we denote the equivalence classes of γ_1 by E_{i*}, and the equivalence classes of γ_2 by E_{*j}, then the equivalence classes of γ are $E_{ij} = E_{i*} \cap E_{*j}$. (Some of these E_{ij} may be empty.) The corresponding representation g^γ gives then L as a connected 2-isometric irredundant subspace of a product $M = \square_{i,j} M_{ij}$. The representations g^{γ_1} and g^{γ_2} are contained in this representation by Lemma 6.4, with the partition of factors for the containment consisting of factors $i*$ and $*j$ respectively. The containment mappings are isometric by Lemma 6.7, and their ranges are L_{i*} in M_{i*} and L_{*j} in M_{*j} respectively. Therefore the two retract representations g^{γ_1} and g^{γ_2} give L as a retract of $\square_i L_{i*}$ and of $\square_j L_{*j}$ respectively. We can then use Lemma 6.39 to conclude that L is a retract of M, so that g^γ is a retract representation.

There is therefore a retract representation containing all retract representations, namely the representation g^ρ where ρ is the intersection of all the equivalence relations γ such that g^γ is a retract representation. The fact that all representations contained in g^ρ are retract representations follows from Lemma 6.34, and the fact that these are precisely the representations with $\rho \subseteq \gamma^*$ follows from Lemma 6.4 and Theorem 6.9 □

This relation ρ relates two edges in $G(L)$ if and only if the two edges belong to the same factor in every retract representation of $G(L)$. It can be shown, along the lines of Lemma 6.32 and Corollary 6.36, that the problem of deciding whether two given edges are related by ρ is co$\mathcal{NP}$-complete. Thus the problem of finding the canonical retract representation is $\mathcal{NP}$-hard, but can be solved by making simultaneous queries to an $\mathcal{NP}$ oracle, one query for each pair of of edges in $G(L)$.

6.7 Dynamic Search in Graphs

Chung, Graham, and Saks [20, 21] studied a dynamic location problem in a finite connected unweighted graph G. Given a *request* sequence $Q = (q_1, q_2, \ldots)$ of vertices $q_i \in V(G)$, a *pebbling* sequence $P = (p_0, p_1, p_2, \ldots)$ with $p_i \in V(G)$ has an associated cost

$$c_N(Q, P) = \sum_{1 \le i \le N} (d(p_{i-1}, p_i) + d(p_i, q_i)),$$

where d is the distance function $d_{M(G)}$. There are therefore two costs associated with a pebbling sequence, the cost of moving the pebble from p_{i-1} to p_i and the cost of satisfying the request q_i from the pebble p_i.

A sequence P is Q-*optimal* if

$$\sup_N (c_N(Q, P) - c_N(Q, \widehat{P}))$$

is bounded for all pebbling sequences $\widehat{P}$. An algorithm A which produces a Q-optimal pebbling sequence $A(Q)$ for each request sequence Q is said to be an *optimal* algorithm for G. Of particular interest are algorithms that produce Q-optimal pebbling sequences even though at any time only a finite portion of Q can be seen by A. A graph G is said to have *windex* k if there is an optimal algorithm A for G with the property that A always determines p_i with knowledge of q_j only for $j < i + k$. If there is no such k, then G is said to have infinite windex. The name windex, a shortened form of window index, refers to the fact that one can think of A as having a window through which exactly k future request symbols of Q can be seen.

It can be shown, using the König infinity lemma, that if an algorithm on a graph G can produce, using a window of size k, a pebbling sequence P for each request sequence Q so that

$$\limsup_{N \to \infty} (c_N(Q, P) / c_N(Q, \widehat{P})) \le 1$$

for all pebbling sequences $\widehat{P}$, then there is an algorithm that can produce a Q-optimal pebbling sequence so that G has windex k. In other words, the weaker condition stating that the ratio tends to 1 implies the stronger condition requiring the difference to be bounded for some other algorithm, so there is a discrete jump on the kind of behavior that the best sequences on a given finite graph can exhibit. This kind of 'gap' seems to arise often in the

CHAPTER 6. METRIC NETWORKS AND PRODUCT GRAPHS

study of infinite processes on finite domains. See the notion of asymptotic convergence for circuits in section 3.4.4 for another example of this phenomenon.

Chung et al. gave a complete characterization of the graphs of windex at most k. First, they observed that if two graphs G_1 and G_2 have windex at most k, then so does the product $G_1 \square G_2$. Intuitively, this holds because we can treat the request and pebbling sequences separately in the two factors, and use the fact that the distance function in the product is the sum of the distance functions in each factor. Second, they observed that if a graph H has windex at most k, then so does every retract G of H. The main reason for this is that a pebbling sequence in H can be mapped to a pebbling sequence in G without increasing the cost, using the corresponding retraction mapping. Third, they showed that a clique on k vertices has windex k. As a result, the retracts of products of cliques on at most k vertices have windex at most k. The most involved part of their characterization consists of showing that these are *all* the graphs of windex at most k.

In order to establish this characterization, they defined the imprint function for products of cliques. It is this notion that motivated our imprint function for general metric spaces. They showed then that the retracts of product of cliques are the same as the connected induced subgraphs closed under imprints. In our terminology, by Theorem 6.22, this is equivalent to the statement that retracts and connected 2-isometric subspaces coincide, for products of cliques. Furthermore, by Theorem 6.23 and Theorem 6.35, this can be proved by considering the product of just two cliques, where a characterization of connected 2-isometric subspaces and retracts is not hard to obtain. The connected 2-isometric irredundant subspaces of the product of two cliques are the subspaces defined by a condition $(x_1 = a) \vee (x_2 = b)$, and the whole product space. The second case clearly defines a retract; the first case also defines a retract, under the retraction mappping f defined by $f(x) = x$ if $x = x_1 x_2$ satisfies the stated condition, and $f(x) = ab$ otherwise. The mapping f can be viewed as a *generalized* X *gate*: The usual X gate from the boolean case is obtained by letting $ab = 00$.

Therefore retracts and connected 2-isometric irredundant subspaces of products of cliques coincide. Given a graph G, we can determine whether G is the graph of a connected 2-isometric irredundant subspace of a product of cliques by finding its canonical 2-isometric representation and verifying that the factors in the representation are cliques. The number of vertices in the largest clique that occurs as a factor will then be the windex of G. As mentioned in section 6.4.3, this representation can be found in $O(nm)$ time and $O(m)$ space, so the windex of a graph can be determined in $O(nm)$ time and $O(m)$ space.

An open question is the study of this problem under different cost measures, say of the form
$$c_N(Q, P) = \sum_{1 \leq i \leq N} (d(p_{i-1}, p_i) + \lambda d(p_i, q_i)),$$
for some $\lambda > 0$. Here again, the class of graphs of windex at most k is closed under products and retracts, so the general structure of the retracts of products may be useful. It is however

unclear whether the graphs of finite windex will again be the retracts of products of some simple family of basic graphs.

6.8 Network Stability

The framework for the study of stability in general metric networks is essentially the same as in the boolean case, as presented in section 2.1. A *gate* is a mapping g from a metric space $M_I = \prod_{i \in I} M_i$ to a metric space $M_O = \prod_{i \in O} M_i$. The coordinates in $I = I(g)$ and $O = O(g)$ are called *inputs* and *outputs* of g. The gate g is *nonexpansive* if it satisfies $d(g(x), g(y)) \leq d(x, y)$ for all points x, y in M_I.

A *network* is a set of gates that do not share any inputs and do not share any outputs. This means that if N is a network and g, g' are distinct gates in N, then $I(g) \cap I(g') = \emptyset$ and $O(g) \cap O(g') = \emptyset$. The *transition function* of a network N is a single gate f that describes all the gates in N. The gate f has $I(f) = \bigcup_{g \in N} I(g)$ and $O(f) = \bigcup_{g \in N} O(g)$, and satisfies $y = f(x)$ if and only if $y_{O(g)} = f(x_{I(g)})$ for all gates $g \in N$. Given a network N with transition function f, the set $R(N) = I(f) \cup O(f)$ is the set of *coordinates* of the network N, and consists of three parts: the set of *links* $L(N) = I(f) \cap O(f)$, the set of *inputs* $I(N) = I(f) \backslash O(f)$, and the set of *outputs* $O(N) = O(f) \backslash I(f)$ of the network. A link $i \in L(N)$ belongs to $I(g) \cap I(g')$ for some $g, g' \in N$; for such a link, the metric space M_i must be the same as an input to g and as an output to g'.

A *configuration* of a network N is a point x in $M = M_{R(N)}$, and consists of an input assignment $x_{I(N)}$ in $M_{I(N)}$, an *output assignment* $x_{O(N)}$ in $M_{O(N)}$, and an *internal assignment* $x_{L(N)}$ in $M_{L(N)}$. A network N can be used to define an associated mapping on the configurations of N. Given two configurations x and y of a network N with transition function f, we write $y = N(x)$ if $y_{I(N)} = x_{I(N)}$ and $y_{O(N) \cup L(N)} = f(x_{I(N) \cup L(N)})$. A configuration x is *stable* if $N(x) = x$. A configuration x is therefore stable if it satisfies $f(x_{I(f)}) = x_{O(f)}$ for the transition function f, or equivalently $g(x_{I(g)}) = x_{O(g)}$ for each gate $g \in N$. A configuration x is *periodic* if $N^{(p)}(x) = x$ for some $p \geq 1$; the least such p is the *period* of x.

Relabellings are used as in section 2.1.1 to create and to break links, and therefore to build new networks from given networks.

The following sections use the structural theory developed in this chapter to extend most of the results for nonexpansive networks from the boolean case the the general metric case. Section 6.8.1 looks at convergent networks; section 6.8.2 is concerned with the properties of fixed points and retracts; section 6.8.3 studies the isomorphism on the set of periodic configurations; section 6.8.4 proves a fixed product theorem for metric spaces; and section 6.8.5 looks a the iterative behavior of metric networks on non-periodic points.

CHAPTER 6. METRIC NETWORKS AND PRODUCT GRAPHS   197

6.8.1 Convergent Networks

A network N is said to be *convergent* if for every input assignment x there exists an output assignment y such that every configuration consistent with x maps to a configuration consistent with y under sufficiently many iterations of N. More formally, for every configuration t consistent with x, there must exist an integer k_0 such that $N^{(k)}(t)$ is consistent with y for all $k \geq k_0$. If a network N is convergent, then for every input assignment x there is a unique corresponding output assignment y, and we say that N *converges* to the gate g with $I(g) = I(N)$ and $O(g) = O(N)$ that computes $g(x) = y$.

The basic results on convergent networks in the boolean case from section 3.1 extend to the metric case without substantial modifications. We summarize these results.

Theorem 6.41 *All metric networks of nonexpansive gates are convergent, and converge to a nonexpansive gate.*

The iterative behavior of networks can again be used to determine how fast they converge.

Lemma 6.42 *Given a nonexpansive metric network N that converges to a gate g, and an input assignment x, let $y^1, y^2, \ldots$ be the output sequence obtained by iterating N, using the input assignment x and an arbitrary initial internal assignment z^0, and let $\widehat{z^0}$ be a periodic internal assignment of the internal function for input x. Then*

$$\sum_{0 < i \leq k} d(y^i, g(x)) \leq d(z^0, \widehat{z^0}) - d(z^k, \widehat{z^k}) \leq |L(N)|,$$

where the z^i and the $\widehat{z^i}$ are the internal assignments obtained by iterating N on input x and initial internal assignments $z^0, \widehat{z^0}$.

The properties of minimum distance points and the substitutivity property for convergent networks generalize to the metric case as well. This makes it once again possible to reduce the problem of finding stable configurations to finding such configurations for various projections. Recall that (N, S) is essentially the network obtained from N by breaking the links in S. Given a network $N = \{f\}$ and a set $S \subseteq L(N)$, we let the *projection* of f under S be the gate f' such that (N, S) converges to (f', S). The projection f' is denoted by f_S.

Theorem 6.43 *Let $N = \{f\}$ be a nonexpansive network, and consider two sets $S, T \subseteq L(N)$. Then $\{f_{S \cup T}\}$ has a stable configuration consistent with a given input assignment if and only if both $\{f_S\}$ and $\{f_T\}$ have such a stable configuration. If these equivalent conditions hold, then the stable configurations of $\{f_S\}$ are precisely the configurations consistent with stable configurations of $\{f_{S \cup T}\}$.*

Lemma 6.42 makes it possible to evaluate the gate defined by a convergent network. If L is a metric space, let $d_{\min}(L)$ denote the minimum distance between distinct points in L (set to 0 if L is a single point), and let $d_{\max}(L)$ denote the maximum distance between points in L. Let N by a network on an underlying metric space $M = M_{R(N)}$ with transition function f, and suppose that N converges to a gate g. Let $r = \lfloor d_{\max}(M_{L(N)})/d_{\min}(M_{O(N)}) \rfloor$. Lemma 6.42 indicates that at most r of the $d(y^i, g(x))$ can be nonzero; if we let $k = 2r+1$, then the majority of the y^i must equal $g(x)$. Therefore g can be evaluated with at most k queries to the transition function of N, each taking $O(m)$ time for $m = |R(N)|$.

We can now use Theorem 6.43 to find a stable configuration in a nonexpansive metric network, if one exists, for a fixed input assignment. First we discard the fixed inputs and the outputs, so that the network has no inputs or outputs. Then we find a fixed point a of f_i for some coordinate $i \in L(N)$, by trying all $|V(M_i)|$ possible values. Finally, we recursively apply this procedure to the network $(N, \{i\})$ obtained by breaking link i, with the input assignment $x_i = a$.

Theorem 6.44 *Suppose that we are given a network N on a metric space $M = M_{R(N)}$, and consider the three parameters $k = 2\lfloor d_{\max}(M_{L(N)})/d_{\min}(M_{L(N)}) \rfloor + 1$, $l = \sum_{i \in L(N)} |V(M_i)|$, and $m = |R(N)|$. Then one can find a stable configuration for N, if one exists, with kl queries to the transition function of N, in $O(klm)$ time.*

This gives a polynomial time algorithm for stability, provided that $d_{\max}/d_{\min}$ is polynomially bounded. If M_i is large, but a minimum distance point in M_i can be found in t_i queries, then a fixed point in $\Box_i M_i$ can be found in $\prod_i t_i$ queries. First a fixed point a_1 of the projection f_1 is found with t_1 queries for a minimum distance point a_2 of the projection f_2' for some $f' = f_{x_1, L \setminus \{1\}}$ (using substitutivity), and so on, giving eventually a fixed point a of f if one exists. For example, if M_i is a path on $2^r - 1$ points, then a binary search gives $t_i = r$. In $M_i \Box M_i$, Yannakakis [117] has observed that the direction $(x, f(x))$ restricts the search for a fixed point to a halfspace through x, so the complexity is still r.

Once again, Lemma 6.42 and Theorem 6.43 reduce stability as a decision problem to the evaluation question, since the existence of a fixed point for f is equivalent to the existence for each f_i, and the evaluation of f_i can be viewed as circuit evaluation.

6.8.2 Fixed Points and Retracts

For nonexpansive mappings in the hypercube, the fixed points set and the periodic set were studied using the median function and the notion of a median set. For nonexpansive mappings in the cartesian product of metric spaces, the imprint function and the 2-isometric subspaces play the same role.

Theorem 6.45 *Let f be a nonexpansive mapping on a cartesian product $M = \Box_{i \in S} M_i$, let L be the periodic space of f, and let L' be the fixed point space of f. Then L is a retract of M and hence a connected 2-isometric subspace of M, and L' is a 2-isometric subspace of $M' = \Box_{i \in S} L_i'$ (not necessarily connected).*

Proof. The mapping $g = f^{(n!)}$, where $n = |V(M)|$, is a retraction on M with L as retract, since g maps every point in M to a point in L, and leaves every point in L fixed because its period divides $n!$. Therefore L is a retract and hence a connected 2-isometric subspace of M.

The mapping f is an isomorphism on L (see Lemma 3.2). Given a fixed point x and a set of fixed points U, the imprint $I_L(x, U)$ must also be a fixed point, because the definition of imprints is invariant under isomorphisms. The space L is a connected 2-isometric irredundant subspace of $M'' = \square_{i \in S} L_i$, so the imprint function $I_{M''}$ coincides with the imprint function I_L within L. Therefore $I_{M''}(x, U)$ is in L' for a point x in L' and a set of points U in L'.

Given a point z in L' and a nonempty set of points V in L', suppose that z is not a maximal lower bound of V in M'. Then some point $t \neq z$ is a lower bound of V with respect to z in M'. Pick a coordinate i such that $t_i \neq z_i$. Note that t_i is a lower bound of V_i with respect to z_i in $M'_i = L'_i$, and hence also in $M''_i = L_i$. Let W be the set of points w in L' such that $w_i = t_i$. For each point v in V, the point $y = I_{M''}(z, W \cup \{v\})$ is in L' by the above argument. Furthermore, $y_i = I_{M''_i}(z_i, \{t_i, v_i\}) = t_i$ since $t_i \leq_{z_i} v_i$, so y is actually in W. It follows, by the definition of $I_{M''}$, that two points y, y' corresponding to two points v, v' in in V must satisfy $y \leq_z y'$ and $y' \leq_z y$, so $y = y'$. Therefore the point y is independent of the choice of v in V, and so $y \leq_z v$ for all v in V holds in M'' and as result in M' as well. Since $y_i = t_i \neq z_i$, we have $y \neq z$, proving that z is not a maximal lower bound of V in L' either. Therefore L' is a 2-isometric subspace of M'. □

It follows from this lemma and from Theorem 6.23 that a point x is in L if and only if each x_T with $|T| \leq 2$ is in L, and that a point x is in L' if and only if each x_T with $|T| \leq 2$ is in L'. By Theorem 6.43, if L' is nonempty, then L'_T is precisely the fixed point space of f_T; furthermore, if each L'_T with $|T| = 1$ is nonempty, then L' is nonempty. This gives:

Theorem 6.46 *A point x in the product $M = \square_{i \in S} M_i$ is a fixed point of a nonexpansive mapping f on M if and only if x_T is a fixed point of f_T for each $T \subseteq S$ with $|T| \leq 2$.*

The sets L_T for the periodic space L are harder to find; we shall examine this question in the next section.

We have seen that both the fixed point set and the periodic point set are 2-isometric irredundant subspaces of an appropriate product, and are therefore characterized by their projections under sets T with $|T| \leq 2$. Let L is a 2-isometric irredundant subspace of a product M. We do not known whether a point in L can be found from the sets L_T with $|T| \leq 2$ in $\mathcal{NC}^1$ as in the boolean case; we give an algorithm that works in $\mathcal{NC}$. The construction maintains a partition of the set of coordinates S, and a point in $L_{S'}$ for each S' in the partition. Initially, each set in the partition contains a single coordinate i, and a point in L_i is known. We show that given two coordinates sets S_1 and S_2 and a point from each of the sets L_{S_1} and L_{S_2}, one can find in parallel a point in $L_{S_1 \cup S_2}$. This makes it possible to repeatedly combine the sets of the partition in pairs, until the partition consists of the single set S. The number of times that such a pairing must be performed is $\lg |S|$.

Suppose then that a is in L_{S_1} and b is in L_{S_2}. Let x be an unknown point in L such that $x_{S_1} = a$, and let U be the set of points y in L such that $y_{S_2} = b$. We claim that the point $z_{S_1 \cup S_2}$ for $z = I_M(x, U)$ is easy to find. Since L is a 2-isometric subspace of M, the imprint z must also be in L, so the point $z_{S_1 \cup S_2}$ is in $L_{S_1 \cup S_2}$ as required. To find this point, it is sufficient to find $z_i = I_M(x_i, U_i)$ for each $i \in S_1 \cup S_2$. Note that U_i is L_i-closed, since U is L-closed as the intersection with the 2-isometric subspace L of the set of points in the M-closed set defined by the points y satisfying $y_{S_2} = b$. Therefore z_i is the point in U_i closest to x_i. If $i \in S_2$, then $U_i = \{b_i\}$, so $z_i = b_i$. If $i \in S_1 \backslash S_2$, then $x_i = a_i$, so in order to find z_i it is sufficient to find U_i. To achive this, we must be able to test whether a point c is in U_i, or equivalently, whether $w = bc$ is in $L_{S_2 \cup \{i\}}$. Since this space is a 2-isometric subspace of $M_{S_2 \cup \{i\}}$, the test reduces to whether w_T is in L_T for all $T \subseteq S_2 \cup \{i\}$ with $|T| \leq 2$. In fact, it is sufficient to test those subsets T that have $i \in T$, since b_T is in the remaining subsets by assumption. We can perform these tests in parallel for each T to test c, and for each c to determine U_i, then use the U_i to determine the point z_i in U_i closest to a_i in the case $i \in S_1 \backslash S_2$, and set $z_i = b$ for each $i \in S_2$, also in parallel for all $i \in S_1 \cup S_2$.

This construction provides an $\mathcal{NC}$ reduction from the stability question to the evaluation of convergent networks, since the characterization of the L_T can be obtained by evaluating the projections f_T as indicated by the preceding theorem. The evaluation of convergent networks consists in itself of the evaluation of a circuit consisting of repeated copies of f, as discussed following Theorem 6.43 above. Therefore the stability question as a search problem reduces to the evaluation question under $\mathcal{NC}$ reductions.

6.8.3 Periodic Points and Isomorphisms

Given a nonexpansive mapping f on a cartesian product $M = \square_{i \in S} M_i$, let L be the periodic space of f, and let $M'' = \square_{i \in S} L_i$. Each L_i has a canonical (isometric, 2-isometric, or retract) representation as a subspace of a product $\widetilde{M}_{S_i} = \square_{j \in S_i} \widetilde{M}_j$, where the S_i are disjoint sets whose union is a set T. These representations give a canonical (isometric, 2-isometric, or retract) representation for L as a subspace of the product $\widetilde{M} = \square_{i \in S} \widetilde{M}_{S_i} = \square_{j \in T} \widetilde{M}_j$.

Recall that the mapping g_σ on a product $\widetilde{M} = \square_{j \in T} \widetilde{M}_j$ is defined by a permutation σ on pairs (j, a) with j in T and a in $\widetilde{M}_j$. The permutation has $\sigma(j, g_j(b)) = (\nu(j), b)$, where ν is a permutation on T and each g_j is an isomorphism from $\widetilde{M}_{\nu(j)}$ to $\widetilde{M}_j$. The mapping g_σ is given then by $g_\sigma(x) = y$ if and only if $g_j(x_{\nu(j)}) = y_j$.

Consider the canonical (isometric, 2-isometric, or retract) representation for the periodic space L of f in a product $\widetilde{M}$ obtained above. If we replace each point x in L with its image $f(x)$ in $L = f(L)$, we obtain a new isometric, 2-isometric, or retract representation for L in $\widetilde{M}$, since f is an isomorphism on L. Since the original representation was the canonical representation, it must contain this new representation. Furthermore, since the number of factors in all isometric representations is determined by the number of equivalence classes of edges, and both representations have the same number of factors, it follows by containment that both representations must have the same equivalence classes, and hence the new representation is also a canonical representation and contains the original representation. Since both representations contain each other, one can obtain one from the other by means of an isomorphism g_σ on $\widetilde{M}$, as observed in section 6.2. This gives an isomorphism g_σ that extends the isomorphism f on L to the product $\widetilde{M}$.

Theorem 6.47 *Given a nonexpansive mapping f on a cartesian product $M = \square_{i \in S} M_i$ with L as its periodic space, represent each L_i as a subspace of a product $\widetilde{M}_{S_i} = \square_{j \in S_i} \widetilde{M}_j$, where the S_i are disjoint, by means of a canonical isometric, 2-isometric, or retract representation. Then f restricted to L coincides with an isomorphism g_σ on $\widetilde{M} = \square_{i \in S} \widetilde{M}_{S_i} = \square_{j \in T} \widetilde{M}_j$.*

Therefore, by decomposing the factors in the product appropriately, we obtain a simple representation for the behavior of f restricted to its periodic space.

In order to find the above representation, one needs to find the L_i. Unlike the case of the fixed point space, where a nonempty L' gives L_i' as the fixed point space of the projection f_i, there is no obvious connection between the fixed points of f and the fixed points of f_i. For example if we consider the mapping on the 2-cube defined by $f(00) = 00$, $f(01) = 10$, $f(10) = 01$, $f(11) = 00$, then 10 is a periodic point of f, but the mapping f_1, given by $f_1(0) = 0$, $f_1(1) = 0$ does not have a corresponding periodic point at 1. If we consider the mapping on the product of a path $0, 1, 2$ of length two and a path $0, 1$ of length 1 defined by $f(00) = 20$, $f(01) = 10$, $f(10) = 10$, $f(11) = 11$, $f(20) = 11$, $f(21) = 01$, then the mapping f_1, given by $f_1(0) = 2$, $f_1(1) = 1$, $f_1(2) = 0$ has a periodic point 0, but neither 00 nor 01 are periodic points of f. We give below a procedure for testing membership in L_i that depends on an a priori bound k such that $f^{(k)}(x)$ is a periodic point for all x in M; such bounds will be discussed in section 6.8.5.

Suppose that there exists an unknown periodic point t such that $t_i = b$ for some given b. Find a periodic point $y = f^{(k)}(x)$. Let $z = f^{(r)}(y)$, where r will be specified later; if $z_i = b$ then we are done, so we assume that $z_i \neq b$, so that $z \neq t$. Let w be the point $w_i = b$ and $w_{S \setminus \{i\}} = z_{S \setminus \{i\}}$, not necessarily a periodic point. Since $w_i \neq z_i$, we have $w \neq z$. Observe that $d(z, t) = d(z, w) + d(w, t)$; therefore $d(z', t') = d(z', w') + d(w', t')$ for $z' = f^{(k)}(z)$, $t' = f^{(k)}(t)$, and $w' = f^{(k)}(w)$, with $w' \neq z'$, by nonexpansiveness and the fact that f preserves distances on periodic points. Since z' and w' are periodic, there exists a periodic point $\widetilde{z}'$ adjacent to z' in the periodic space L such that $d(z', w') = d(z', \widetilde{z}') + d(\widetilde{z}', w')$. Since u' is periodic, we have $u' = f^{(r+k)}(v)$ for some periodic point v. Furthermore, v must be adjacent to y in M, because a point between y and v in M would map to a periodic point between $z' = f^{(r+k)}(y)$ and $u' = f^{(r+k)}(v)$, contradicting the adjacency of z' and u'. If we test all the points v' adjacent to y in M, some such point must have $f^{(r+k)}(v') = \widetilde{z}'$; this enables us to find periodic points $\widetilde{y} = f^{(k)}(v')$ and $\widetilde{z} = f^{(r-k)}(\widetilde{y})$ such that $f^{(k)}(\widetilde{z}) = \widetilde{z}'$. Furthermore, $d(z, w) = d(z, \widetilde{z}) + d(\widetilde{z}, w)$ since f preserves distances on periodic points, so

$d(z,t) = d(z,\tilde{z}) + d(\tilde{z},t)$, with $z \neq \tilde{z}$ since the images of these two points under $f^{(k)}$ are two adjacent yet different points. This implies that $d(\tilde{z},t) \leq d(z,t)\backslash d_{\min}(M)$. We have therefore replaced the original periodic points y and $z = f^{(r)}(y)$ with periodic points $\tilde{y}$ and $\tilde{z} = f^{(r-k)}(\tilde{y})$ such that $d(\tilde{z},t)$ is smaller than $d(z,t)$ by at least $d_{\min}(M)$. Repeating this at most $l = \lfloor d_{\max}(M)/d_{\min}(M) \rfloor$ times, we must obtain at some point a periodic point $\hat{z}$ with $\hat{z}_i = b$, because otherwise we have periodic points $\hat{y}$ and $\hat{z} = f^{(r-lk)}(\hat{y})$ with $d(\hat{z},t) \leq d_{\max}(M) - ld_{\min}(M) < d_{\min}(M)$, so that $\hat{z} = t$ and $\hat{z}_i = t_i = b$. The choice of r must be such that $r \geq lk$, and the total number of calls to f is then $O(lrd)$, where d is the maximum degree of a point in $G(M)$, equal to the sum over all factors M_i of the maximum degree of a point in M_i.

6.8.4 Unstable Mappings and Fixed Products

We now generalize Bandelt and Vel's fixed cube theorem on the hypercube to cartesian products of metric spaces. Let L be the periodic space of a nonexpansive mapping f on a product $M = \square_{i \in S} M_i$. Since L is a retract of M, it has no holes in M, so by Theorems 6.28 and 6.31, its distance center and the L-closure U of its distance center are subproducts of M. Since f is an isomorphism on L, and the definitions of distance center and closure are invariant under isomorphisms, it follows that $f(U) = U$. A set of points U is a *fixed product* in M if U is a subproduct of M and $f(U) = U$. The preceding discussion gives the following.

Theorem 6.48 *Every nonexpansive mapping f on a product M with periodic space L has an L-closed fixed product in M.*

A factor M_i of a product M is *trivial* for U if $|U_i| = 1$. Once again, we can show that all fixed products, under a minimality and also a closure condition, involve the same factors.

Theorem 6.49 *All minimal L-closed fixed products in M of a nonexpansive mapping f have the same trivial factors.*

Proof. Let U and V be two minimal L-closed fixed products in M. Suppose that $|U_i| = 1$, so that $U_i = \{u_i\}$ for some u_i. Define $d(x',U')$ to be the minimum of $d(x',u')$ for $u' \in U'$. Let W be the set of points x in V that minimize $d(x,U)$. Since f is nonexpansive and $f(U) = U$, it follows that if x is a minimizing point then so is $f(x)$, and therefore $f(W) = W$. The points x in V that minimize $d(x,U)$ are those points for which x_i is in V_i and minimizes $d(x_i, V_i)$; since this minimization can be performed independently for each factor, it follows that W is a subproduct of M. Since V is L-closed, each V_j is L_j-closed, so $W_i = \{w_i\}$, where w_i is the unique point in V_i closest to u_i. Therefore $|W_i| = 1$. Since L is a 2-isometric subspace of M, the L-closure W' of W is also a subproduct of M. Furthermore, since the set of points y in M such that $y_i = w_i$ is M-closed, the intersection of this set with L must be L-closed, since this intersection contains W, it must also contain its L-closure W'. It follows that all points z in W' also satisfy $z_i = w_i$ and therefore has $|W'_i| = 1$. Since

V is L-closed and contains W, it must also contain its L-closure W'. We have thus shown that W' is an L-closed fixed product contained in V, so by minimality $V = W'$. Therefore $|V_i| = 1$ as well. □

The last two theorems give a different proof of the fact that f has a fixed point if and only if each f_i has a fixed point. The 'only if' direction is obvious. For the 'if' direction, suppose that each f_i has a fixed point a_i. The mapping f can be written as $f(xy) = g_x(y)h_x(y)$, with $x, g_x(y)$ in M_i and $y, h_x(y)$ in $M_{S\setminus\{i\}}$. Let $x = a$ be a fixed point of f_i, and let $h = h_a$. Then h is a nonexpansive mapping on $M_{S\setminus\{i\}}$, and therefore has a fixed product U by Theorem 6.48. Since all the points in U are periodic points of h, we have $g_a(u) = f_i(a) = a$, so $f(au) = ah_a(u)$ for all u in U, so U can be viewed as a fixed product V of f in M consisting of the points v with $v_i = a$ and $z_{S\setminus\{i\}} \in U$. Therefore f has a fixed product V with $|V_i| = 1$. It follows, as in the proof of the last theorem, that f has an L-closed fixed product V' with $|V'_i| = 1$, and thus a minimal such V'. By the last theorem, all minimal L-closed fixed products V' of f must have $|V_i| = 1$. Applying this argument to all i, we concude that every minimal L-closed fixed product V' of f has $|V_i| = 1$ for all i, and therefore V consists of a single fixed point. Since f has an L-closed fixed products, it follows that f has a fixed point.

We can even go further and give a precise characterization of the trivial factors of minimal L-closed fixed products in terms of the isomorphism g_σ on the product $\widetilde{M} = \square_{j \in T} \widetilde{M}_j$ from the canonical representation for the periodic space L, where g_σ coincides with f in the periodic space L. The permutation σ is defined by $\sigma(j, g_j(b)) = (\nu(j), b)$, where ν is a permutation on T and g_j is an isomorphism from $\widetilde{M}_{\nu(j)}$ to $\widetilde{M}_j$; then $g_\sigma(x) = y$ if and only if $g_j(x_{\nu(j)}) = y_j$ for all $j \in T$. A cycle C of σ is a *stable* cycle if it contains at most one pair of the form (j, b) for each $j \in T$. Equivalently, C must contain exactly one pair of the form (j, b) for some $j \in T$. This condition is clearly necessary for a cycle to be stable; it is also sufficient, for if C contains two distinct pairs (j', c) and (j', c') for some other $j' \in T$, then the fact that $j = \sigma^{(k)}(j')$ for some k and σ is one-to-one gives two pairs $\sigma^{(k)}(j', c)$ and $\sigma^{(k)}(j', c')$ of the form (j, b). A factor $\widetilde{M}_j$ is a *stable* factor if σ has some stable cycle C that contains a pair of the form (j, b). For the following theorem, we shall assume that the canonical representation for L in $\widetilde{M}$ is the canonical retract representation.

Theorem 6.50 *Let f be a nonexpansive mapping on a product M, and let L be the periodic space of f. Then the minimal L-closed fixed products of f in M are precisely the minimal L-closed fixed products of f in $\widetilde{M}$. The trivial factors of the minimal L-closed fixed products of f in $\widetilde{M}$ are precisely the stable factors of σ.*

Proof. The property of being L-closed and fixed is the same in M and in $\widetilde{M}$. The subproducts of $\widetilde{M}$ contained in L are clearly subproducts of M, since M combines factors of $\widetilde{M}$. In order to establish the first claim, It is then sufficient to show that every minimal L-closed fixed product of f in M is in fact a subproduct of $\widetilde{M}$. Let L' be the subspace of L induced by the points in a minimal L-closed fixed product of f in M. Since L' is L-closed, it follows that L' is a retract of L by Lemma 6.30; but L is a retract of $\widetilde{M}$ by the choice of representation, so L' is a retract of $\widetilde{M}$. We can then extend the mapping f on L' to a mapping g on $\widetilde{M}$ with $f(\widetilde{M}) = L'$, by pre-composing it with a retraction on $\widetilde{M}$ with L' as retract. The mapping g must have an L'-closed fixed product U. Since L' is L-closed, it follows that U is L-closed. Therefore U is an L-closed fixed product of f in $\widetilde{M}$, and hence U is an L-closed fixed product in M. By the minimality condition in the choice of L', the set U must consist of all the points in L', so the fact that U is a subproduct implies that L' is a subproduct.

We now prove the second claim in the statement of the theorem. Suppose that $\widetilde{M}_j$ is a trivial factor of a fixed product U of f in $\widetilde{M}$. Then $U_j = b$ for some point b in $\widetilde{M}_j$; let x be a point in U such that $x_j = b$. The pair (j, b) belongs to some cycle C of σ. To show that $\widetilde{M}_j$ is a stable factor, it is sufficient to show that C is a stable cycle. To show that C is a stable cycle, it is sufficient to show that C contains no pair (j, b') with $b' \neq b$. But if C contains such a pair, so that $(j, b) = \sigma^{(k)}(j, b')$ for some k, then $y = f^{(k)}(x) = g_\sigma^{(k)}(x)$ has $y_j = b'$, because $x_j = b$. Since x is in U, so is y, contradicting the fact that $b' \notin U_j$.

In the other direction, suppose that $\widetilde{M}_j$ is a stable factor of σ. Then there is a stable cycle C of σ that contains a pair (j, b). Suppose that U is a minimal L-closed fixed product of f. Then each $U_{j'}$ is $L_{j'}$-closed by Lemma 6.20. For every pair $(j', v_{j'})$ in C, let $u_{j'}$ be the point in $U_{j'}$ closest to $v_{j'}$. Since $f = g_\sigma$ is an isomorphism on U, the isomorphism $g_{j'}$ used in defining σ must be a one-to-one correspondence from $U_{\nu(j')}$ to $U_{j'}$. Furthermore, $g_{j'}(v_{\nu(j')}) = v_{j'}$ for all j' that have a pair in C. Since $g_{j'}$ is an isomorphism from $L_{\nu(j')}$ to $L_{j'}$, it must map the point closest to $v_{\nu(j')}$ in $U_{\nu(j')}$ to the point closest to $v_{j'}$ in $U_{j'}$, so $g_{j'}(u_{\nu(j')}) = u_{j'}$. Let W be the subproduct of U consisting of the points x such that $x_{j'} = u_{j'}$ for j' that have a pair in C. Since each condition $x_{j'} = u_{j'}$ defines an $\widetilde{M}$-closed set, it also defines an L-closed set inside L, and so W is an intersection of L-closed sets, hence L-closed. Furthermore, if x is in W, then $y = f(x) = g_\sigma(x)$ has $y_{j'} = g_{j'}(x_{\nu(j')}) = g_{j'}(u_{\nu(j')}) = u_{j'}$ for all j' that have a pair in C, so y is in W. Therefore W is an L-closed fixed product contained in U, and so by minimality of U we have $U = W$. As a result, $U_{j'} = W_{j'} = \{u_{j'}\}$ for all j' that have a pair in C, and in particular for $j' = j$, so $|U_j| = 1$ and $\widetilde{M}_j$ is a trivial factor of U. □

Theorems 6.48 and 6.50 imply that f has a fixed point if and only if all factors of $\widetilde{M}$ are stable factors.

One drawback of the notion of an L-closed fixed product is that, in order to test that a subproduct U is L-closed, we must test that each U_i is L_i-closed, so explicit knowledge of the L_i for the periodic space L is required. Two other drawbacks are (a) the fact that the canonical retract representation is used in the characterization of the last theorem, and this representation is $\mathcal{NP}$-hard to find, and (b) the fact that the proof of the existence of fixed products uses the distance center, which is $\#\mathcal{P}$-hard to find as we saw in section 3.4. A fixed product can be found in polynomial time as follows. First, we find the isometric (or 2-isometric) representation for the periodic set. In this representation, the original coordinates form a partition $E_1, E_2, \ldots, E_k$ of the new coordinates. We partition the sets E_i further as follows. In the new representation, the isomorphism on the periodic set permutes the coordinates via a permutation ν, so if two coordinates j, j' in the same E_i are mapped to two different $E_{i'}$, then E_i is partitioned depending on which $j \in E_i$ map to each different $E_{i'}$. Continuing this process, we end up with a refinement $E'_1, E'_2, \ldots, E'_{k'}$, such that ν is a permutation on the E'_i. This partition defines a representation contained in every isometric representation where ν permutes the factors, and in particular, in the retract representation. So a fixed cube must exist in this representation. Let L be the periodic space. Find a minimal L_1-closed set U_1 of points in the first factor, with corresponding sets U_i in the factors that belong to the same cycle of ν as the first factor, and such that every combination of values from these U_i occurs for some periodic point (it is sufficient to test pairwise consistency of values). This defines a subproduct that must contain a fixed product, and we proceed similarly to find sets U_i in the factors corresponding to the remaining cycles of ν. In the end, the product of the U_i is a fixed product.

6.8.5 Non-Periodic Points and Iterates

Some of the problems on metric networks, such as the characterization of the L_i for the periodic set L discussed in section 6.8.3, require finding a periodic point. For the convergence question, we actually want to find a periodic point attained when f is iterated starting with a specific point. For this reason, it is useful to give a bound on the number of iterations of f required to reach a periodic point. We concentrate on the unweighted case, where all the edges of the graphs associated with metric spaces have weight 1. The result proved below seems to extend in a staightforward manner to the weighted case to give a polynomial bound involving $d_{\max}(M)/d_{\min}(M)$. The advantage of the unweighted case is that it forces nonexpansive mappings to map adjacent points to adjacent or equal points, so that they can be viewed as mappings on edges; this makes the construction intuitively simpler.

We identify a metric space M with its corresponding graph $G(M)$. Given a graph G with edges of weight 1, let $l(G) = \max_{x,y \in V(G)} d(x,y)$ denote the diameter of G. Given a product $H = \Box_{i \in S} H_i$, with edges of weight 1, let $l = l(H) = \sum_{i \in S} l(H_i)$, let $n = \sum_{i \in S} (|V(H_i)| - 1)$, and let $m = \sum_i |E(H_i)|$.

Theorem 6.51 *Let f be a nonexpansive mapping on an unweighted product $H = \Box_{i \in S} H_i$. If x is a vertex in H at distance d from the periodic set of f, then $f^{(k)}(x)$ is a periodic point for all $k \geq 5d(l+1)nm$, where $d \leq l \leq n \leq m$.*

Proof. It is sufficient to consider the case $d = 1$, because we can prove that if $f^{(k)}(x)$ is a periodic point for all x at distance 1 from the periodic set, then $f^{(kd)}(x)$ is a periodic point for all x at distance d from the periodic set. The inductive proof of this fact considers a point x is at distance $d \geq 1$ from the periodic set U. The point x is at distance $d-1$ from some point y at distance 1 from U, so $f^{(k)}(x)$ is at distance at most $d-1$ from the periodic point $f^{(k)}(y)$ and hence from the set U, and by inductive hypothesis $f^{(kd)}(x) = f^{(k(d-1))}(f^{(k)}(x))$ is a periodic point, completing the induction.

The basic idea of the proof for $d=1$ is to define a *potential* for points that are adjacent (possibly equal) to periodic points, and show that this potential decreases quite rapidly as the mapping f is iterated. Given such a point x, we consider two cases. If there exists a pair of periodic points y, z with y adjacent or equal to x, such that $d(x,z) \leq d(y,z)$, then we let the potential of x be the minimum of $d(y,z)$ over all such pairs. If no such pair exists, then we shall choose an arbitrary point y adjacent to x anyway; the potential of x is set to $l+1$. Note that the potential of x is 0 if and only if $x = y = z$ so that x is a periodic point; the potential of x is between 1 and l for nonperiodic points satisfying the first case; and the potential of x is $l+1$ for points in the second case. The potential of $f(x)$ is also no greater than the potential of x, because given the appropriate y and z for x, we have $d(f(x), f(z)) \leq d(x,z) \leq d(y,z) = d(f(y), f(z))$. The main tool that enables us to show that potentials decrease is given by the following two lemmas. Let G be the the graph of the periodic space of f. We say that two edges e and e' in G iterate in *different factors* with respect to f if $f^{(r)}(e)$ and $f^{(r)}(e')$ belong to different factors in H for some $r \geq 0$. We say that an edge e'' in G iterates in factor i with respect to f if $f^{(r)}(e'')$ belongs to factor H_i for all $r \geq 0$.

Lemma 6.52 *Suppose that an edge e'' does not iterate in factor i'' with respect to f. Then for every $a \geq 0$, there exists an r with $a \leq r < a+n$ such that $f^{(r)}(e'')$ does not belong to factor i''. Suppose that two edges e and e' in G iterate in different factors with respect to f. Then for every $a \geq 0$, there exists an r with $a \leq r < a+n$ such that $f^{(r)}(e)$ and $f^{(r)}(e')$ belong to different factors.*

Proof. Consider the canonical isometric (alternatively, 2-isometric or retract) representation for G as a subgraph of a product $\tilde{H} = \square_{j \in T} \tilde{H}_j$, where T is the disjoint union of sets S_i with $i \in S$, and the edges of G that belong to factor j in $\tilde{H}$ belong to the factor i in H such that $j \in S_i$. Therefore the value of j determines the value of i. There is an isomorphism g_σ on $\tilde{H}$ that coincides with f on G. The mapping g_σ is defined by a permutation σ with $\sigma(j, g_j(b)) = (\nu(j), b)$; then $g_\sigma(x) = y$ if and only if $g_j(x_{\nu(j)}) = y_j$. The factor $j''(r)$ to which an edge $f^{(r)}(e'')$ belongs in $\tilde{H}$ is thus periodic in r; the period is the length of the corresponding cycle of ν. It follows that the corresponding factor $i''(r)$ is also periodic, and its period divides the period of $j''(r)$. The period of a cycle in ν is at most $|T|$. Recall from section 6.2 that $|S_i| \leq |V(G_i)| - 1$; therefore $|T| \leq n$, proving the first claim in the lemma.

Consider now two edges e, e' with the corresponding factors $j(r), j'(r)$ in $\tilde{H}$ and $i(r), i'(r)$ in H for $f^{(r)}(e), f^{(r)}(e')$. The periods q, q' of $j(r), j'(r)$ satisfy $q, q' \leq |T| \leq n$; furthermore, if $q \neq q'$, then the two families of edges belong to different cycles of ν, and therefore $q + q' \leq |T| \leq n$. Since $j(r), j'(r)$ determine $i(r), i'(r)$, the periods of the latter must divide q, q' respectively. Let R be the range of values for r with $a \leq r < a+n$. To prove the lemma, it is sufficient to show that $i(r) \neq i'(r)$ for some $r \in R$. Suppose towards a contradiction that $i(r) = i'(r)$ for all $r \in R$. Consider first the case where the period of the $i(r)$ divides q'. Then the period of both $i(r)$ and $i'(r)$ divides q', and since $i(r) = i'(r)$ for all r with $a \leq r < a + q' \leq r + n$, it follows that $i(r) = i'(r)$ for all r, contrary to the assumption that e and e' iterate in different factors. If the period of the $i(r)$ does not divide q', then it follows from the fact that it divides q that $q' \neq q$, so that $q + q' \leq n$. Since the period of $i(r)$ does not divide q', it follows that $i(r) \neq i(r + q')$ for some r. Since the period of $i(r)$ divides q, this inequality holds for congruent values of r modulo q, and in particular for some r with $a \leq r < a + q$. Then $a \leq r < r + q' < a + q + q' \leq a + n$, so both r and $r + q'$ are in R and therefore $i'(r) = i(r)$ and $i'(r + q') = i(r + q')$. But this means that $i'(r) \neq i'(r + q')$, contradicting the fact that the period of the $i'(r)$ divides q'. □

The fact that certain edges iterate in different factors will be used to decrease potentials by means of the following lemma.

Lemma 6.53 *Suppose that x is adjacent to a periodic point y, and that the periodic points z, z' satisfy $d(x, z) \leq d(y, z)$ and $d(x, z') \leq d(y, z')$. Suppose further that P and P' are shortest paths in G from y to z and from y to z' respectively, and that if e and e' are edges on P and P' respectively other than the first p edges on these paths, then e and e' iterate in different factors. Then $f^{(ns)}(x)$ has potential at most p for $t = \max(0, |P| + |P'| - 2p - 1)$.*

Proof. The proof is by induction on s. If $s = 0$, then either $|P| \leq p$ or $|P'| \leq p$. If $|P| \leq p$, then $d(y, z) \leq p$ and $x = f^{(0)}(x)$ has potential at most p; the case where $|P'| \leq p$ is similar. If $s \geq 1$, assume that the lemma holds for $s-1$. If $|P| \leq p$ or $|P'| \leq p$, then we just saw that x has potential at most p, so that $f^{(ns)}(x)$ also has potential at most p. We can therefore assume that $|P|, |P'| > p$. Let e and e' be the last two edges of the paths P and P' respectively. Then e and e' iterate in different factors by assumption, so $f^{(r)}(e)$ and $f^{(r)}(e')$ belong to different factors for some r with $0 \leq r < n$ by the previous lemma. In particular, one of these two edges must belong to a factor different from the factor to which the edge $\{\tilde{x}, \tilde{y}\} = .\{f^{(r)}(x), f^{(r)}(y)\}$ belongs. Say it is the edge $f^{(r)}(e) = \{\tilde{w}, \tilde{z}\}$, where $\tilde{z} = f^{(r)}(z)$. Since these two edges belong to different factors, we have

$$d(\tilde{x}, \tilde{w}) - d(\tilde{y}, \tilde{w}) = d(\tilde{x}, \tilde{z}) - d(\tilde{y}, \tilde{z}) \leq d(x, z) - d(y, z) \leq 0.$$

Letting $\tilde{z}' = f^{(r)}(z')$, and taking images of the path P without its last edge and of the path P', we have paths $\tilde{P}$ and $\tilde{P}'$ from $\tilde{y}$ to $\tilde{w}$ and from $\tilde{y}$ to $\tilde{z}'$ respectively with $|\tilde{P}| = |P| - 1$, $|\tilde{P}'| = |P'|$, and also $d(\tilde{x}, \tilde{w}) \leq d(\tilde{y}, \tilde{w})$, $d(\tilde{x}, \tilde{z}') \leq d(\tilde{y}, \tilde{z}')$. Since $\tilde{s} = |\tilde{P}| + |\tilde{P}'| - 2p - 1 = s - 1$, and the images under $f^{(r)}$ of two edges that iterate in different factors also iterate in different factors by the previous lemma, we can use the inductive hypothesis to infer that $f^{(n(s-1))}(\tilde{x}) = f^{(n(s-1)+r)}(x)$ has potential at most p, and since $r < n$, the point $f^{(ns)}(x)$ has potential at most p as well. □

To carry out the proof of the theorem, we shall start with a point x of potential $p + 1$, and show that after a relatively small number of iterations the scenario from Lemma 6.53 arises, so that after an additional ns iterations the potential must decrease to at most p. To achieve this, we need to define several points related to x that will eventually give points satisfying the conditions of the lemma.

Suppose then that x is a nonperiodic point of potential $p + 1$, so that either there is a pair of periodic points y, z with x adjacent to y and $d(x, z) \leq d(y, z) = p + 1$, or there is no such pair, in which case $p+1 = l+1$ and y is an arbitrary periodic point adjacent to x. Note that if r is sufficiently large, then $f^{(r)}(x)$ is a periodic point adjacent or equal to $f^{(r)}(y)$. We let u be the periodic point adjacent or equal to y such that $f^{(r)}(u) = f^{(r)}(x)$ for such an r.

We shall be interested in points obtained from these points by iterating f. The case where $f^{(2m)}(x)$ is a periodic point and hence has potential 0 is a good case in the sense that the potential decreases within only $2m$ iterations. We therefore consider the case where $f^{(r)}(x)$ is nonperiodic for all $0 \leq r \leq 2m$. These nonperiodic points $f^{(r)}(x)$ are adjacent to periodic points $f^{(r)}(y)$ for all such r, and the directed edge $e = e(r)$ joining them belongs to some factor $i = i(r)$. The corresponding edges e_i in factor $i = i(r)$ cannot all be distinct for $0 \leq r \leq 2m$, because there are $2m+1$ such r but only $2m$ possible e_i, namely the edges in the graphs H_i with the two possible ways of orienting them. Therefore two of these e_i projections coincide, say those corresponding to values r_1 and r_2 of r with $0 \leq r_1 < r_2 \leq 2m$. We denote by $\hat{\imath}$ the common value $i(r_1) = i(r_2)$. For every point v, we introduce the notation $v^r = f^{(r_1 + r(r_2 - r_1))}(v)$. In particular, for $v = x, y$ and $r = 0, 1$, we have edges $\{x^0, y^0\}$ and $\{x^1, y^1\}$ that belong to factor $\hat{\imath}$, with $x_{\hat{\imath}}^0 = x_{\hat{\imath}}^1$ and $y_{\hat{\imath}}^0 = y_{\hat{\imath}}^1$.

Two special values $r = a$ and $r = b$ will give particularly useful pairs of points x^r, y^r, because they will make it possible to satisfy the conditions of Lemma 6.53. The value $r = a$ is the least $r \geq 0$ such that the edge $\{x^r, y^r\}$ does not belong to factor $\hat{\imath}$. If no such r exists, we let $a = \infty$. Note that $a \geq 2$. The value $r = b$ is chosen so as to satisfy certain properties, mentioned in the following lemma.

Lemma 6.54 *If the value a is finite, then there exists a value b with $1 \leq b \leq a$ and $b \leq 2n$, such that $d(y^b, y^a) \leq d(y^b, x^a)$.*

Proof. Consider again the canonical isometric (alternatively, 2-isometric or retract) representation for G in a product $\widetilde{H} = \square_{i \in S} \widetilde{H}_{S_i} = \square_{j \in T} \widetilde{H}_j$, as in the proof of Lemma 6.43. We first wish to show that there exists a value b with $1 \leq b \leq a$ and $b \leq 2n$ such that $d(y^b, y^a) \leq d(y^b, u^a)$. This is clear if $y^a = u^a$, where we can set $b = 1$. We therefore assume $y^a \neq u^a$, and let $\hat{\jmath}$ be the factor in $\widetilde{H}$ to which the edge $\{y^a, u^a\}$ belongs. Just as for the mapping f, the mapping $f' = f^{(r_2 - r_1)}$ restricted to G coincides with an isomorphism g_σ on $\widetilde{H}$, where σ is a permutation on pairs (j, b) with $j \subset T$ and $b \in V(\widetilde{H}_j)$. We claim that the number of such pairs, and hence the period of a pair under σ, is at most $2n$. This number $\sum_{j \in T} |V(\widetilde{H}_j)|$, or equivalently the sum over all i of $\sum_{j \in S_i} |V(\widetilde{H}_j)|$. If we consider the graph G_i embedded in H_{S_i}, then a single vertex of G_i contributes $|S_i|$ to this sum, or one vertex for each factor $j \in S_i$, but subsequent vertices of G_i are adjacent to earlier vertices, so they each contribute at most one vertex to one factor, hence at most $|V(G_i)| - 1$ all together. Using the fact that the number of factors satisfies $|S_i| \leq |V(G_i)| - 1$, we obtain a total sum of at most $2 \sum_{i \in S}(|V(G_i)| - 1) \leq 2n$.

Thus every pair has period at most $2n$ in σ. In particular, the pair $(\hat{\jmath}, y_{\hat{\jmath}}^a)$ has some period $q \leq 2n$, implying that $y_{\hat{\jmath}}^{a-rq} = y_{\hat{\jmath}}^a$ for all r. There is therefore a value b with

CHAPTER 6. METRIC NETWORKS AND PRODUCT GRAPHS

$1 \leq b \leq q \leq 2n$ and $b \leq a$ such that $y_{\hat{j}}^b = y_{\hat{j}}^a$. Then $d(y^b, y^a) \leq d(y^b, u^a)$, because y^a and u^a only differ in coordinate $\hat{j}$, and y^b agrees with y^a in this coordinate, proving the claim.

Since $x^{a+r} = u^{a+r}$ for some sufficiently large r, we have $d(y^b, y^a) = d(y^{b+r}, y^{a+r}) \leq d(y^{b+r}, u^{a+r}) = d(y^{b+r}, x^{a+r}) \leq d(y^b, x^a)$. □

In order to prove that potentials decrease, we shall need one more distance property.

Lemma 6.55 *The inequality $d(x^0, x^r) \leq d(y^0, y^r)$ holds for all $0 \leq r < a$, where the value a may be infinite. If a is finite, then $d(x^b, x^a) \leq d(y^b, y^a)$.*

Proof. The proof is by induction on r. The base case $r = 0$ is vacuously true. Suppose that $d(x^0, x^{r-1}) \leq d(y^0, y^{r-1})$, with $1 \leq r < a$. Then $d(x^1, x^r) \leq d(x^0, x^{r-1}) \leq d(y^0, y^{r-1}) = d(y^1, y^r)$. By the definition of a, the edges $\{x^0, y^0\}$, $\{x^1, y^1\}$ and $\{x^r, y^r\}$ all belong to factor $\hat{i}$, so $d_{H_i}(x_i^1, x_i^r) = d_{H_i}(y_i^1, y_i^r)$ and $d_{H_i}(x_i^0, x_i^r) = d_{H_i}(y_i^0, y_i^r)$ for $i \neq \hat{i}$. This implies that $d_{H_{\hat{i}}}(x_{\hat{i}}^1, x_{\hat{i}}^r) \leq d_{H_{\hat{i}}}(y_{\hat{i}}^1, y_{\hat{i}}^r)$, and so $d_{H_{\hat{i}}}(x_{\hat{i}}^0, x_{\hat{i}}^r) \leq d_{H_{\hat{i}}}(y_{\hat{i}}^0, y_{\hat{i}}^r)$ because $x_{\hat{i}}^0 = x_{\hat{i}}^1$, $y_{\hat{i}}^0 = y_{\hat{i}}^1$, proving that $d(x^0, x^r) \leq d(y^0, y^r)$ and completing the induction.

In particular, since $1 \leq b \leq a$, we have $d(x^b, x^a) \leq d(x^0, x^{a-b}) \leq d(y^0, y^{a-b}) = d(y^b, y^a)$.
□

Before proving the main lemma, we examine the factors where certain edges iterate. This will be needed to satisfy the conditions of Lemma 6.53.

Lemma 6.56 *For all $r', r'' \geq 0$, the edges on a shortest path from $y^{r'}$ to $y^{r''}$ in G do not iterate in factor $\hat{i}$ with respect to $f' = f^{(r_2 - r_1)}$*

Proof. The edges on a shortest path from y^0 to y^1 in G do not belong to the factor $\hat{i}$ because $y_{\hat{i}}^0 = y_{\hat{i}}^1$. Therefore these edges do not iterate in factor $\hat{i}$. The edges on a shortest path from y^r to y^{r+1} in G map to edges on a shortest path from y^0 to y^1 after some number of iterations of f', since y^0 is periodic, so these edges do not iterate in the factor $\hat{i}$ either. If $r' < r''$, then there is a path P from $y^{r'}$ to $y^{r''}$ in G that goes through $y^{r'}, y^{r'+1}, \ldots, y^{r''-1}, y^{r''}$, joining two consecutive points among these by shortest paths in G. None of the edges on this path P iterate in the factor $\hat{i}$. In the canonical isometric (or 2-isometric, or retract) representation for G in a product $\widetilde{H}$, every edge on a shortest path Q from y^r to $y^{r'}$ visits a factor j that is visited by some corresponding edge on P as well. The factors in which these two associated edges iterate are determined by the permutation ν used in defining σ in the isomorphism g_σ that coincides with f in G, so the edge in Q iterates in the same factors as the corresponding edge in P, inside $\widetilde{H}$. Since the factor j in $\widetilde{H}$ that an edge visits determines the corresponding factor i in H uniquely, it follows that the edge in Q iterates in the same factors as the corresponding edge in P inside H has well. Therefore the edges on the shortest path Q do not iterate in the factor $\hat{i}$ either. □

The main lemma showing that potentials decrease is the following.

Lemma 6.57 *If x has potential $p + 1$, then $f^{(k)}(x)$ has potential at most p for $k = 5nm$.*

Proof. The statement of the lemma clearly holds if $f^{(2m)}(x)$ is a periodic point, so we can assume as we have so far that $f^{(r)}(x)$ is not a periodic point for $0 \leq r \leq 2m$ and use the definitions and properties obtained under this assumption up to this point. We consider three cases, depending on the value of a.

Suppose that $2n < a < \infty$. Then $b < a$ by Lemma 6.54, so by the definition of a, the edge $\{x^a, y^a\}$ does not belong to factor $\hat{\imath}$, but the edge $\{x^b, y^b\}$ belongs to factor $\hat{\imath}$. Therefore $d(x^b, y^a) - d(y^b, y^a) = d(x^b, x^a) - d(y^b, x^a)$. Recall that $d(x^b, x^a) \leq d(y^b, y^a) \leq d(y^b, x^a)$ by Lemmas 6.54 and 6.55. Therefore $d(x^b, y^a) \leq d(y^b, y^a)$. This shows that the potential of x^b is given by the first case in the definition of potential, and is therefore at most l; if the potential of x is given by the second case in the definition of potential, then it equals $l+1$, so $f^{(k)}(x) = x^b$ has smaller potential for $k = r_1 + b(r_2 - r_1) \leq 2mb \leq 4nm$, satisfying the condition of the lemma. We can now assume that the potential of x is given by the first case in the definition of potential, so that $d(x, z) \leq d(y, z) = p + 1$ for some pair of periodic points y, z, with y adjacent to x in H. The edges on a shortest path P from y^b to y^a in G do not iterate in factor $\hat{\imath}$ with respect to $f' = f^{r_2 - r_1}$ by the previous lemma. On the other hand, we have $d(x^b, z^b) \leq d(y^b, z^b) = p + 1$, with a path P' of length $p + 1$ from y^b to z^b in G that has some edge $e' = \{w^b, z^b\}$ as its last edge. Suppose that e' iterates in factor $\hat{\imath}$ with respect to f'. Then e' and e'' iterate in different factors with respect to f', for each edge e'' in P, and therefore e' and e'' iterate in different factors with respect to f, for each edge e'' in P. By Lemma 6.53 we can conclude that $f^{(nt)}(x^b)$ has potential at most p for $t = |P| + |P'| - 2p - 1 = |P| - p \leq l \leq m$. Therefore $f^{(k)}(x) = f^{(nt)}(x^b)$ has potential smaller than that of x for $k = r_1 + b(r_2 - r_1) + nt \leq 2mb + nt \leq 4nm + nm = 5nm$, and so the statement of the lemma holds in this case. Suppose instead that e' does not iterate in factor $\hat{\imath}$ with respect to f'. By Lemma 6.52, the edge $\{w^r, z^r\}$ does not belong to factor $\hat{\imath}$ for some r with $0 \leq r < n$. But $\{x^r, y^r\}$ belongs to factor $\hat{\imath}$ since $r < a$, using the definition of a, so $d(x^r, w^r) - d(y^r, w^r) = d(x^r, z^r) - d(y^r, z^r) \leq 0$. This shows that $d(x^r, w^r) \leq d(y^r, w^r) = p$, so x^r has potential at most p. Since $f^{(k)}(x) = x^r$ for $k = r_1 + r(r_2 - r_1) \leq 2mr < 2nm$, the statement of the lemma holds in this case as well.

Consider now the case $a \leq 2n$. Note that $\{x^{a-1}, y^{a-1}\}$ belongs to factor $\hat{\imath}$, but $\{x^a, y^a\}$ does not, so that $d(x^{a-1}, y^a) + d(y^{a-1}, x^a) = d(x^{a-1}, x^a) + d(y^{a-1}, y^a)$. Furthermore, $d(x^0, x^1) \leq d(y^0, y^1)$ by Lemma 6.55, so $d(x^{a-1}, x^a) \leq d(y^{a-1}, y^a)$. It follows that one of the two alternatives $d(x^{a-1}, y^a) \leq d(y^{a-1}, y^a)$ and $d(x^a, y^{a-1}) \leq d(y^a, y^{a-1})$ must hold. In either case we have a point $f^{(k)}(x)$, either x^{a-1} or x^a, of potential at most l, thus decreasing the potential of x if this potential is given by the second case of the definition of potential, with $k \leq r_1 + a(r_2 - r_1) \leq 2ma \leq 4nm$. We therefore assume that the potential of x is given by the first case of the definition, so that $d(x, z) \leq d(y, z) = p + 1$ for some periodic points y, z, with y adjacent to x in H. Back to the two alternatives above, suppose that the first alternative holds (resp. the second alternative holds). If the last edge $\{w^{a-1}, z^{a-1}\}$ (resp. $\{w^a, z^a\}$) on a shortest path in G from y^{a-1} to z^{a-1} (resp. from y^a to z^a) iterates in different factors when compared with the edges on a shortest path from y^{a-1} to y^a (resp. from y^a to y^{a-1}), with respect to f, then we can again use Lemma 6.53 to show that the potential of $f^{(nt)}(x^{a-1})$ (resp. $f^{(nt)}(x^a)$) is at most p, with $t \leq m$, so that $f^{(k)}(x)$ has potential at most p for $k \leq r_1 + a(r_2 - r_1) + nt \leq 4nm + nm = 5nm$. If $\{w^{a-1}, z^{a-1}\}$ (resp. $\{w^a, z^a\}$) iterates in the same factors as the edges from y^{a-1} to y^a, then $\{w^0, z^0\}$ (resp. $\{w^1, z^1\}$) iterates in the same factors as the edges from y^0 to y^1, so this edge does not belong to factor $\hat{\imath}$ because $y_i^0 = y_i^1$, and since $d(x^0, z^0) \leq d(y^0, z^0)$ (resp. $d(x^1, z^1) \leq d(y^1, z^1)$) we can infer that $d(x^0, w^0) \leq d(y^0, w^0) = p$ (resp. $d(x^1, w^1) \leq d(y^1, w^1) = p$), using the fact that $\{x^0, y^0\}$ and $\{x^1, y^1\}$ belong to factor $\hat{\imath}$. Therefore $f^{(k)}(x)$ has potential at most p for some $k \leq r_2 \leq 2m$.

If $a = \infty$, then $d(x^0, x^r) \leq d(y^0, y^r)$ for all $r \geq 0$ by Lemma 6.55. If we choose $r \geq 1$ so that $y^0 = y^r$ (since y is periodic), then this inequality implies that $x^0 = x^r$ as well, so x^0 is a periodic point, decreasing the potential once again. □

This lemma can be used to complete the proof of the theorem. Since the potential decreases by at least 1 within $5nm$ iterations of f, and the potential does not exceed $l + 1$, it must decrease to 0 within $5(l+1)nm$ iterations, i.e., $f^{(k)}(x)$ is a periodic point for $k = 5(l+1)nm$ in the case $d = 1$, and therefore $f^{(k)}(x)$ is a periodic point for $k = 5d(l+1)nm$ for arbitrary d. □

As a consequence of this theorem, the convergence question can be answered by just iterating f a polynomial number of times, therefore reducing the convergence question to the evaluation question.

6.9 Discussion

The structure of metric nonexpansive networks seems to be due to some extent to two complementary aspects of the axioms of the model. Nonexpansiveness, on the one hand, implies that certain distances, namely those between images of points, cannot be too large. On the other hand, the choice of a product metric that is additive, rather than simply subadditive, with respect to the metrics in each of the factors, implies that certain distances are as large as possible, while still satisfying the triangle inequality. These two underlying assumptions are present in the reduction of the existence of fixed points for nonexpansive mappings on products to the existence of fixed points in each coordinate separately. It is this reduction that leads to a polynomial time algorithm for network stability.

The efficiency of the algorithms for stability depends on the assumption that the relative ratio between minimum and maximum distances is not too large. An open question is whether a polynomial time algorithm can be found when this ratio is exponentially large. A seemingly harder scenario allows factors with an exponential number of points but with a simple stucture, e.g., exponentially long paths with edges of unit weight. For a product of m paths on $2^r - 1$ points, bounds on the order of $2^{2r}m^2$ and $r^{\lceil m/2 \rceil}$ on the number of queries are known (section 6.8.1), with a lower bound of about rm (extending a construction from section 3.1.7).

Many of the structural results seem to extend to the case of infinite metric spaces. For instance, convergent networks can be defined using limits, assuming that the metric spaces are compact. The proof of the fixed product theorem can be carried out by using the distance center of the attractor set (the set of points that remain present after arbitrarily many iterations of the given mapping), provided that a notion of measure that is well-behaved with respect to distances is used for the integration required in the proof. The 2-isometric subspaces can also be studied; their connection with the attractor set of nonexpansive mappings is unclear, since it is no longer possible to transform these mappings into retractions by iteration. The existence of canonical representations depends on the finiteness of the space. For instance, a unit interval can be represented isometrically inside a product of arbitrarily many factors. On the other hand, canonical isometric representations are made possible by the fact that any two isometric representations have a third isometric representation as a common refinement, in the sense that the third representation contains the first two. This also holds for all other canonical representations. It may well be that this refinement property can be extended to infinite domains; this is indeed the case for the cartesian factorization and the isometric representation of possibly infinite, connected graphs, as shown by Walker [114].

As an example, we mention a stability problem suggested by Condon [24] in the study of probabilistic game automata. All coordinates take values in $[0, 1]$. There are three two-input gates, *min*, *max*, *average*, which on inputs x, y give $min(x, y)$, $max(x, y)$, and $(x + y)/2$ as outputs, with unlimited fanout. Since all three gates are monotone, a network always has a stable configuration, and the least stable configuration is sought. The problem of comparing some coordinate of this configuration with some given constant is in $\mathcal{NP}$ and in co$\mathcal{NP}$, but is not known to be in $\mathcal{P}$. We may obtain a nonexpansive version of this problem by using just two gates, namely the comparator that outputs both $min(x, y)$ and $max(x, y)$ on inputs x, y and an *average* gate that outputs exactly two copies of $(x + y)/2$, with no copy gates. A discrete version of this problem has all coordinates varying over integers in some finite range $\{0, 1, \ldots, r\}$, and the *average* gate is replaced by the *discrete average* gate that outputs $\lfloor (x+y)/2 \rfloor$, $\lceil (x+y)/2 \rceil$, on inputs x, y, or more generally, an *average-of-k* gate that on k inputs x_i produces k outputs $\lfloor (j + \sum x_i)/k \rfloor$, for $0 \leq j < k$. This problem is not known to be in $\mathcal{P}$ because r may be written in binary and hence exponential in the input size, so the associated metric space is an exponentially long path, as considered above. We may also include a generalization of the negation gate, which in this case outputs $y = r - x$; the network is now no longer monotone and may no longer have a stable configuration, but the output values, which are independent of the choice of a stable or periodic configuration, can still be tested by guessing a fixed product in $\mathcal{NP}$ and in co$\mathcal{NP}$.

The main notion used to study the structure of fixed points and periodic points is that of a 2-isometric subspace. This notion was introduced as an intermediate link between isometric representations and retracts, but seems to be of interest in itself. The interaction between 2-isometric subspaces and other elements of the structure is not fully understood. For instance, it may be that 2-isometric subspaces maintain enough properties of the subspaces without holes to characterize their distance center as a product.

Bibliography

[1] A. V. Aho, J. E. Hopcroft, and J. D. Ullman, "The Design and Analysis of Computer Algorithms," Addison-Wesley, Reading, MA (1974).

[2] R. K. Ahuja, J. B. Orlin, and R. E. Tarjan, "Improved time bounds for the maximum flow problem," SIAM J. Comput., 18-5 (1989), 939–954.

[3] D. Angluin and L. Valiant, "Fast probabilistic algorithms for Hamiltonian circuits and matchings," J. Comp. Syst. Sci. 19 (1979), 155–193.

[4] S. Arora, C. Lund, R. Motwani, M. Sudan, and M. Szegedy, "Proof verification and the hardness of approximations," Proc. 33rd IEEE Symp. on Foundations of Computer Science (1992), 14–23.

[5] F. Aurenhammer and J. Hagauer, "Computing equivalence classes among the edges of a graph with applications," Discrete Math. 109 (1992), 3–12.

[6] F. Aurenhammer, J. Hagauer, and W. Imrich, "Factoring cartesian product graphs at logarithmic cost per edge," Arbeitsbericht 2/1990, Montanuniversitaet Leoben, Austria (1990).

[7] S. P. Avann, "Metric ternary distributive semi-lattices," Proc. Amer. Math. Soc. 12 (1961) 407–414.

[8] B. Banaschewski and G. Bruns, "Categorical characterization of the MacNeille completion," Archiv. der Math. Basel, 18 (1967), 369–377.

[9] H. J. Bandelt, "Retracts of Hypercubes," J. Graph Theory, 8 (1984), 501–510.

[10] H.-J. Bandelt and J. Hedlíková, "Median algebras," Discrete Math. 45 (1983) 1–30.

[11] H. J. Bandelt and M. Vel, "A fixed cube theorem for median graphs," Discr. Math., 67 (1987), 129–137.

[12] R. Bar-Yehuda and S. Even, "A linear time approximation algorithm for the weighted vertex cover problem," J. Algorithms, vol. 2 (1981), 198–203.

[13] R. Bar-Yehuda and S. Even, "A local ratio theorem for approximating the weighted cover problem," Mathematical Studies 109, Annals of Discrete Math. 25, North Holland, Amsterdam (1985).

[14] R. Beigel, personal communication.

[15] G. Birkhoff, "Lattice Theory," American Mathematical Society Colloquium Publlications, vol. 25, Amer. Math. Soc., Providence, RI (1967).

[16] C. Blair, "Every finite distributive lattice is a set of stable matchings," J. Comb. Theory (A), 37 (1984), 353–356.

[17] A. Borodin, J. von zur Gathen and J. Hopcroft, "Fast parallel matrix and GCD computations," Inform. and Control, 52 (1982) 241–256.

[18] I. Borosh and L. B. Treybig, "Bounds on positive integral solutions of linear Diophantine equations," Proc. Amer. Math. Soc. 55 (1976), 299–304.

[19] H. Chernoff, "A measure of asymptotic efficiency for tests based on the sum of observations," Annals of Mathematical Statistics, vol. 23 (1952), 493–509.

[20] F. R. K. Chung, R. L. Graham and M. E. Saks, "Dynamic search in graphs," Discr. Algorithms and Complexity (1987), 351–387.

[21] F. R. K. Chung, R. L. Graham and M. E. Saks, "A dynamic location problem for graphs," Combinatorica, 9 (1989), 111–132.

[22] M. J. Chung and B. Ravikumar, "Strong nondeterministic reduction – a technique for proving intractability," J. Comput. Syst. Sci. 39 (1989), 2–20.

[23] K. Clarkson, "A modification of the greedy algorithm for vertex cover," Inform. Process. Lett., 16 (1983), 23–25.

[24] A. Condon, "Computational Models of Games," MIT Press, Cambridge, Mass., 1989.

[25] S. A. Cook and M. Luby, "A simple parallel algorithm for finding a satisfying truth assignment to a 2-CNF formula," Inform. Proc. Letters 27 (1988), 141–145.

[26] R. P. Dilworth, "A decomposition theorem for partially ordered sets," Annals of Math., 51 (1950), 161–166.

[27] E. A. Dinic, "Algorithm for solution of a problem of maximum flow in a network with power estimation," Soviet Math. Dokl., 11 (1970), 1277–1280.

[28] J. Edmonds and R. M. Karp, "Theoretical improvements in algorithmic efficiency for network flow problems," J. Assoc. Comput. Mach., 19-2 (1972), 248–264.

[29] S. Even, A. Itai, and A. Shamir, "On the complexity of timetable and multicommodity flow problems," SIAM J. Comput., 5 (1976), 691–703.

[30] T. Feder, "A new fixed point approach for stable networks and stable marriages," Proc. 21st ACM Symp. on Theory of Computing (1989), 513–522. Full version in J. Comp. Syst. Sci. 45 (1992), 233–284.

[31] T. Feder, "Product graph representations," J. Graph Theory 16 (1992), 467–488.

[32] T. Feder, "Network flow and 2-satisfiability," Algorithmica, to appear.

[33] T. Feder and M. Vardi, "Monotone monadic SNP and constraint satisfaction," Proc. 25th ACM Symp. on Theory of Computing (1993), to appear.

[34] J. Feigenbaum, "Product graphs: some algorithmic and combinatorial results," doctoral dissertation, Stanford University (1986).

[35] J. Feigenbaum, J. Hershberger, and A.A. Schäffer, "A polynomial time algorithm for finding the prime factors of cartesian-product graphs," Discrete Appl. Math., 12-2 (1985), 123–138.

[36] L. R. Ford, Jr., and D. R. Fulkerson, "Maximal flow through a network," Can. J. Math., 8 (1956), 399–404.

[37] D. Gale and L. S. Shapley, "College admissions and the stability of marriage," Amer. Math. Monthly, 69 (1962), 9–15.

[38] D. Gale and M. Sotomayor, "Some remarks on the stable matching problem," Discrete Appl. Math., 11 (1985) 223–232.

[39] M. R. Garey and D. S. Johnson, "Computers and Intractability: A Guide to the Theory of NP-Completeness," W. H. Freeman and Company, NY (1979).

[40] A. V. Goldberg, S. A. Plotkin, D. B. Shmoys, and E. Tardos, "Using interior-point methods for fast parallel algorithms for bipartite matching and related problems," SIAM J. Comput. 21 (1992), 140–150.

[41] A. V. Goldberg, E. Tardos, and R. E. Tarjan, "Network flow algorithms," in Paths, Flows, and VLSI-Layout, B. Korte, L. Lovász, H. J. Prömel, and A. Schriver, eds., Springer-Verlag, Berlin, (1990), 101–164.

[42] A. V. Goldberg and R. E. Tarjan, "Finding minimum-cost circulations by successive approximation," Math. of Oper. Res., 15-3 (1990), 430–466.

[43] L. M. Goldschlager, "The monotone and planar circuit value problems are logspace complete for P," SIGACT News, 9-2 (1977), 25–29.

[44] L. M. Goldschlager and I. Parberry, "On the construction of parallel computers from various bases of boolean functions," Theoret. Comput. Sci., 43 (1986), 43–58.

[45] R. L. Graham, "A mathematical study of magnetic domain interactions," Bell Sys. Tech. J., 49-8 (1970), 1627–1644.

[46] R. L. Graham and P. M. Winkler, "On isometric embeddings of graphs," Trans. Amer. Math. Soc., 288-2 (1985), 527–536.

[47] D. Gusfield, personal communication.

[48] D. Gusfield, "Three fast algorithms for four problems in stable marriage," SIAM J. Comput. 16-1 (1987), 111–128.

[49] D. Gusfield, "The structure of the stable roommate problem: efficient representation and enumeration of all stable assignments," SIAM J. Comput. 17-4 (1988), 742–769.

[50] D. Gusfield and R. W. Irving, "The parametric stable marriage problem," Inform. Proc. Lett., 30 (1989), 255–259.

[51] D. Gusfield and R. W. Irving, "The Stable Marriage Problem: Structure and Algorithms," MIT Press Series in the Foundations of Computing, MIT Press (1989).

[52] D. Gusfield, R. Irving, P. Leather, and M. Saks, "Every finite distributive lattice is a set of stable matchings for a small stable marriage," J. Combin. Theory, Ser. A 44 (1987), 304–309.

[53] D. Gusfield and L. Pitt, "Equivalent approximation algorithms for node cover," Inform. Process. Lett, 22-6 (1986), 291–294.

[54] D. Gusfield and L. Pitt, "A bounded approximation for the minimum cost 2-SAT problem," Technical Report CSE-89-4, University of California, Davis (1989).

[55] F. Harary, "Graph Theory," Addison-Wesley, Reading, MA (1969).

[56] P. Hell, "Absolute retracts of graphs," Lecture notes 406 (1974), 292–301.

[57] D. S. Hochbaum, "Approximation algorithms for the set covering and vertex cover problems," SIAM J. Computing, 11-3 (1982), 555–556.

[58] D. S. Hochbaum, "Efficient bounds for the stable set, vertex cover and set packing problems," Discrete Appl. Math., 6 (1983), 243–254.

[59] B. Hochstrasser, "A note on Winkler's algorithm for factoring a connected graph," Discrete Math. 109 (1992), 127–132.

[60] J. E. Hopcroft and R. M. Karp, "An $n^{5/2}$ algorithm for maximum matching in bipartite graphs," SIAM J. Comput., 2 (1973), 225–231.

[61] N. Immerman, "Nondeterministic space is closed under complementation," SIAM J. Comput., 17-5 (1988), 935–938.

[62] R. W. Irving, "An efficient algorithm for the stable room-mates problem," J. Algorithms 6 (1985), 477–595.

[63] R. W. Irving, "On the stable room-mates problem," Technical Report CSC/86/R5, University of Glasgow (1986).

[64] R. W. Irving and P. Leather, "The complexity of counting stable marriages," SIAM J. Comput. 15-3 (1986), 655–667.

[65] R. W. Irving, P. Leather, and D. Gusfield, "An efficient algorithm for the optimal stable marriage," J. Assoc. Comput. Mach. 34-3 (1987), 532–543.

[66] J. R. Isbell, "Median algebra," Trans. Amer. Math. Soc. 260 (1980), 319–362.

[67] E. Jawhari, D. Misane, and M. Pouzet, "Retracts: graphs and ordered sets from the metric point of view," in Contemporary Mathematics, vol. 57, Amer. Math. Soc. (1986).

[68] D. S. Johnson, "The NP-completeness column: an ongoing guide," J. Algorithms, 3-3 (1982), 298–300.

[69] D. S. Johnson, "The NP-completeness column: an ongoing guide," J. Algorithms, 8-3 (1987), 438–448.

[70] D. S. Johnson, C. H. Papadimitriou, and M. Yannakakis, "How easy is local search?," J. Comput. Syst. Sci., 37 (1988), 79–100.

[71] N. D. Jones, Y. E. Lien, and W. T. Laaser, "New problems complete for nondeterministic log space," Mathematical Systems Theory 10 (1976), 1–17.

[72] R. M. Karp, "Reducibility among combinatorial problems," in R. E. Miller and J. W. Thatcher (eds.), Complexity of Computer Computations, Plenum Press, New York (1972), 85–103.

[73] V. King, S. Rao, and R. Tarjan, "A faster deterministic maximum flow algorithm," Proc. 3rd ACM-SIAM Symp. on Discrete Algorithms (1992), 157–164.

[74] D. E. Knuth, personal communication.

[75] D. E. Knuth, "Mariages Stables et leur Relations avec d'autres Problèmes Combinatoires," Les Presses de l'Université de Montréal (1976).

[76] D. E. Knuth, R. Motwani, and B. Pittel, "Stable husbands," Random Structures and Algorithms, 1-1 (1990), 1–14.

[77] R. E. Ladner. The circuit value problem is logspace complete for P. SIGACT news, 7-1 (1975), 18–20.

[78] E. Mayr and A. Subramanian, "The complexity of circuit value and network stability," in Fourth Annual Conference on Structure in Complexity Theory (1989), 114–123. Full version in J. Comp. Syst. Sci. 44 (1992), 302–323.

[79] R. D. C. Monteiro and I. Adler, "Interior path following primal-dual algorithms. Part I: Linear programming," Math. Programming 44 (1989), 27–41.

[80] M. Mulder, "n-cubes and median graphs," J. Graph Theory 4 (1980), 107–110.

[81] H. M. Mulder and J. Schrijver, "Median graphs and Helly hypergraphs," Discrete Math., 25 (1979), 41–50.

[82] K. Mulmuley, "A fast parallel algorithm to compute the rank of a matrix over an arbitrary field," in Proc. 18th Symp. on Theory of Computing (1986), 338–339.

[83] M. Naor, personal communication.

[84] G.L. Nemhauser and R.E. Trotter, "Vertex packing structural properties and algorithms," Math. Programming, 8 (1975), 232–248.

[85] C. Ng, "An $O(n^3\sqrt{\log n})$ algorithm for the optimal stable marriage problem," Technical Report 90-22, University of California, Irvine (1990).

[86] C. Ng and D. S. Hirschberg, "Lower bounds for the stable marriage problem and its variants," SIAM J. on Comput. 19-1 (1990), 71–77.

[87] J. Nieminen, "Distance center and centroid of a median graph," J. Franklin. Inst., 323 (1987), 89–94.

[88] R. Nowakowski and I. Rival, "Retract rigid cartesian products of graphs," Discete Math. 70 (1988), 169–184.

[89] J.B. Orlin, "A faster strongly polynomial minimum cost flow algorithm," Proc. 20th ACM Symp. on Theory of Computing (1988), 377–387.

[90] C. H. Papadimitriou, "On selecting a satisfying truth assignment," Proc. 32nd. IEEE Symp. on Foundations of Computer Science (1991), 163–169.

[91] C. H. Papadimitriou, A. A. Schäffer, and M. Yannakakis, "On the complexity of local search," Proc. 22nd ACM Symp. on Theory of Computing (1990), 438–445.

[92] B. Pittel, "The average number of stable matchings," SIAM J. Discrete Math. 2 (1989), 530–549.

[93] B. Pittel, "On likely solutions of a stable matching problem," Proc. 3rd ACM-SIAM Symp. on Discrete Algorithms (1992), 10–15

[94] G. Pólya, R.E. Tarjan, and D.R. Woods, "Notes on Introductory Combinatorics," Birkhäuser-Verlag (1983).

[95] J.S. Provan and M.O. Ball, "The complexity of counting cuts and of computing the probability that a graph is connected," SIAM J. Comput. 12-4 (1983), 777–788.

[96] A. Quillot, "A retraction problem in graph theory," Discrete Math., 54 (1985), 61–71.

[97] A. Quillot, "On the Helly property working as a compactness criterion on graphs," J. Comb. Theory, Ser. A, 40 (1985), 186–193.

[98] A. Quillot, "Homomorphismes, points fixes, rétractions et jeux de poursuite dans les graphes, les ensembles ordonnés et les espaces métriques," Thèse de doctorat d'Etat, Univ. Paris VI (1983).

[99] V. Ramachandran and L.-C. Wang, "Parallel algorithms and complexity results for telephone link simulation," Proc. 3rd IEEE Symp. on Parallel and Distributed Processing (1991), 378–385.

[100] E. Ronn, "NP-complete stable matching problems," Journal of Algorithms 11-2 (990), 285–304.

[101] E. Ronn, "On the complexity of stable matchings with and without ties," doctoral dissertation, Yale University (1986).

[102] A. E. Roth and M. Sotomayor, "Two-Sided Matching: A Study in Game-Theoretic Modeling and Analysis," Econometric Society Monographs, 18 (1990).

[103] G. Sabidussi, "Graph multiplication," Math. Zeitschr. 72 (1960), 446–457.

[104] T. J. Schaefer, "The complexity of satisfiability problems," Proc. 10th ACM Symp. on Theory of Computing (1978), 216–226.

[105] D. D. Sleator and R. E. Tarjan, "A data structure for dynamic trees," J. Comput. Syst. Sci., 26 (1983), 362–391.

[106] D. D. Sleator and R. E. Tarjan, "Self-adjusting binary search trees," J. Assoc. Comput. Mach 32 (1985), 652–686.

[107] A. Subramanian, personal communication.

[108] A. Subramanian, "A new approach to stable matching problems," Technical Report STAN-CS-89-1275, Stanford University (1989).

[109] A. Subramanian, "The complexity of circuit value and network stability," doctoral dissertation, Stanford University (1989).

[110] A. Tamura, "A property of the divorce digraph for a stable marriage," Research Report B-234, Tokyo Institute of Technology (1990).

[111] R. E. Tarjan, "Depth-first search and linear graph algorithms," SIAM J. Comput., 1 (1972), 146–160.

[112] R. E. Tarjan, "Data structures and network algorithms," Society for Industrial and Applied Mathematics, Philadelphia, PA (1983).

[113] V. G. Vizing, "The cartesian product of graphs," Vychislitel'nye Sistemy 9 (1963), 30–43.

[114] J. W. Walker, "Strict refinement for graphs and digraphs," J. Combin. Theory, Ser. B, 43 (1987), 140–150.

[115] P. M. Winkler, "Factoring a graph in polynomial time," European J. Combinatorics 8 (1987), 209–212.

[116] P. M. Winkler, "The metric structure of graphs: theory and applications," Eleventh British Combinatorial Conference, in Surveys in Combinatorics 1987, London Math. Soc. Lecture Note Series, 123 (1987) 197–221.

[117] M. Yannakakis, personal communication.

[118] A. C. Yao, "Probabilistic complexity: towards a unified measure of complexity," Proc. 18th IEEE Symp. on Foundations of Computer Science (1977), 222–227.

Index

$S(x)$, 8
x_T, 8
xy, 8
$I(g)$, 9
$O(g)$, 9
$g_{x,T}$, 9
$I(N)$, 10
$L(N)$, 10
$O(N)$, 10
$R(N)$, 10
$N(x)$, 10
(x, α), 11
(T, α), 11
$(N, S, \alpha, \beta, \gamma)$, 11
(N, S), 12
(N, α, β), 12
$(N_1, \ldots, N_k)$, 12
(N, k), 12
$x \leq y$, 18
$y = x \oplus z$, 20
e^i assignment, 20
X gate, 22, 65, 138
X_k gate, 22, 24, 42, 136
X_{ij}^{ab} gate, 65
Δ_4 gate, 23, 42
k-sorter, 27, 56
2SAT instance, 61
2SAT optimization, 123
2SAT enumeration, 128
t_{pi}, 134
$G(M)$, 160
$M(G)$, 160
ρ^*, 162
g^ρ, 162
g_σ, 162, 200

θ, 165
θ_1, 166
τ, 168
κ, 182
$y \leq_x z$, 171
x-closed in M, 171
x-closure in M, 172
M-closed, 172
M-closure, 172
q-isometric subspace, 174
2-isometric representation, 179

absorption gate, 9
acyclic 2SAT instance, 61
adjacent assignments, 20
adjacent points, 160
arrangement, 134
assignment, 8, 161
asymptotic convergence, 91

balanced bipartite independent set, 27
balanced stable matching, 155
basis, 9
between points, 160
bipartite 2SAT instance, 63
bipartite matching with bounds, 154
bipartite stable matching, 135
blocking flow, 113
breaking links, 11

canonical isometric representation, 166
canonical 2-isometric representation, 182
canonical retract representation, 193
capacitated flow problem, 109
cartesian product of metric spaces, 161
certificate, 85

INDEX

circuit, 9
closed under imprints, 175
comparator, 21, 65, 138
compatibility graph, 63
complete gate, 92
composition of networks, 12
configuration of a network, 10
connected median set, 60
connected subspace, 179
consistency condition, 135
consistent assignments, 8
constant gate, 9
containment among representations, 162
convergence question, 15
convergent network, 32, 197
convergence to a gate, 32
coordinate, 8, 161
coordinates contained in a subcube, 79
coordinates of a network, 10
counting problem, 15
creating links, 11
cut width, 112

decoder, 22
degree of 2SAT instance, 129
demand vertex, 110
depth-first searches, 121
distance, 31
distance center, 79, 184
distributive lattice, 59
dual gates, 92
dynamic search, 194
dynamic trees, 114

edge belongs to factor, 161
enumeration problem, 15
equivalent variables, 62
evaluation question, 15
explicit width, 110

feedback arc set, 40
fixed cube, 79
fixed point, 11, 58, 183
fixed product, 202

forest solution, 116
forcer, 26

gate, 9
gatewidth of a network, 28

heart, 81
Helly property, 178
homomorphism, 31
hypercube, 30

identity gate, 9
implication graph, 61
implicit width, 110
imprint function, 172
inference graph of 2SAT instance, 62
input assignment, 10
inputs of a network, 10
internal assignment, 10
internal function, 11
irredundant subspace, 161
isometric representation, 165
isomorphism, 31, 162
iterate of a network, 11

layered graph, 115
lexicographic stable matching, 154
linear basis, 18
linear programming, 94, 144
links of a network, 10
local search, 138
lower bound, 171

max-flow min-cut theorem, 109
maximal lower bound, 171
median, 58
median algebra, 60
median graph, 58
median of arrangements, 141
median semilattice, 60
median set, 59
metric space, 159
minimal fixed cube, 81
minimal vertex cover, 124

INDEX

minimum distance point, 33
minimum regret stable matching, 154
monotone 2SAT instance, 63
monotone basis, 18

near-sorter, 21
network, 9
network over a basis, 10
network converges to gate, 32
non-sorter, 26
nonexpansive gate, 20
nonexpansive mapping, 31, 162

optimization problem, 15
oracle model, 28
output assignment, 10
output function, 11
outputs of a network, 10

partial nonexpansive mapping, 187
partial transitive closure, 156
partnership condition, 134
partnership instability, 134
path cover, 110
periodic configuration, 11
periodic point, 11
preference condition, 134
preference instability, 134
preference list, 134
prime factorization, 168
projection of a mapping, 34, 187
proper restriction, 21

relabelling, 9
representation for a metric space, 162
residual graph, 109
restriction in the weak sense, 23
restriction of an assignment, 8
restriction of a gate, 9
retract, 60, 183
retract representation, 184
retraction, 60, 183
rotations, 143

scatter-free gate, 21

signed coordinate, 68
signed permutation, 68
simulation of a gate, 11
size of network, 28
sorter, 21
sorting network, 27
stability question, 15
stable arrangement, 134
stable arrangements and convergence, 140
stable configuration, 10
stable cycle, 81
stable marriage, 135
stable matching, 135
stable matching and counting, 152
stable matching enumeration, 155
stable matching optimization, 152
stable matching with ties, 147
stable roommates, 135
stretching a coordinate, 14
strict representation, 169
subcube, 79
subproduct, 161
subspace, 160
subspace without holes, 184
substitutivity property, 37
supply vertex, 110

tags, 11
transition function of a network, 10
transitive closure of 2SAT instance, 62
transitively closed 2SAT instance, 61
trivial gate, 69
trivial variable, 62

uncapacitated flow problem, 110
union of assignments, 8
universal basis, 16
unravelling a network, 12
unstable cycle, 82
useful gate, 69

weak convergence, 91
width in stable matching, 149

Editorial Information

To be published in the *Memoirs*, a paper must be correct, new, nontrivial, and significant. Further, it must be well written and of interest to a substantial number of mathematicians. Piecemeal results, such as an inconclusive step toward an unproved major theorem or a minor variation on a known result, are in general not acceptable for publication. *Transactions* Editors shall solicit and encourage publication of worthy papers. Papers appearing in *Memoirs* are generally longer than those appearing in *Transactions* with which it shares an editorial committee.

As of March 31, 1995, the backlog for this journal was approximately 4 volumes. This estimate is the result of dividing the number of manuscripts for this journal in the Providence office that have not yet gone to the printer on the above date by the average number of monographs per volume over the previous twelve months, reduced by the number of issues published in four months (the time necessary for preparing an issue for the printer). (There are 6 volumes per year, each containing at least 4 numbers.)

A Copyright Transfer Agreement is required before a paper will be published in this journal. By submitting a paper to this journal, authors certify that the manuscript has not been submitted to nor is it under consideration for publication by another journal, conference proceedings, or similar publication.

Information for Authors and Editors

Memoirs are printed by photo-offset from camera copy fully prepared by the author. This means that the finished book will look exactly like the copy submitted.

The paper must contain a *descriptive title* and an *abstract* that summarizes the article in language suitable for workers in the general field (algebra, analysis, etc.). The *descriptive title* should be short, but informative; useless or vague phrases such as "some remarks about" or "concerning" should be avoided. The *abstract* should be at least one complete sentence, and at most 300 words. Included with the footnotes to the paper, there should be the 1991 *Mathematics Subject Classification* representing the primary and secondary subjects of the article. This may be followed by a list of *key words and phrases* describing the subject matter of the article and taken from it. A list of the numbers may be found in the annual index of *Mathematical Reviews*, published with the December issue starting in 1990, as well as from the electronic service e-MATH [**telnet e-MATH.ams.org** (or **telnet 130.44.1.100**). Login and password are **e-math**]. For journal abbreviations used in bibliographies, see the list of serials in the latest *Mathematical Reviews* annual index. When the manuscript is submitted, authors should supply the editor with electronic addresses if available. These will be printed after the postal address at the end of each article.

Electronically prepared manuscripts. The AMS encourages submission of electronically prepared manuscripts in $\mathcal{A}_{\mathcal{M}}\mathcal{S}$-TeX or $\mathcal{A}_{\mathcal{M}}\mathcal{S}$-LaTeX because properly prepared electronic manuscripts save the author proofreading time and move more quickly through the production process. To this end, the Society has prepared "preprint" style files, specifically the amsppt style of $\mathcal{A}_{\mathcal{M}}\mathcal{S}$-TeX and the amsart style of $\mathcal{A}_{\mathcal{M}}\mathcal{S}$-LaTeX, which will simplify the work of authors and of the production staff. Those authors who make use of these style files from the beginning of the writing process will further reduce their own effort. Electronically submitted manuscripts prepared in plain TeX or LaTeX do not mesh properly with the AMS production systems and cannot, therefore, realize the same kind of expedited processing. Users of plain TeX should have little difficulty learning $\mathcal{AMS}$-TeX, and LaTeX users will find that $\mathcal{AMS}$-LaTeX is the same as LaTeX with additional commands to simplify the typesetting of mathematics.

Guidelines for Preparing Electronic Manuscripts provides additional assistance and is available for use with either $\mathcal{AMS}$-TeX or $\mathcal{AMS}$-LaTeX. Authors with FTP access may obtain *Guidelines* from the Society's Internet node e-MATH.ams.org (130.44.1.100). For those without FTP access *Guidelines* can be obtained free of charge from the e-mail address guide-elec@math.ams.org (Internet) or from the Customer Services Department, American Mathematical Society, P.O. Box 6248, Providence, RI 02940-6248. When requesting *Guidelines*, please specify which version you want.

At the time of submission, authors should indicate if the paper has been prepared using $\mathcal{AMS}$-TeX or $\mathcal{AMS}$-LaTeX. The *Manual for Authors of Mathematical Papers* should be consulted for symbols and style conventions. The *Manual* may be obtained free of charge from the e-mail address cust-serv@math.ams.org or from the Customer Services Department, American Mathematical Society, P.O. Box 6248, Providence, RI 02940-6248. The Providence office should be supplied with a manuscript that corresponds to the electronic file being submitted.

Electronic manuscripts should be sent to the Providence office immediately after the paper has been accepted for publication. They can be sent via e-mail to pub-submit@math.ams.org (Internet) or on diskettes to the Publications Department, American Mathematical Society, P.O. Box 6248, Providence, RI 02940-6248. When submitting electronic manuscripts please be sure to include a message indicating in which publication the paper has been accepted.

Two copies of the paper should be sent directly to the appropriate Editor and the author should keep one copy. The *Guide for Authors of Memoirs* gives detailed information on preparing papers for *Memoirs* and may be obtained free of charge from the Editorial Department, American Mathematical Society, P.O. Box 6248, Providence, RI 02940-6248. For papers not prepared electronically, model paper may also be obtained free of charge from the Editorial Department.

Any inquiries concerning a paper that has been accepted for publication should be sent directly to the Editorial Department, American Mathematical Society, P.O. Box 6248, Providence, RI 02940-6248.

Editors

This journal is designed particularly for long research papers (and groups of cognate papers) in pure and applied mathematics. Papers intended for publication in the *Memoirs* should be addressed to one of the following editors:

Ordinary differential equations, partial differential equations, and applied mathematics to JOHN MALLET-PARET, Division of Applied Mathematics, Brown University, Providence, RI 02912-9000; e-mail: am438000@brownvm.brown.edu.

Harmonic analysis, representation theory, and Lie theory to ROBERT J. STANTON, Department of Mathematics, The Ohio State University, 231 West 18th Avenue, Columbus, OH 43210-1174; electronic mail: stanton@function.mps.ohio-state.edu.

Ergodic theory, dynamical systems, and abstract analysis to DANIEL J. RUDOLPH, Department of Mathematics, University of Maryland, College Park, MD 20742; e-mail: djr@math.umd.edu.

Real and harmonic analysis and elliptic partial differential equations to JILL C. PIPHER, Department of Mathematics, Brown University, Providence, RI 02910-9000; e-mail: jpipher@gauss.math.brown.edu.

Algebra and algebraic geometry to EFIM ZELMANOV, Department of Mathematics, University of Wisconsin, 480 Lincoln Drive, Madison, WI 53706-1388; e-mail: zelmanov@math.wisc.edu

Algebraic topology and differential topology to MARK MAHOWALD, Department of Mathematics, Northwestern University, 2033 Sheridan Road, Evanston, IL 60208-2730; e-mail: mark@math.nwu.edu.

Global analysis and differential geometry to ROBERT L. BRYANT, Department of Mathematics, Duke University, Durham, NC 27706-7706; e-mail: bryant@math.duke.edu.

Probability and statistics to RICHARD DURRETT, Department of Mathematics, Cornell University, White Hall, Ithaca, NY 14853-7901; e-mail: rtd@cornella.cit.cornell.edu.

Combinatorics and Lie theory to PHILIP J. HANLON, Department of Mathematics, University of Michigan, Ann Arbor, MI 48109-1003; e-mail: phil.hanlon@math.lsa.umich.edu.

Logic and universal algebra to GREGORY L. CHERLIN, Department of Mathematics, Rutgers University, Hill Center, Busch Campus, New Brunswick, NJ 08903; e-mail: cherlin@math.rutgers.edu.

Algebraic number theory, analytic number theory, and automorphic forms to WEN-CHING WINNIE LI, Department of Mathematics, Pennsylvania State University, University Park, PA 16802-6401; e-mail: wli@math.psu.edu.

Complex analysis and complex geometry to DANIEL M. BURNS, Department of Mathematics, University of Michigan, Ann Arbor, MI 48109-1003; e-mail: burns@gauss.stanford.edu.

Algebraic geometry and commutative algebra to LAWRENCE EIN, Department of Mathematics, University of Illinois, 851 S. Morgan (MIC 249), Chicago, IL 60607-7045; email: u22425@uicvm.uic.edu.

All other communications to the editors should be addressed to the Managing Editor, PETER SHALEN, Department of Mathematics, Statistics, and Computer Science, University of Illinois at Chicago, Chicago, IL 60680; e-mail: shalen@math.uic.edu.

Recent Titles in This Series

(*Continued from the front of this publication*)

526 **Shiro Goto and Koji Nishida,** The Cohen-Macaulay and Gorenstein Rees algebras associated to filtrations, 1994
525 **Enrique Artal-Bartolo,** Forme de Jordan de la monodromie des singularités superisolées de surfaces, 1994
524 **Justin R. Smith,** Iterating the cobar construction, 1994
523 **Mark I. Freidlin and Alexander D. Wentzell,** Random perturbations of Hamiltonian systems, 1994
522 **Joel D. Pincus and Shaojie Zhou,** Principal currents for a pair of unitary operators, 1994
521 **K. R. Goodearl and E. S. Letzter,** Prime ideals in skew and q-skew polynomial rings, 1994
520 **Tom Ilmanen,** Elliptic regularization and partial regularity for motion by mean curvature, 1994
519 **William M. McGovern,** Completely prime maximal ideals and quantization, 1994
518 **René A. Carmona and S. A. Molchanov,** Parabolic Anderson problem and intermittency, 1994
517 **Takashi Shioya,** Behavior of distant maximal geodesics in finitely connected complete 2-dimensional Riemannian manifolds, 1994
516 **Kevin W. J. Kadell,** A proof of the q-Macdonald-Morris conjecture for BC_n, 1994
515 **Krzysztof Ciesielski, Lee Larson, and Krzysztof Ostaszewski,** $\mathcal{I}$-density continuous functions, 1994
514 **Anthony A. Iarrobino,** Associated graded algebra of a Gorenstein Artin algebra, 1994
513 **Jaume Llibre and Ana Nunes,** Separatrix surfaces and invariant manifolds of a class of integrable Hamiltonian systems and their perturbations, 1994
512 **Maria R. Gonzalez-Dorrego,** $(16,6)$ configurations and geometry of Kummer surfaces in $\mathbb{P}^3$, 1994
511 **Monique Sablé-Tougeron,** Ondes de gradients multidimensionnelles, 1993
510 **Gennady Bachman,** On the coefficients of cyclotomic polynomials, 1993
509 **Ralph Howard,** The kinematic formula in Riemannian homogeneous spaces, 1993
508 **Kunio Murasugi and Jozef H. Przytycki,** An index of a graph with applications to knot theory, 1993
507 **Cristiano Husu,** Extensions of the Jacobi identity for vertex operators, and standard $A_1^{(1)}$-modules, 1993
506 **Marc A. Rieffel,** Deformation quantization for actions of R^d, 1993
505 **Stephen S.-T. Yau and Yung Yu,** Gorenstein quotient singularities in dimension three, 1993
504 **Anthony V. Phillips and David A. Stone,** A topological Chern-Weil theory, 1993
503 **Michael Makkai,** Duality and definability in first order logic, 1993
502 **Eriko Hironaka,** Abelian coverings of the complex projective plane branched along configurations of real lines, 1993
501 **E. N. Dancer,** Weakly nonlinear Dirichlet problems on long or thin domains, 1993
500 **David Soudry,** Rankin-Selberg convolutions for $SO_{2\ell+1} \times GL_n$: Local theory, 1993
499 **Karl-Hermann Neeb,** Invariant subsemigroups of Lie groups, 1993
498 **J. Nikiel, H. M. Tuncali, and E. D. Tymchatyn,** Continuous images of arcs and inverse limit methods, 1993
497 **John Roe,** Coarse cohomology and index theory on complete Riemannian manifolds, 1993

(See the AMS catalog for earlier titles)